CLIMATOLOGY: FUNDAMENTALS AND APPLICATIONS

McGRAW-HILL SERIES IN GEOGRAPHY
Edward J. Taaffe and John W. Webb, *Consulting Editors*

CLIMATOLOGY: FUNDAMENTALS AND APPLICATIONS

JOHN R. MATHER
Professor and Chairman
Department of Geography
University of Delaware

McGRAW-HILL BOOK COMPANY
New York St. Louis San Francisco Düsseldorf Johannesburg
Kuala Lumpur London Mexico Montreal New Delhi
Panama Paris São Paulo Singapore Sydney Tokyo Toronto

CLIMATOLOGY:
FUNDAMENTALS AND APPLICATIONS

1 2 3 4 5 6 7 8 9 0 M A M M 7 9 8 7 6 5 4

Library of Congress Cataloging in Publication Data

Mather, John Russell, date
      Climatology: fundamentals and applications.

      (McGraw-Hill series in geography)
      Bibliography: p.
        1. Climatology. I. Title. [DNLM: 1. Climate.
2. Weather. QC981 M427c 1974]
QC981.M43   551.5   73-23082
ISBN 0-07-040891-2

This book was set in Theme by Scripta Technica, Inc.
The editors were Janis Yates and M. E. Margolies;
the designer was Nicholas Krenitsky;
the production supervisor was Thomas J. LoPinto.
The drawings were done by Danmark & Michaels, Inc.
The Maple Press Company was printer and binder.

To
RUDOLF GEIGER

and in memory of
C. WARREN THORNTHWAITE

CONTENTS

CHAPTER 8

CHAPTER 9

CHAPTER 10

CHAPTER 11

PREFACE

Many current textbooks in climatology provide broad surveys of the whole field or deal in depth with some limited aspect of the subject matter. It might seem that another textbook, especially at the introductory level, would hardly be necessary. This volume exists not because of the author's desire to discuss anew the basic facts of climatology in a slightly different format but rather because of the author's strong conviction that the present approach to the field of climatology at the introductory level is wrong for many of today's students.

There is no question that a sound foundation in the elements of any subject matter is absolutely essential for comprehension of that field. Many older textbooks in climatology attempt such an approach; as a result they often include "dry as dust" discussions of the nature of the basic weather elements, their distribution both in space and time, and their relation to the distribution of soils and vegetation. The approach undoubtedly has turned off many bright students and left them with the feeling that climatology is far from dynamic, less than relevant, and of little use except to those interested in static distributions and statistical displays. The result has been apathy or even rejection at a time when the field should have been rapidly developing a new role for itself in a world concerned with the interrelation of man and his environment.

This volume represents an attempt to reverse old approaches to introductory climatology and to present the field in a way that will, it is hoped, challenge a number of today's students to delve further into the field. It should, at least, encourage all students to seek new ways to live rationally within their climatic environment. The present approach seeks to provide enough discussion of the basic elements of climate and weather to provide an understanding of their nature. However, it concentrates its main thrust on the uses of the climatic information in various aspects of our lives—hydrology, agriculture, health, comfort, clothing, architecture, commerce, and industry. It is felt that if beginning students in climatology (or even those with a sound background in the basic elements of weather and climate) can be shown how knowledge of climate permits us to live and to work more comfortably and economically, many will want to pursue the field in greater depth. Today's student wants to see the need, the application, and the value of a subject before expending significant effort in its study. Applied climatology has much to offer in this respect if only we will introduce it to the student before he or she is overwhelmed by pages of descriptions about climatic distributions or the seasonal changes in conditions in remote areas of the earth.

The book requires no background in climatology, physics, or mathematics, although clearly if the student has such background he will be able to achieve a greater depth of understanding. The first chapter describes briefly the history and applications of climatology in general and provides some specific comments on types or classes of problems that require climatic information for their solution. Since heat- and water-budget approaches to climatology are emphasized, the second chapter (on basic climatic elements) devotes most of its attention to the factors of radiation, temperature, atmospheric moisture, and evapotranspiration. The third chapter describes the water-budget approach to climatology in some detail. The level of generalization shifts from the annual global balance, to the balance of the Northern Hemisphere, to the balance over North America, and ultimately to the water budget at a particular place. Chapter 4 utilizes the water-budget approach in the field of climatic classification and shows how the factors of the water budget correlate with the distribution of vegetation in North America.

The remaining seven chapters in the book describe current and past work in the field of applied climatology. The effort here is to show how climate affects nearly all aspects of our lives and how, through an understanding of climate, we can better adjust our activities to our atmospheric environment. Chapter 5 describes how the water budget relates to problems in hydrology and water resources, while Chapters 6 and 7 discuss the role of climate in agriculture. Both heat- and water-budget approaches are fundamental in agricultural relations. Chapter 8 focuses on problems of clothing and human comfort. The approach examines the heat fluxes between the body and the outside environment and the problems of maintaining a nearly constant body temperature. Chapter 9 describes our present understanding of the effect of climate on human health and includes discussions of air pollution as well as urban heat-island problems. Chapter 10 introduces the role of climate and building design, pointing out how most current developments in building have produced houses that are at odds with their environments; indoor comfort can thus be achieved only through mechanical devices at additional expense to the homeowner. Finally, Chapter 11 considers the role of climate in industrial production, commerce, business, and engineering. Just the normal day-to-day changes in weather can result in significant economic gains or losses to the businessman because of weather effects on both production schedules and purchasing habits.

Each of these last seven chapters deserves a book in its own right; at the same time many books have already been written covering just the materials in Chapters 2 and 4. Thus much has had to be eliminated or greatly condensed to keep within the bounds of a practical length. The text abounds in references to other works, and many of these should be utilized to amplify the brief discussions included in the book. If the book is used in a beginning course in climatology, Chapter 2 might be expanded by reference to other basic texts. If it is used in a more advanced course in which the students have had some background in weather or climate, Chapter 2 can be passed over rapidly and more time spent on the applications of climatic data to specific problems.

I would like to express my appreciation to all the many individuals and groups who have helped in the preparation of this book in any way. Special thanks must be extended to those who have given permission to use material from other books, journals, or reports. Some may not be acknowledged in the text, but only through oversight on my part. To these, I extend my sincere apologies, for without the help and cooperation of many, this work would not have been possible. To the staff of C. W. Thornthwaite Associates Laboratory of Climatology must go especial appreciation not only for permission to use many diagrams and other materials but also for their great help over the years in assisting in the collection of many of the basic data used here. And to Dr. F. Kenneth Hare of Environment Canada and the University of Toronto as well as to the other reviewers, my thanks for very sympathetic and helpful suggestions. Finally, to my wife must go my special appreciation for her patience and understanding through unnumbered months while this book was in preparation; and to my secretaries Mrs. S. Helen Bano, who typed the first draft, and Mrs. Marie Donaghay, who cheerfully typed later drafts and revisions, I extend my thanks for their great help.

JOHN R. MATHER

CLIMATOLOGY— HISTORY, TOOLS, AND TECHNIQUES OF PROBLEM SOLVING

INTRODUCTION

Most scientific undertakings are concerned with the useful application of the results. Seldom is scientific work attempted without at least some thought as to its ultimate utility to the investigator. New and unexpected knowledge can, of course, be obtained during a scientific investigation and results emerge that are entirely unanticipated. No immediate application of the results may be visualized at the time. While some of our great scientific discoveries result in this way, it is apparent that the great bulk are the product of directed research—studies undertaken with a purpose, seeking results that can be applied to some practical problem confronting mankind.

The results of scientific work in one field are sometimes applied to problems in entirely different fields. Such investigations are often undertaken by teams of scientists from different backgrounds. By working together, they are able to focus their knowledge on a particular problem that may be influenced by factors in one or more neighboring disciplines.

Climatology, an old field of scientific investigation in its own right, has many contacts with other fields. Climate influences agricultural and industrial operations, hydrology, geology, health and disease, forestry, transportation, and many other fields of human endeavor. Because of this interdisciplinary influence, climatology becomes a prime field for (1) the application of its knowledge to problems in other fields and (2) the application of a team approach to problem solving.

It is clear that weather data are indispensable in planning many agricultural, industrial, engineering, or resource-use activities. Most of these applications of meteorological data do not necessarily involve any short-range weather forecast but rather some knowledge of past and existing weather elements in an area, including the understanding of frequencies and extremes of weather conditions. Such knowledge can be used to evaluate the likelihood of some future event, but a short-range, specific forecast of the actual occurrence of certain events is not the major aim of the applied climatologist. Rather, his goal is to use his knowledge of the influence of different weather elements on various environmental factors to provide information on how individuals can better adapt their lives and activities to the given or expected climatic conditions for more comfortable, efficient, and inexpensive living.

If weather is considered to be the instantaneous picture of various atmospheric elements such as radiation, moisture, pressure, wind, and temperature, then climate is defined by some to be the statistical summary, the averaging of the individual weather elements over a period of time. Such a definition of climate is too restricted for it emphasizes only the statistical approach and infers that it is nothing more than average weather. Weather is the condition of the atmosphere at a given moment at a given place. Climate, therefore, is the characteristic weather considered for longer periods—days, months, seasons, years—and for larger areas. Climate does refer to average conditions, but it also refers to the extremes of weather and to the probability or likelihood of different events occurring at a place or over a region. Climatology does involve statistics, but the statistics should be used only as an aid in describing the fluctuations or changing nature of climate, its long-term variability or trends, and the interrelated influence of all the weather elements.

Applied climatology can be considered to be the analysis and application of climatic data for some specific operational purpose (Landsberg and Jacobs, 1951). Table 1-1 lists just a few of the many operational purposes to which climatic data are now being applied. It suggests something of the wide range of interests facing an applied climatologist.

RECENT HISTORY

By the beginning of the nineteenth century, when climatological records were becoming more commonplace, great interest was focused on the influence of climate on particular operations or undertakings. Thomas Jefferson, Alexander von Humbolt, and others called for the collection of systematic climatological records for the express purpose of answering specific questions involving agriculture, crop introduction, health, and soil conservation. Before 1820, several orders of the Surgeon General of the United States directed that weather observations were to be taken by the Post Surgeons to provide data which could be used in answering questions about the effect of climate on the health of troops in remote outposts, on "medical topography . . . prevalent regional complaints . . . change of climate . . . cultivation of soil . . . density of population . . ." (U.S. Weather Bureau, 1955, p. 2).

In the years following, however, developments in synoptic meteorology which followed the introduction of the telegraph, in weather forecasting, and later, in aeronautical meteorology overshadowed the original interest in applied climatology. About the only phase of applied climatology that was able to develop was agricultural climatology. The Federal Weather Service had been established in the Signal Office of the U.S. Army in 1870. Recognizing the great potential value of weather and climatology to agriculture, the Weather Service was transferred from the Army to the Department of Agriculture in 1891 and became the U.S. Weather Bureau. Within the first few years of its existence the Weather Bureau published a number of important works on various phases of agricultural climatology. Those by Hilgard (1892), Whitney (1892), Mell (1893), and Abbe (1905) are classic studies well worth reading even today. These works charted lines for later, more detailed agricultural climatology studies. The Climate and Crop Weather Division was established in the Weather Bureau in the mid-1890s, and its work has continued to this day.

As the twentieth century started, the airplane became a reality and the Weather Bureau found a new and challenging field for its services—

TABLE 1-1

Applied climatology in action

COMMUNITY ACTIVITIES	HUMAN ACTIVITIES
Climatic effects on	Climatic effects on
Air pollution abatement and regulation	Physiological responses
Water supply planning	Health, morbidity, and mortality relations
City design and layout	Acclimatization
Land-use planning	Clothing requirements
Traffic hazards	House design and characteristics
Severe storm hazards	Heating and cooling requirements
Highway, bridge, culvert design	Sports activities

INDUSTRIAL AND COMMERCIAL ACTIVITIES	AGRICULTURAL ACTIVITIES
Climatic effects on	Climatic effects on
Plant operations, water supply, and pollution dispersion	Land-use planning
	Agricultural pests, diseases, viruses
Product design and engineering	Yield
Product sales planning	Freeze probabilities, frost-free periods
Weather insurance	Rainfall frequencies, irrigation needs
Transportation, storage of raw or manufactured materials	Crop scheduling
	Weather insurance
Plant absenteeism, accidents	Forest fire probabilities
Utility operations	Soil tractionability
Aircraft, ship, rail systems	Coastal fisheries activities

ogy. By 1926, the Weather Bureau became officially responsible for weather
services to civil aviation, and by 1940, the Weather Bureau was transferred
from the Department of Agriculture to the Department of Commerce. Both
surface- and upper-air studies, largely for the purposes of aviation or weather
modification, became increasingly significant.

Applied climatology, both agricultural and industrial, began to
redevelop a role for itself after World War II. First, during the war, the need
to have climatic information from many different parts of the globe and to
apply these data to military operations or to the supplying and equipping of
troops brought to the fore the vital role that the applied climatologist could
play in an increasingly complex world. Second, the development of
high-speed electronic computers that could assimilate, analyze, and present in
tabular form vast quantities of climatic data made it feasible to apply some of
our large store of weather data to the practical problems of weather
influence. Third, the large supply of trained meteorologists, needed during the
war effort but who were then being released into civilian life, provided the
nucleus of skilled workers to begin the attack on problems of applied
climatology.

Woodrow Jacobs (1947) published a classic monograph entitled
"Wartime Developments in Applied Climatology" in which he detailed a
number of different types of problems that applied climatologists faced
during the war and indicated the kinds of solutions that were possible. This
monograph undoubtedly had an impact on some of the meteorologists just
being released from military service, and a number of them moved into the
industrial applied climatology field.

The stage is now set for a redevelopment of applied climatology. While
many recent developments in the industrial and consulting aspects of weather
have related to weather forecasting or weather modification, there are other
movements toward the development of useful applications of climatology for
operational purposes. In the long run, these developments may be of as
far-reaching significance as the work on forecasting and weather modification.

THE TOOLS OF APPLIED CLIMATOLOGY

What are the tools with which the applied climatologist works? Quite clearly
they are, in part, the meteorological data collected daily and hourly by
observing stations around the world. The amount of material so available is
staggering in quantity, yet the standard summaries based on these data leave
much to be desired as far as the applied climatologist is concerned.
Mean-value climatologies of temperature, precipitation, or other basic
meteorological factors often do not provide summaries that are particularly
pertinent or useful. The applied climatologist might, in a given study, need
information on the occurrence of dew, the composition of precipitation, the
weight of snow, the rate of snowmelt, the rate of icing on exposed surfaces,
the depth of freezing of the ground, the soil moisture content, the
evapotranspiration, the wind shear, the distribution of vertical winds, or on
the factors of the heat balance, just to name a few possible items. Information

on these and other similar factors is not generally available. Here the applied climatologist must (1) interpolate data between more distant observing points, (2) substitute some available observations for the required but not observed data, or (3) supply missing data based on past experience or known data from similar environmental situations. To be successful the applied climatologist must always be ready to bridge the gap between available and required observation.

Early Instrumental Observations in the United States

Rev. John Campanius took the first known surface weather observations in America at Swedes Fort, Delaware, in 1644–1645. The first use of the Fahrenheit thermometer to record air temperature in America probably dates back to Dr. John Lining of Charleston, South Carolina, in 1738. Dr. Lining took systematic observations under the sponsorship of the local medical society from 1738 to 1750. Weather diaries became more prevalent in the early 1800s as interest in climate and in its effect on the health and activities of man, especially in the more remote areas of the West, gave rise to the rapid collection of instrumental observations of climate. Dr. Samuel Forry attempted the first comprehensive summarization of available records in 1842 in a publication entitled "The Climate of the United States and its Endemic Influences." Lorin Blodget (1857) published the first maps of seasonal and annual precipitation of the United States in 1857. Based on data from about 400 weather stations, Blodget's map only approximates the areal distribution of precipitation as we understand it today, but his contribution to climatology and his insight into the fundamental facts of climate mark his work as one of the foremost in the field of climatology for half a century. Our own map of precipitation may indeed seem crude in comparison with later maps to be produced from more advanced sensors possibly carried over both land and water areas of the globe by satellites a few years hence.

At about the same time, J. H. Coffin (1853, and later in 1875) published comprehensive memoirs on winds over the Northern Hemisphere, while C. A. Schott (1872, 1876) prepared tables and maps of precipitation and temperature over the United States and other areas of North and South America. Blodget, Coffin, and Schott, like Dr. A. H. Guyot who authored the Smithsonian Meteorological Tables and prepared extensive instructions on meteorological instrumentation, its calibration, and its installation, all carried out their pioneering work under the direction of Joseph Henry, first Secretary of the Smithsonian Institution. High on Henry's list of research objectives was the desire to extend the network of meteorological observations for the purpose of understanding American storms. As a result, he pushed the Institution into the very forefront of American climatology, establishing observation networks, collecting, tabulating, and analyzing data, and publishing scientific memoirs and reports of highest quality.

The number of precipitation stations in the conterminous United States increased from about 400 in Blodget's time to 3,000 in 1891 when the

Weather Bureau was established, to 4,500 in 1930 and to about 10,000 by the 1940s. There has not been any great increase since that time. If all 50 of the United States as well as island possessions are considered, the observing network includes some 12,000 to 15,000 stations measuring at least precipitation on a daily basis. About 5,000 of these stations are designated as part of the permanent climatic network, taking, in addition, observations of maximum and minimum temperature each day as well as notes on frosts, thunderstorms, and other limiting weather conditions.

Upper-air observations, as valuable to the meteorologist as surface weather observations, developed only slowly. Originally only mountain stations were able to provide records of conditions at elevations much above sea level but even these stations were not truly upper air. Benjamin Franklin used kites for upper-air investigations, but it was not until 1893 that the Chief of the Weather Bureau first asked Congress to appropriate some money for upper-air kite observations. In the same year, Hargreave in Australia used a series of box kites to obtain observations to over 9,000 m (30,000 ft)!

Some observations had been made before that time by observers in manned balloon flights, but no systematic observations were possible. Pilot balloon observations (following the flight of a free balloon) to obtain winds aloft developed during World War I as did the direct measurement of upper-air pressure, temperature, and humidity by airplanes.

The first successful radiosonde ascent (a radiosonde is a lightweight meteorograph and radio transmitter attached to a small balloon, capable of sending signals of pressure, temperature, and humidity to a ground observatory) was made in 1928 by the Russian meteorologist Moltchanoff. The first United States ascent did not occur until 1935 but with it came the end of the airplane observing program, which was limited by weather conditions and expense. The number of radiosonde observing stations operated by the various United States weather services increased from 6 in 1937 to a peak of 335 in 1945 and has leveled off at about 200 in more recent years. The impact of World War II on the weather observation program of the United States cannot be overemphasized.

During the Fourth World Meteorological Congress in Geneva in 1963, the concept of a World Weather Watch was approved. Actual plans were adopted at the Fifth Congress in 1967. The purpose of the Watch was to encourage all member nations of the World Meteorological Organization (WMO) to join in the collection and analysis of meteorologic data on a worldwide basis. Such a program would provide an unprecedented opportunity for all nations to contribute to the solution of pressing observational and analytical problems as well as to derive increased benefits from the improved meteorological services such a cooperative effort would make possible. One basic aspect of the World Weather Watch was to be a Global Observing System (GOS) optimizing the density of stations for sampling purposes. Table 1-2 provides a tabulation of the number of stations in the GOS, by continent, as of the end of December 1970. Other aspects of the Watch were improved data transmission, storage and retrieval systems, as well as greatly expanded observations from the ocean areas of the globe by means of ships, buoys, and satellites.

Data Processing

Before 1940 there was essentially no centralized weather data processing facility in existence—certainly a necessary adjunct to the establishment of any large-scale climatic observing network. The British, Dutch, and Czechoslovakian meteorological organizations had used punched cards in the production of climatic summaries of marine meteorological data in the 1920s, and a WPA project established in New Orleans in 1934 had begun to explore the utility of weather data processing, but the great explosion of interest in the use of high-speed data processing waited for (1) the development of the need for a large number of weather summaries from many different places and (2) the development of equipment to do the job rapidly and accurately.

Actually, H. Hollerith, working in the U.S. Bureau of the Census in 1890, was possibly the first to employ punched cards quite similar to those now in use in the tabulation of data—in this case census data. He first had to develop equipment that could be used to read the data from the cards. Five years later, Commander C. D. Sigsbee, the Hydrographer of the United States, pointed out the value to climatology of the electrical tabulating system devised by Hollerith, but developments along these lines moved exceedingly slowly and for many years thereafter, hand-tabulated summaries of data were still the only climatic records available. With the great increase in the number of observing stations, the problem of data evaluation finally became so formidable that increased use of electrical data tabulating and processing equipment became necessary.

TABLE 1-2

Number of surface and upper air stations, Global Observing System, World Weather Watch, as of Dec. 31, 1970*

AREA	SURFACE		UPPER AIR (RADIOSONDE/RAWINSONDE)
	0000†	1200†	0000†
Africa	511	644	74/88
Asia	875	885	188/200
South America	298	315	35/32
North America	444	447	147/147
South Pacific	314	259	68/91
Europe	835	849	142/141
Antarctic	24	24	11/11
Total	3,301	3,423	665/710

* Adapted from WMO, 1970, pp. I-11 to I-26.
† Observation hour.

By 1941, some 20 million meteorological observations taken during the 1928–1941 period had been entered onto punched cards and were available at the New Orleans Tabulating Unit (NOTU), which had grown out of the earlier WPA project. These, and later data entered onto punched cards, were used extensively during World War II in various studies of applied climatology.

Major achievements during the war years, beyond the establishment of climatology as a significant factor for planning purposes, included (1) the standardization of observation techniques and sites, instrumentation, record keeping, and the beginning of a centralized records collection for the military as a supplement to the National Weather Service; (2) the use of the punched card for data processing purposes; and (3) the collection of weather data on a global basis (Barger, 1960).

From the end of the war in 1946 until 1951 the processing of all United States weather records was centralized in New Orleans. Quality control procedures were introduced to ensure accuracy and reproducibility in the records being tabulated. Data processing equipment improved rapidly from the original machines designed mainly for accounting operations to the more sophisticated equipment for carrying out statistical manipulations. The first machine to use electronic principles as opposed to the earlier electromechanical machines appeared in 1950.

By 1951, the NOTU had outgrown its available space, and, in September of that year, a transfer of all records and equipment from New Orleans to Asheville, North Carolina, was begun (completed January 1952). The new facility, a multistory building covering a full city block and designated the National Weather Records Center (now called the National Climatic Center), was established to handle the weather data tabulating, checking, storage, and processing needs of the Weather Bureau, the Army, Air Force, Navy, and the public in general. The Air Force and Navy continued separate Control Divisions or Liaison offices in Asheville to handle their own needs, but the Joint Punched Card Library served the climatic data needs for all groups.

As of 1970, the National Climatic Center (NCC) held approximately 515 million punched cards and microfilmed card images on which weather data (condensed by many different numerical codes) had been entered. The files were growing at the rate of nearly 8 million cards per year. Other forms of data held at Asheville and their rate of accession are listed briefly in Table 1-3. The NCC has also been designated an extension of the National Archives; this means it must attempt to collect and file a copy of each original weather record, recording chart, or observation form. Even with the large floor space available at Asheville, space is once more at a premium, for, as more records are accumulated and stored, as increasing demands are made for the data, more electronic processing equipment must also be added to handle the needs.

Several developments are now under way. First, the punched-card library is being continually reduced by use of the FOSDIC[1] technique which

[1] For Film Optical Sensing Device for Input to Computers.

provides a microfilm of the punched card for more convenient storage. The punched card can be easily re-created from the microfilm whenever necessary. Second, some of the weather data are being entered onto magnetic tapes which can store vast amounts of data in much less space than the cards themselves (some reels can store the equivalent of 6,000 standard 20-drawer punch-card cabinets). Third, work is continuing so that bulkier paper records can be destroyed once authority has been obtained. Fourth, ever faster data processing equipment is being installed and utilized to speed the process of data retrieval and summarization.

    One new requirement for space which threatens to outstrip all previous demands involves the handling of satellite records including photographs. The problems of identifying such photographs for easy retrieval later as well as the physical problems of storing vast new amounts of material have created difficult problems for the NCC.

Published Data Sources

The National Weather Service (formerly the U.S. Weather Bureau) publishes an extremely large number of climatological summaries, often providing the same type of data (temperature or precipitation, for example) at a station in several different ways to make it more useful to a larger number of individuals. Before 1963, it was difficult even for a trained climatologist to be aware of all the different weather summaries and tabulations prepared by the Weather Bureau. In that year, however, the Weather Bureau (Swartz, 1963) published a most valuable "Selective Guide to Published Climatic Data Sources," designed to inform potential users of climatic data of the availability of such data in some published form. The report not only lists the type of data, the form of the summary, and the name and number of the

TABLE 1-3

Information bank at National Climatic Center, Asheville, N.C.*

FORM OF DATA	ON HAND (1970)	ANNUAL ACCESSION RATE
Original manuscripts and autographic traces (sheets)	71,400,000	2,500,000
Microfilms of original manuscripts and autographic traces (100-ft reels)	71,500	3,000
Punched cards including microfilmed card images	515,000,000	7,800,000
Magnetic tapes (2,400-ft reels)	31,200	5,000
Radar films (100-ft reels)	12,700	1,150
Satellite films (100-ft reels)	9,800	1,000
Climatological publications (back editions)	173,800	7,000
Unpublished tabulations and analyses	17,500	1,200

* Adapted from U.S. Department of Commerce, 1970, NOAA.

publication in which the information appears but also includes information on cost, availability, and gives shortened examples of the tables to aid the user in understanding the exact nature of the published information.

The different components of basic climatic parameters on which some data are available include

Barometric pressure	Snowfall and snow cover
Ceiling, sky condition	Soil moisture
Dew point	Soil temperature
Evaporation	Solar radiation
Frost	Sunshine
Heating degree-days	Temperature: surface and air
Humidity	Visibility
Ozone	Weather and/or obstruction
Precipitation	to vision
River gage data	Wind direction and speed

It might appear that practically all data needed for most applied climatological problems must be available in some form but this is not so. First, the data on certain parameters such as temperature, wind, and precipitation are much more extensive than those on evaporation, soil temperature, or ozone. Not only are more station records available for the former factors but the data may also be given in hourly values as opposed to daily, weekly, or individual spot values which may be the only information available on the latter factors. Second, the form of the summary may have little value for applied climatology. Total depth of snow cover is important, but it is possible that snow cover in a certain time period, snow amounts associated with different wind velocities, or weight of snow cover would be more useful for particular climatological problems. Third, the length of record at a particular station may be quite short and may not reveal the true expectancies of conditions over time. Fourth, the published data are for particular points, and often data are needed for other points more or less removed from the observation point. Microclimatic differences between the two places may make it quite impossible to extrapolate from one to another with any degree of confidence. Fifth, the reliability of the data in each of the component categories is not in every case the same. Problems of exposure of sensors, their accuracy and sensitivity, and the precision with which they can be read all influence the usefulness of the different data tabulations.

As a result, the applied climatologist must somehow provide the needed meteorologic information. In certain cases, direct extrapolation or interpolation of data is possible. For example, extrapolation of temperature taken at one height to the temperature at another height is quite possible within reasonable limits. Extrapolation of precipitation data, even on a monthly basis between two stations just a few miles apart, is a very questionable procedure, however. When daily precipitation is extrapolated, little reliability can be placed in the results especially in the summer period of the year when shower activity predominates.

Some derivation of new records must also be attempted. For example, if evaporation records are needed but not available, it is possible to obtain an estimate of such data from air temperature using any one of several empirical relationships. Less direct is the derivation of surface temperatures from available shelter temperatures in the case of frost determination. Many other environmental factors such as slope and aspect, nature of the vegetation, soil moisture condition, sky cover, and wind may make the relation between temperature at two different levels (especially near the surface) quite variable. The experience and the skill of the climatologist becomes paramount in any such undertaking.

THE ROLE OF THE APPLIED CLIMATOLOGIST

In an assessment of applied meteorologists and climatologists, Jacobs (1959) notes that those who have been successful have often broken away from the traditional data summary or mean-value climatologies of the descriptive climatologists and have innovated new methods of analysis, of summary, and of presentation of meaningful data. These applied climatologists have become involved in the problems of the operator so that the search for a solution has become of personal importance; it is no longer an impersonal piece of scientific research. No matter who the client is—an industry, a utility, a municipality, a farmer, a medical group, or any one of a number of other possible users—the applied climatologist must be able to understand and talk intelligently with him. He must recognize the user's problem, and how climate and weather affect his operation, in order to bring to bear on the question the correct and most meaningful climatological information. It is almost axiomatic that the user himself does not understand the real influence of climate and weather on his operation. Users often approach the applied climatologist with the request for a certain piece of climatic information; in other words, they present what they think are their weather problems rather than just outlining their total operational program. In many cases, a more detailed study of the operation reveals an entirely different weather or climate problem which is more significant to the user's operations although unsuspected by him.

The applied climatologist must put himself in the role of a diagnostician who does not accept the statement of the ills of the user but who studies and probes in detail, until he is able to uncover the real ills for himself. In doing so, the applied climatologist (1) may come up with other, more significant problems which must be solved and (2) may be able to frame the problem in such a way that a helpful solution within the limits of available, supplied, interpolated, or derived data can be achieved. The applied climatologist must always be ready to remove himself from his possibly more familiar frame of

reference as a forecaster or an atmospheric scientist and involve himself in the
details of the user's operations.

ORGANIZATION FOR PROBLEM SOLVING

Increasing requirements for the application of climatic information to both
military and civilian problems have resulted in the need for functional
agencies capable of accepting many problems of a diverse nature, rapidly
determining the proper meteorological information needed to solve the
problem, and providing answers in terms of the problem itself that can be
clearly understood and immediately used by the initial requester.

The Air Weather Service has established a Climatic Center, USAF
(1210th Weather Squadron), as its climatic analysis agency (U.S. Air Force,
1961). This unique group has the flexibility to investigate a wide range of
climatological problems and the capability to apply almost immediately,
climatic data to the solution of military planning and operational problems.
When requests for information are received by the Climatic Center, they pass
through a number of definite steps or stages within the Center on their way
toward analysis and solution. Let us consider what might happen to a
hypothetical request for information about the frequency of ice storms at
northern coastal stations made by the U.S. Army Engineers.[1] The Engineers'
request is prompted by the need to design and construct power transmission
lines at these stations.

The request appears fairly routine, although a conference between the
requester and the consultant group is desirable to outline the problem in
more detail and to fix the limits of the study. The conference clarifies the
need for information on both the frequency and the amount of ice
accumulation on wires in terms of radial thickness and weight in order to aid
in the design and construction of the power lines. The consulting climatologists point out that ice accumulation is possible under conditions of
freezing rain as well as with long-duration fog or low clouds under the proper
temperature and wind conditions. Such a conference establishes the extent of
the study, informs the user of the nature of the parameters to be investigated,
and provides sufficient liaison to ensure that the final product is directly
applicable to the user's problem and couched in terms that can be
immediately understandable.

A routine search of the literature and available data sources reveals the
existence of only hourly weather data. There are no existing methods for
estimating ice accumulation from available data in a manner directly usable
by the Engineers. Thus a team made up of consultants and analysts must be
formed to develop an estimation procedure. Once this is successfully
completed, a detailed data survey is undertaken, and data in the required
format are requested from the Joint Punched Card Library in Asheville,
North Carolina. When these data become available, the estimation procedure
is used to determine the thickness and weight of ice that would have
accumulated in past years on transmission wires. This provides estimates of

[1]   This material is adapted from an example in the AWSM 105-3 with permission.

frequencies and design values for various life expectancies in the future. The final step is the preparation of a report not in meteorological terms but in terms of the problem itself (transmission line design), its review, and transmission to the requester.

CATEGORIES OF PROBLEMS

Landsberg and Jacobs (1951) have suggested four major classes of problems in applied climatology. They are

1   The design and specification of equipment or material. This is the simplest type of problem in applied climatology for the needed data are usually the frequencies of severe (but possibly less than extreme) meteorological events. The problem might be the design of a storm water runoff culvert, which would require information on storm frequencies and intensities of precipitation, or it might be the specification of the type of clothes for individuals working in a limiting environment, which would require information on the nature of the limiting conditions.

2   The location and use of a facility or equipment, for example, a manufacturing or hydroelectric facility. Generally, several meteorological factors are involved and more than just frequencies of events must be provided to allow the most reasonable choice among several alternatives. Usually, climatic factors are not the only factors involved in the selection of the location, and so the analysis will have to set forth the various possibilities presented by the climatic environment to permit evaluation of different courses of action.

3   The planning of an operation. These more complicated problems in applied climatology usually make the greatest demands on the applied climatologist. Once the industrial plant site has been located (item 2 above), another series of weather-related problems must be considered as plans are made for the construction of the buildings, and later, for the operation of the plant under different conditions through the year. Selecting the time of year for construction so that weather delays are minimized, specifying the type of building materials and insulation so that heat exchanges are controlled, selecting the construction materials so that weathering and air pollution effects will be least damaging once the plant is built—these and many more concerns must involve the applied climatologist in the planning of an operation.

Even after construction is completed and production begins, weather-related problems involving the transportation of raw materials to the plant and finished products from the plant, the health, morale and absenteeism of workers, the availability of water supplies for production or for dilution of wastes, the occurrence of limiting air pollution episodes, the deterioration of stockpiled materials, all must involve the applied climatologist in the operation. Simple weather summaries are no longer helpful since many complex interrelationships are involved. Many separate studies involving the influence of weather or climate on materials, humans, transportation, and water supplies are necessary. The climatic data are usually expressed in terms of probabilities, but, since a number of elements influence each phase of the operation, it is necessary to provide some composite frequency analysis based generally on the techniques of synoptic climatology (Jacobs, 1947).

4   Climatic influences on biologic activities. Even the fairly well-defined problem of climate in relation to crop development and production involves not only the climatic variables of temperature, radiation, duration of sunlight, evapotranspiration, precipitation, and soil moisture storage, but also biologic variables such as seed selection, fertility, timing of the various stages in plant development, resistance to disease, and food value or palatability of the final product. Secondary influences of climate on plant diseases, fungi, molds, and insect pests must also be considered.

More complicated problems such as the influence of climate on human health and comfort are only just beginning to be studied with the fairly limited data and analytical techniques at hand. Human response to weather conditions is so varied and poorly understood that the applied climatologist who seeks to investigate these relationships is faced at every turn with a myriad of problems.

The range of applications under each of these four categories is almost limitless and the techniques that the climatologist must use to solve them are certainly not well understood at the present time.

Recognizing that the three factors of climate, space, and time can all enter into problems in applied climatology, these same authors have provided a concise table (Table 1-4) which shows possible combinations of these three factors.

The authors suggest problems that are fairly typical to illustrate each of the above eight classes. Some of these are listed in Table 1-5 along with

TABLE 1-4

Combinations of parameters in climatic problems*

SPACE	TIME	CLIMATE
Single point	Simple series	Single element Multiple elements
	Complex relation	Single element Multiple elements
Multiple point or area	Simple series	Single element Multiple elements
	Complex relation	Single element Multiple elements

* With permission of the American Meteorological Society, from Landsberg and Jacobs, 1951, table IV, p. 981.

TABLE 1-5

Examples of problems in applied climatology*

PARAMETERS	TYPICAL PROBLEMS
Single point, simple time series, single climatic element	Insurance against adverse weather (hail insurance) Probability of wind damage
Single point, simple time series, multiple climatic elements	Clothing and housing designs and specifications Airport construction for optimum use Plant or house site selection
Single point, complex time relation, single climatic element	Design of heating system for melting snow on pavement Evaluation of frost hazard to tender fruits and plants Design of window surfaces, blinds, and awnings for protection against solar radiation
Single point, complex time relation, multiple climatic elements	Deterioration of materials under weather influences Design of irrigation layout Air pollution control work at particular point
Multiple point, simple time relation, single climatic element	Distributing and marketing "weather goods" on regional or national scale Demand for electrical power in large geographic areas under single, sudden weather event such as incoming cloud deck or thunderstorm activity
Multiple point, simple time relation, multiple climatic elements	Planning aerial mapping of large geographic areas Use of the synoptic-climatic methods of weather forecasting (objective forecasting)
Multiple point, complex time relation, single climatic element	Probable maximum floodwater runoff from areal patterns of intensity duration of rainfall Assessing frost hazard or frost damage over large geographic area of diverse topography
Multiple point, complex time relation, multiple climatic elements	Agricultural land usage in relation to climate Fire-weather hazards to forests over large geographic area Establishment of time schedule for aircraft flights optimally adapted to weather conditions Air pollution control work for city or area

* With permission of the American Meteorological Society, adapted from Landsberg and Jacobs, 1951, pp. 981–988.

additional examples to indicate a few of the varied problems that have climatic influences.

Techniques used in solving applied climatology problems are as many and varied as the problems to be solved. The technique used may involve some fairly simple summary of data, the expression of probabilities or likelihoods of certain events happening, or the computation of a correlation coefficient with a regression equation that can be used to predict one variable from knowledge of another. However, as the problems to be solved become more theoretical or the number of variables involved increase, it may become necessary to originate different problem-solving techniques. Here, the experience of the applied climatologist and the fertility of his imagination set the only limits to techniques that may be applied. But knowing the proper technique to utilize is not always the only problem. As we have seen, having the right tools with which to start the analysis is sometimes a difficult problem. Before the climatologist can begin to apply a problem-solving technique, he must often first collect, synthesize, interpolate, or otherwise derive the basic information, the climatic data.

BASIC CLIMATIC ELEMENTS 2

This book is concerned with the application of climatic data to fields such as agriculture, industry, health, clothing, and architecture. A full discussion of the various weather elements, their distribution, and measurement, is beyond our scope. The reader is referred to more basic texts in meteorology or climatology (Sellers, 1965; Longley, 1970; Landsberg, 1960; for example) or in instrumentation (Middleton and Spilhaus, 1953) for fuller discussions of the climatic elements and their measurement.

   For those with little background in meteorology or climatology, however, some general discussion of the more common climatic elements—radiation, temperature, atmospheric moisture and precipitation, evaporation, and wind—would be helpful in understanding later sections. Thus this chapter serves as a brief introduction to the major climatic elements while not attempting to be complete or definitive. It should provide, at least, some of the background needed for an understanding of the applications of the climatic elements to follow.

RADIATION

1 Basic Relations
Radiation, conduction, and convection are the three avenues by which energy can be transferred from one place to another. Radiation transfer occurs with

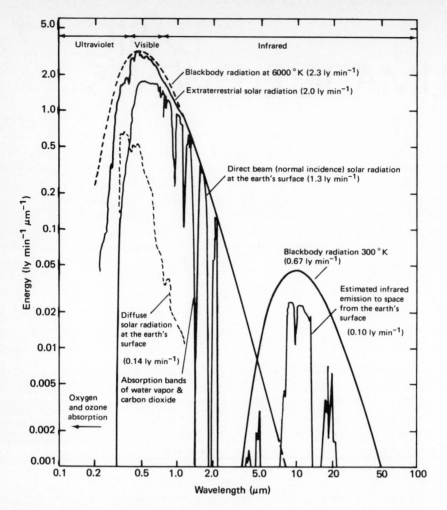

Spectra of solar and terrestrial radiation. Note use of logarithmic scale for   FIG. 2-1
both energy and wavelength. Blackbody radiation at 6000 K has been
reduced by square of ratio of sun's radius to average earth-sun distance in
order to provide flux incident on top of atmosphere. *(Copyright 1965, by the
University of Chicago Press, and reproduced with permission from Sellers,
1965, Fig. 6, p. 20.)*

the speed of light and can take place without the presence of matter between
the radiator and the receptor. In contrast, both convection and conduction
result in a much slower transfer of energy and require the presence of some
intermediate substances such as air. Radiation transfers energy by means of
electromagnetic waves having an extremely wide range of wavelengths, from
the very high-frequency cosmic waves—wavelength 10^{-9} μm (1 μm = 10^{-6}
m)—to the low-frequency alternating current—wavelength 3×10^6 μm.

　　As far as applied climatology is concerned, the wavelengths of
importance are those in which the sun, the earth, and the atmosphere radiate.

Solar radiation occurs in the so-called "short" wavelengths (99 percent in the range 0.1 to 4.0 μm) while radiation of terrestrial or atmospheric origin is in the so-called "long" wavelengths (99 percent in the range from 4.0 to about 50 μm) (Fig. 2-1). There is essentially no overlap in wavelengths between solar and terrestrial radiation.

Radiation is most conveniently measured in terms of the gram calorie or the British thermal unit. The gram calorie (cal) is the amount of heat necessary to raise the temperature of a cubic centimeter of water one degree Celsius from 14.5° to 15.5°C (also known as the 15°C g cal). Radiation in the cgs system is usually expressed in calories per square centimeter per some period of time. The unit cal/cm^2 is also defined as a langley (ly). The British thermal unit (Btu) is the amount of heat needed to raise the temperature of one pound of water one degree Fahrenheit. In English units, radiation rates are usually given in Btu/ft^2 hr. For conversion purposes 252 g cal = 1 Btu, or (ly/min) $\times$ 221 = Btu/ft^2 hr.

All substances having any heat will radiate at a rate proportional to the fourth power of their absolute temperature. A so-called "blackbody" (a body that absorbs perfectly all wavelengths falling on its surface) will also radiate perfectly and can, thus, provide a theoretical maximum rate of energy radiated:

$$F = \sigma K^4$$

where F = radiation flux per unit time and area from a blackbody
$\quad$ σ = Stefan-Boltzmann constant (8.132 $\times$ 10^{-11} ly/K^4 min)
$\quad$ K = absolute temperature (°C + 273) of the blackbody

A blackbody at room temperature (294 K) radiates 0.61 ly/min while the sun at 6000 K radiates 10^5 ly/min. A doubling of the absolute temperature, of course, will increase the radiation flux by 16 times.

Wien has shown that the wavelength of maximum flux from a blackbody (λ_{max}) depends only on temperature (λ_{max} = 2,897/K). The wavelength of maximum emission from the sun is about 0.5 μm. Using the expression, the mean absolute temperature of the sun is found to be approximately 6000 K.

Most bodies absorb only a percentage of the radiant energy falling on them. This percentage, known as the "absorptivity," varies from 0 for a perfect reflector to 1 for a perfect absorber (or blackbody). The sum of the fraction reflected (the reflectivity) and the absorptivity must equal 1 for opaque materials.

Different materials vary in their ability to absorb different wavelengths. A material may absorb well in one wavelength but poorly in another (e.g., new snow absorbs poorly in the short wavelengths but is a good absorber of the long or terrestrial wavelengths; iron absorbs short wavelengths well but not the longer wavelengths; moist earth is a good absorber of both long and short wavelengths). Thus, absorptivity and reflectivity can be defined for a particular wavelength, a portion of the spectrum, or for all-wave radiation.

The term "albedo" is usually defined as the fraction or percent of the incoming solar radiation that is reflected from a surface. Thus it usually refers to the reflectivity of a particular band or portion of the spectrum. The radiation that is reflected does not enter into the heat balance of the body. The albedo of the whole earth and atmospheric system approximates 35 percent, while for particular surfaces it has quite widely varying values, some of which are included in Table 2-1.

A number of substances are transparent (transmissivity is defined as the fraction of incident radiation that is passed through an object) to certain wavelengths and yet will absorb other wavelengths. Glass is one such material; crown glass readily transmits radiation of 0.3 to 2.0 μm wavelengths while quartz is transparent to wavelengths between 0.18 and 4.0 μm. Thus solar radiation (short-wave) will pass through these substances readily, but the long-wave or terrestrial radiation will be absorbed by these materials. Polyethylene plastic material (0.1 mm thick) is relatively transparent in all wavelengths except for narrow bands at 3.5, 6.9, and 14 μm. Thus it makes a good material for use with radiation instruments that are designed for all wavelength use.

All natural bodies will absorb and later emit a portion of the energy falling on them. Emissivity and absorptivity are directly related; a loose interpretation of the Kirchhoff law suggests that good radiators are good absorbers of energy. Without proper modification, this can cause confusion for we have just seen that snow is a poor absorber at short wavelengths but a good absorber (and, hence, a good emitter) at long wavelengths. The particular wavelength must be specified for proper application of the Kirchhoff law; bodies that are good absorbers at a particular wavelength are also good emitters at that wavelength and, similarly, poor absorbers at particular wavelengths are poor emitters. Since most bodies on the earth's surface are absorbing short-wave energy from the sun and emitting much longer-wave terrestrial radiation, absorptivity and emissivity are often not similar (unless the body is in thermal equilibrium with all of its surroundings). For example, short-wave absorptance of plowed ground is 0.75 to 0.80 ly/min while long-wave emittance approximates 0.90 ly/min; absorptance for high, dry grass and oak woodland is 0.68 and 0.82 ly/min, respectively, while emittance for both surfaces approximates 0.90 ly/min (Brooks, 1959).

Since radiation from a point source, such as a light or the sun, spreads out uniformly in all directions (assuming a homogeneous, transparent medium around the source), the flux through a unit area on a sphere around

TABLE 2-1

Values of albedo of selected surfaces

SURFACE	ALBEDO, %	SURFACE	ALBEDO, %
Evergreen forests	7	Dry pastures	19
Green fields	11	Clouds	55
Sand	13–18	Snow	80–90

the source must be the same everywhere on that sphere. Spheres with greater radii have greater surface areas, so the flux of radiation through a unit area decreases the farther the surface of the sphere is from the source. The surface area of a sphere is proportional to the square of the radius, so the intensity of radiation or the flux through any unit area will decrease with the square of the radius from the source. Variations in radiation received from the sun are of the order of ± 3 percent because of the elliptical path the earth makes in its annual course around the sun.

When a surface is perpendicular to the beam of radiation, the surface will receive an amount of energy equal to the radiant flux density. However, if the surface is at some angle to the beam of radiation, the surface receives less energy because the beam will be spread over a larger surface area. The actual flux density on a surface not perpendicular to the beam is related to the flux density on a perpendicular surface by the cosine of the angle of incidence of the beam (the angle between the beam and a perpendicular to the surface). The relation can be written

$$I_s = I_0 \cos \beta$$

where I_s = flux density on given surface
 I_0 = flux density perpendicular to surface
 β = angle of incidence of beam (from the perpendicular)

This relation is basic in any discussion of the radiation reaching the earth's surface; due to the spherical shape of the earth, the more poleward surfaces are increasingly inclined away from the perpendicular to the solar beam and, hence, receive smaller and smaller amounts of flux density.

Space does not permit any real discussion of the geometry of the earth's orbit about the sun or the effect of the constant inclination of the polar axis to the plane of revolution. These two factors, however, result in the apparent motion of the sun from $23\frac{1}{2}°$S to $23\frac{1}{2}°$N during the course of the year. (It is the earth, not the sun, that moves but earth-centered people have a tendency to think of the sun moving from Tropic to Tropic every 6 months—the solstices—and being vertically over the equator twice each year—the equinoxes.) As a result of this apparent movement of the sun, the angle with which the solar beam falls on any place on the earth's surface will vary through the day and year as will the radiant flux density. Added to this is the fact that few surfaces are absolutely horizontal, and it is clear that the mathematics of determining the exact flux density on any reasonably sized, undulating surface with a partial cover of vegetation or buildings becomes quite complex. Great differences in radiation receipt can be found between north- and south-facing slopes at the same latitude because of the angle which the sloping surface may make to the solar beam. For example, a south-facing slope on the Tropic of Cancer, $23\frac{1}{2}°$N, having an inclination of $23\frac{1}{2}°$, with the sun over the equator, will receive the sun's rays perpendicularly (Fig. 2-2). The same slope facing to the north will have the sun's rays coming

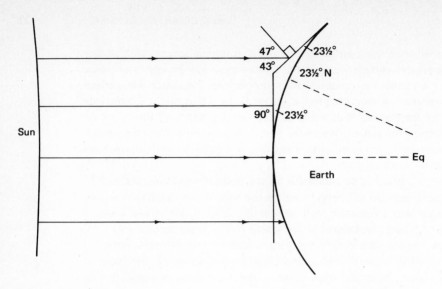

Angle of incidence of solar beam on sloping land surfaces at Tropic of Cancer   FIG. 2-2
$(23\frac{1}{2}°N)$ during equinox.

in at an angle of 43° to the surface or 47° to the perpendicular to the surface. The perpendicular flux density will have to be multiplied by cos β or 0.68. The flux density on the north-facing slope will be about two-thirds of what it is on the south-facing slope. At higher latitudes even greater differences will occur.

2 Instruments to Measure Radiation
Miller (1968, pp. 56–57) has prepared a simple ledger by means of which all components of the radiation balance can be determined. Not all components need be measured; knowing two in any row or column permits determination of the third. The ledger may use either instantaneous, daily, or monthly values (Table 2-2).

Combined values of direct and diffuse solar radiation are often available from pyranometers (the most well-known type in the United States is the Eppley pyranometer), while reflected solar radiation can be measured by an inverted pyranometer. The Eppley pyranometer consists of a thermopile mounted within a glass hemisphere approximately 75 mm in diameter. The thermopile is located immediately beneath pie-shaped disks alternately covered with Parson's optical matt black lacquer and magnesium oxide. The temperature difference between the black and white surfaces generates a current that is related to the energy received directly or diffusely from the sun. Use of a glass hemisphere over the thermopile eliminates the effect of long-wave atmospheric radiation as well as wind or water on the sensor itself. Dust or water on the glass hemisphere will, of course, result in some error.

Earlier models of the Eppley with the thermopile encased in a glass sphere developed problems when the instrument was inverted for use in albedo measurements. Convective currents were established within the sphere, resulting in measurement errors of 4 to 5 percent. The present version of the Eppley pyranometer has largely eliminated this problem.

A pyrheliometer is actually an instrument used to measure just the direct-beam solar radiation while a pyranometer measures both direct and diffuse solar radiation. Over the years, names have been loosely used and the word "pyrheliometer" is sometimes applied to both types of instruments.

A large number of instruments are available to measure net all-wave radiation, reflecting the importance of this quantity in many areas of climatology. The Thornthwaite Associates net radiometer consists of a small blackened disk about 50 mm in diameter with an internal thermopile. The disk is exposed within two 2-mil polyethylene hemispheres mounted on aluminum rings. The hemispheres prevent wind and rain from influencing the readings. Dry air is periodically pumped into the polyethylene sphere, which is inflated to 140 mb pressure to maintain its shape and to keep condensation from forming inside. Response time is fast with a time constant of about 2 sec.

The Schulze radiometer balance meter consists of two total hemispherical radiometers placed back to back. This permits the separate measurement of the downward and upward radiation fluxes. Net radiation can be measured directly if the two internal thermopiles are connected in opposition. Use of a pyranometer with this instrument makes it possible to obtain both the short- and long-wave components of the radiation balance both night and day.

There are several so-called "economical" net radiometers (Suomi and Kuhn, 1958; Lowry, 1957) in which the temperature difference across a small flat plate is read by mercury thermometers. The plates can be protected from wind and moisture by use of polyethylene sheets or bags.

Special instruments to measure long-wave radiation are not entirely necessary; at night, only long-wave radiation occurs and any all-wave radiometer will provide values of the long-wave radiation while, during daytime, the long-wave component may be obtained by subtraction of the short-wave radiation value from the all-wave hemispherical radiation total.

TABLE 2-2

Simple ledger for evaluating components of radiation balance

DIRECTION	SHORT-WAVE RADIATION	LONG-WAVE RADIATION	ALL-WAVE RADIATION
Downward ↓	Direct and diffuse solar	Atmospheric	Total downward
Upward ↑	Reflected solar	Terrestrial	Total upward
Sum	Net short wave	Net long wave	Net all wave

The intensity of direct solar radiation on a clear day depends on latitude and on the amount of surface presented to the direct solar beam. Outside the earth's atmosphere, the receipt of solar energy approximates 2 cal/cm² min (the so-called "solar constant"). Somewhat less than one-half this amount actually passes through the atmosphere to reach the surface annually over the earth as a whole. The actual percentage varies markedly with cloudiness, turbidity of the atmosphere, moisture content, sun angle (thickness of the atmosphere), and other factors. Table 2-3 gives some actual values of radiation received at selected stations. The percentage actually received at the surface at these stations varies from 38 to 66 percent of that theoretically possible (outside the atmosphere) and is strongly influenced by the "clearness" of the atmosphere.

Landsberg (1961) has prepared a map (Fig. 2-3) of the mean global solar radiation, both the direct and diffuse radiation, falling on a horizontal plane at the earth's surface. Along any given meridian some 1440 cal/cm² day are available on a surface perpendicular to the solar beam located outside the atmosphere. At the earth's surface in the equatorial belt approximately 150 kcal/cm² year are received—about 30 percent of the total possible (Table 2-4). Toward the poles the annual percentage decreases as (1) the thickness of the atmosphere through which the solar beam must pass increases and (2) the surface of the earth becomes less perpendicular to the solar beam. (In the solstice months, there is, of course, an increase toward the pole in the summer hemisphere since the day is 24 hr long and the opportunity to receive radiation is maximized).

Landsberg (1960) has provided a useful balance sheet of the average annual short- and long-wave radiation exchanges at the top of the atmosphere, within the atmosphere, and at the surface of the earth (Fig. 2-4). These are only approximate figures; detailed observational networks are not available worldwide. These figures should be considered from a worldwide, annual point of view; they clearly have little significance for a particular place or day.

TABLE 2-3

Observed radiation sums at various places (in ly)*

AREA	LATITUDE NORTH	MEAN DAILY	ANNUAL (× 1000)	AS PERCENT OF RADIATION RECEIVED OUTSIDE OF ATMOSPHERE
San Juan, P.R.	18.5	527	191	64
Honolulu, Hawaii	21.3	513	187	64
New Orleans, La.	29.9	336	126	46
Albuquerque, N. Mex.	35.0	471	171	66
Columbia, Mo.	40.0	369	142	58
Karlsruhe, Germany	49.1	289	106	49
Pavlovsk, USSR	59.7	183	67	38
Fairbanks, Alaska	64.8	222	81	50

* With permission of the Gray Printing Co., from Landsberg, 1960, table 43, p. 129.

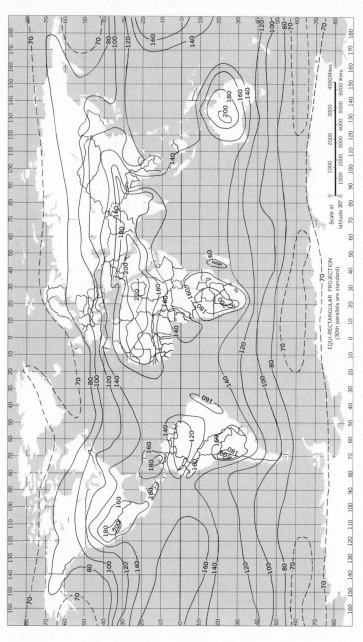

Generalized isolines of annual global radiation on a horizontal surface at the ground. The units are kg cal/cm² year. (Reprinted with permission from Landsberg, 1961, fig. 1, p. 96, copyright, Pergamon Press; also with permission of McGraw-Hill Book Company, from Trewartha, 1968, fig. 1.9.)

FIG. 2-3

If we consider the energy available outside the atmosphere (134×10^{22} cal/year) as 100 percent, we can compare all the other radiation exchanges to this figure. Some 22 percent of this amount (29×10^{22} cal/year) is absorbed within the atmosphere by the clouds, water vapor, ozone, dust, and other constituents of the atmosphere and about 23 percent (31×10^{22} cal/year) comes directly through to the earth's surface. There is appreciable scattering within the atmosphere as well as reflection from cloud and ground surfaces, the downward diffuse radiation equaling about 22 percent (30×10^{22} cal/year) of the value outside the atmosphere and the upward scattering and reflection equaling 33 percent (44×10^{22} cal/year).

The long-wave radiation balance consists of energy flows from the earth to the atmosphere and space and from the atmosphere to the earth as well as to space. Radiation from the earth to the atmosphere (138×10^{22} cal/year or 103 percent of the amount received outside the atmosphere) continues both day and night from the whole earth. It can, therefore, total somewhat more than the energy received from the sun. The atmosphere, in turn, radiates long-wave energy both to the earth's surface (102 percent or 136×10^{22} cal/year) and to space (53 percent or 71×10^{22} cal/year). Finally about 19×10^{22} cal/year pass directly through the atmosphere from earth to space (14 percent of the total received outside the atmosphere).

These radiation totals are not entirely in balance but then the quantities of heat transferred by evaporation and conduction have not been considered. Riehl (1972), based on some new figures by London, suggests that some $40 \times$

TABLE 2-4

Latitudinal distribution of solar energy received on a horizontal plane at the earth's surface in kcal/cm^2 year*

LATITUDINAL ZONE	MEAN	RANGE
90–80°N	65	
80–70	70	68–72
70–60	72	68–85
60–50	85	68–110
50–40	117	68–140
40–30	135	90–200
30–20	159	115–220
20–10	152	120–220
10–0	148	120–170
0–10°S	145	120–180
10–20	150	130–190
20–30	145	120–200
30–40	123	110–170
40–50	100	80–135
50–60	76	65–95
60–70	70	65–75
70–80	72	
80–90	70	

* Reprinted with permission from Landsberg, 1961, table 1,
p. 97, copyright by Pergamon Press.

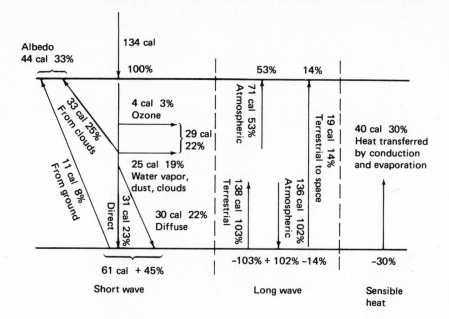

Average annual radiation (and sensible heat) balance of earth and atmosphere   FIG. 2-4
in percent of energy received outside of atmosphere and in 10^{22} cal/year. (*By
permission of the Gray Printing Co., after Landsberg, 1960, table 38, p.
125, modified by data from Riehl, 1972, used with permission of the
McGraw-Hill Book Company.*)

10^{22} cal/year or 30 percent of the incoming solar radiation is transferred
from the ground to the atmosphere over the globe annually by conduction
and evaporation. As a result the receipts and losses of heat at the top of the
atmosphere, within the atmosphere, and at the ground surface all balance.

Various authors (Budyko, 1963; Gates, 1962; Sellers, 1965; Reifsnyder
and Lull, 1965; among others) have prepared maps of the distribution of mean
daily or monthly solar radiation for the world or for different continental
areas similar to Fig. 2-5. Note the seasonal differences in the patterns shown,
for example, in the United States. Latitudinal differences are quite marked in
the late fall and winter, with lines of equal radiation generally oriented
east-west; in summer the isohels of solar radiation across the country show a
predominantly north-south orientation.

Some maps of the distribution of net radiation exist (Gates, 1962)
utilizing data provided largely by Budyko (1955, 1956). These maps are
based on only a very limited amount of actual measured data obtained from a
variety of sensors with different operating characteristics.

Maps of the distribution of sunshine hours (sometimes called "duration
of bright sunshine") are also available and useful in certain studies. The data
can be obtained from relatively simple sunshine recorders. These instruments
indicate the time during which the radiation from the sun is intense enough

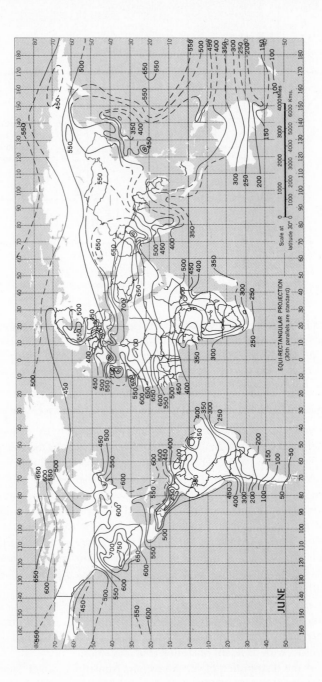

JUNE

EQUI-RECTANGULAR PROJECTION
(30th parallels are standard)

Scale at
latitude 30°

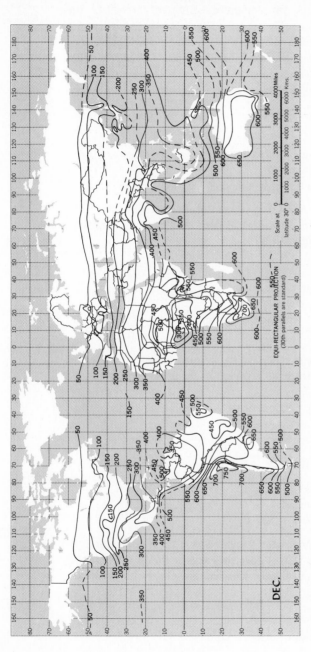

Mean daily solar radiation (ly), June and December. (*From Löf, Duffie, and Smith, 1966, with permission University of Wisconsin Engineering Experiment Station and McGraw-Hill Book Company, from Trewartha, 1968, figs. 1.10, 1.11.*)

FIG. 2-5

to activate a recording device. The great seasonal differences in sunshine hours as well as the areal variation within a particular season are both significant factors in many aspects of applied climatology.

A number of expressions for computing different factors of the radiation exchange from more readily available data have been suggested in order to increase our knowledge of their geographical distribution. For example, Fritz and MacDonald (1949) have suggested the following relation to determine the amount of energy received on any day based on the amount of energy received on a clear day and the ratio of the actual hours of sunshine to those theoretically possible:

$$R = R_0 (0.35 + 0.61S)$$

where R, R_0 = amount of energy received at surface on any day and on a clear day, respectively

S = ratio of sunshine hours described above

The average error of estimating R is 4 to 5 percent of the observed value.

A more basic form of this relation had been developed by Ångström earlier (1922, modified by Savinov, 1933).

$$R = R_0 [1 - (1 - k) C]$$

where R, R_0 = meanings as given above

C = amount of cloud cover in tenths

k = factor that varies with latitude with values of 0.47 at 70°, 0.39 at 60°, 0.32 at 50°, 0.30 at 40°, 0.29 at 20°, and 0.32 at 0°

The value of k is related to cloud type, and the latitudinal differences reflect the fact that some broad latitudinal differences in cloud types can be identified.

Swinbank (1963), using measurements in Australia and the Indian Ocean, suggested that the downward long-wave radiation (clear sky) could be estimated from data of the surface temperature alone with the relation

$$I_0 = \epsilon (0.245 - 0.214 \sigma T^4) \quad \text{ly/min}$$

where ϵ = emissivity of the surface

σ = Stefan-Boltzmann constant

T = absolute air temperature, screen height

Various forms of this equation have been suggested by other investigators.

Selirio, Brown, and King (1971) have reviewed methods of estimating net and solar radiation. For net radiation, they employ a relation of the form

$$R_n = a + bR_s$$

where R_n = net radiation

R_s = direct and diffuse short-wave radiation

a, b = constants

Based on 5 years of data over grass at Guelph, Canada, they found that this relation could be used to estimate R_n satisfactorily for spring, summer, and fall periods; no relationship could be found for the winter period. The relations developed were

Spring: $R_n = 9.6 + 0.49\,R_s$
Summer: $R_n = 7.3 + 0.52\,R_s$
Fall: $R_n = -16.0 + 0.47\,R_s$

When the data for the three seasons were averaged together, they obtained values of $a = -18$ and $b = 0.54$. These values compared quite favorably with the results of others using data from different parts of the world, as shown in Table 2-5.

For estimating the ratio of the direct and diffuse short-wave radiation (R_s) to the total solar radiation at the top of the atmosphere (R_A), these same authors use a relation of the form

$$\frac{R_s}{R_A} = a + b\,\frac{n}{N}$$

where n = hours of bright sunshine
N = daylength

The values of the constants a and b in this relation are found to be 0.23 and 0.57, respectively, at Guelph. Again these values are consistent with the results of others, as shown in Table 2-6 adapted from Selirio et al., 1971.

4 Limitations

Major problems in the field of radiation exchange result from (1) the lack of extensive observation networks, (2) the use of different sensors with different operating characteristics from one place to another, (3) the great variability of the data over just short distances due to cloud or turbidity conditions, and (4) the influence of the spectral characteristics of the plastic or glass used in many radiation sensors and their possible change with time or exposure. Other problems clearly exist, but these four provide significant reasons for

TABLE 2-5

Summary of regression of net radiation (R_n) on global
solar radiation (R_s) from various investigators*

SOURCE	LOCATION	SURFACE	b	a
Davies, 1967	West Africa	Grass, sand	0.61	−28
	Worldwide	Various	0.62	−24
Davies & Buttimor, 1969	Ontario	Crops	0.56	−33
Turner, 1965	Australia	Grass	0.65	13
		Tussock	0.56	4
		Intergrove, etc.	0.46	−16

* With permission of the Agricultural Institute of Canada, from Selirio, Brown, and King, 1971, p. 38.

some of the uncertainty in radiation. Since radiation is a fundamental climatic parameter, it is clear that we need to expand our global coverage and to provide more standardized instruments and exposures. Radiation instruments now being carried on satellites certainly promise a marked increase in the areal extent of our radiation data in the next few years.

Reifsnyder and Lull (1965), in a summary of energy relations in forests, explore the questions of what to measure, where to measure, when to measure, and how to measure. They feel that a lack of real answers to these questions has resulted in many of the limitations in present radiation work. Sometimes the wrong fluxes have been measured at the wrong place, resulting in wasted effort. In some situations, not all wavelengths need to be measured; collecting information on all wavelengths masks the very important influence of particular wavelengths. The measuring instrument is introduced into the environment being sampled, but often not at the particular surface or level of interest. It must be assumed, however, that the sensor does not interfere in any way with the radiative flux, that the actual surface being sampled is completely uniform in terms of radiation characteristics, that there is no absorption or emission of radiation between the sensor and actual surface of interest, and that the variation of absorptivity with the angle of incidence of a sensor is the same as the angular variation of emissivity of the surface being viewed.

Certain of these points are related to the height of the sensor and the so-called "view factor" of the radiometer. The greater the distance from sensor to surface, the greater the area from which most of the energy falling on the sensor comes. Reifsnyder and Lull (1965) have given a simple diagram of the relation between the view factor and the height of the sensor (Fig. 2-6). A flat plate sensor 3 m above a surface receives 95 percent of its energy from a surface area 27 m in diameter.

All objects radiate, and if the temperature and radiation characteristics of the objects in the local environment differ, the total radiation impinging on a surface within that environment will be made up of inputs from many different sources. Hemispherical radiometers integrate all these individual influences, while directional radiometers provide information on input from individual or restricted sources. Decisions on the selection of the instrument

TABLE 2-6

Summary of values of a and b used in the expression
$R_S/R_A = a + b\,(n/N)$ from various investigators*

SOURCE	LOCATION	DATA	b	a
Baier & Robertson, 1965	Ottawa & Edmonton	Daily	0.62	0.25
Smith, 1959	Trinidad	Daily	0.59	0.20
Black et al., 1954	Worldwide	Monthly	0.48	0.23
Davies, 1965	West Africa	Monthly	0.60	0.19
Penman, 1948	Rothamsted	Monthly	0.55	0.18

* *With permission of the Agricultural Institute of Canada, from Selirio, Brown, and King, 1971, p. 38.*

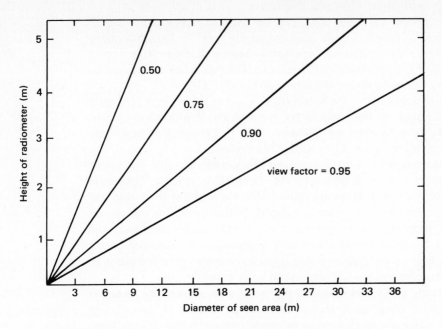

View factor of a radiometer in relation to height. (*With permission U.S.*   FIG. 2-6
Department of Agriculture, Forest Service, from Reifsnyder and Lull, 1965,
fig. 18, *p.* 43.)

to use or the data to be collected, thus, must result from careful analysis of
each particular measurement problem in the light of the special characteristics
of the radiant energy exchanges.

TEMPERATURE

1 Basic Relations

Temperature, next to precipitation, is probably the most talked about
weather element, yet unlike precipitation it is an element which most laymen
have difficulty defining. Even meteorologists question what is meant by "the
air temperature" since air temperature varies from one side of a building to
another, from shaded to nonshaded areas, from grassed to plowed fields or
paved roads. In measuring air temperature we can only hope to obtain a value
which represents some averaged value of the temperature condition of a
heterogeneous mixture of air.

Temperature, to the physicist, can be defined in terms of the movement
of molecules—the more rapid the movement, the higher the temperature. Or
it can be defined in relative terms on the basis of the degree of heat a body
has. Heat moves from a body having a higher temperature to a body with a
lower temperature. The temperature at the earth's surface, determined by the
balance between incoming and outgoing radiation and its transformation into

sensible and latent heat at the earth-air interface, is directly related to the temperature in the layers of air near the surface.

Various scales have been used, over the years, to express temperatures. The Fahrenheit scale has been most frequently used in English-speaking countries. There is now, however, an effort to express all temperature records on the basis of the centigrade or Celsius scale. The Fahrenheit records can be converted to Celsius by means of the relation $C = (F - 32)\frac{5}{9}$.

In some applications, the Kelvin (K) or absolute temperature ($^\circ$A) scale is used. The zero on this scale is the temperature at which a gas would theoretically cease to exert any pressure, -273°C. Centigrade temperatures may be converted to $^\circ$A (or K) by adding 273 to them.

Air temperature is usually measured in a wooden-louvered instrument shelter, which, in most English-speaking countries, is called a "Stevenson screen." There are definite specifications about the height of the screen and its method of exposure to ensure comparable observations from one station to another. In most cases, temperature data are obtained hourly through the day or just the maximum and minimum temperatures are recorded once a day. From these records, the first step is usually to obtain some indication of the daily mean temperature and the diurnal range of temperature.

Landsberg (1960) has discussed the use of such temperature factors as the interdiurnal temperature change (the change in mean air temperature from day to day) and its relation to human responses. He also utilizes "frost days" (daily minimum air temperature less than 0°C), "ice days" (daily maximum less than 0°C), "summer days" (daily maximum greater than 25°C or 77°F), and "tropical days" (daily maximum greater than 30°C or 86°F) in studies of the temperature distribution at particular places. Frequencies of extremes of air temperature as well as the number of consecutive days with air temperatures below or above certain limiting values may all be of significance in applied climatology.

Both surface and atmospheric conditions must be known if air temperature distributions are to be understood. The heat capacity (equals ρC where ρ is the density and C is the specific heat) of water is about five times greater than that of soil. Thus it will take five times more energy to heat a volume of water to the same degree as a similar volume of soil, or put differently, if the same amount of energy is incident on a soil and water surface, the soil surface will heat appreciably more than will the water surface. At the same time, convective currents will be established within the water body so that the heated surface water will be carried downward to mix with lower water layers. Since water is more transparent than soil, the sun's rays will penetrate to a greater depth and heat will be absorbed throughout a deeper profile of water than of soil. All these influences work together to produce more rapid and intense heating at a soil surface than at a water surface. There is a more immediate response to radiation (or to a lack of radiation) on the part of the soil surface than the water surface so that the temperature of the soil surface and of the overlying air layers will rise more in summer or on a sunny day and fall more in winter or on a cloudy day than will the water surface.

It takes time for the radiant energy to warm any surface and to achieve an equilibrium between incoming and outgoing radiation, so that the curve of air temperature lags behind that of solar radiation. This lag is considerably greater over a water surface than over a land surface because of the aforementioned properties of the two surfaces in distributing the heat within their masses. Annual maximum and minimum air temperatures occur 3 to 4 weeks later than the solstices over land and 6 to 8 weeks behind the solstices over water.

Moist or marshy soils are better heat conductors than are dry soils. Cultivated soils or soils with loose litter or grass at the surface, which act as insulating layers, will be particularly poor conductors of heat. The surfaces themselves will heat appreciably and rapidly; likewise they will cool quickly and to a greater degree than will other surfaces.

The wave of temperature moving into the soil is quickly damped out with depth. Observations on a typical December day (without snow) at Seabrook, New Jersey (Fig. 2-7), show a diurnal variation at the surface of some 13°C. At 20 cm depth the diurnal variation was less than 2°C, and at 40 cm depth it was about 0.5°C. No diurnal temperature change could be found at 80 cm depth. Annual variations of temperature are found to somewhat greater depths. Landsberg (1960) reported less than 1°C annual variation at 10 m depth. However, the lag in the wave of temperature reaching this depth is so great that the temperatures may be as much as 6 months out of phase with the surface temperatures.

Air temperature decreases with height above the surface since the surface serves as the direct source of heat for the lower air layers. The normal or average decrease in temperature with height (defined as lapse rate) approximates 0.6°C/100 m. However, due to vertical air movements and the variable nature of moisture, clouds, and particulate matter in the air, the actual lapse rate can differ appreciably from this. At times the lapse rate can even be inverted (a positive lapse rate or an increase of temperature with height), a situation known as a "temperature inversion." When there is no change in temperature with height, the lapse rate is isothermal.

Air is compressible; at any level the air is subjected to a pressure determined by the weight of the column of air over it. Thus, if a parcel of air moves upward in the atmosphere, it will be moving into a region of lower pressure; the parcel of air will expand. As the air expands, its temperature decreases. This rate of decrease of temperature with height by expansion but without condensation, the dry adiabatic rate of cooling, approximates 1°C/100 m.

Comparison of the actual rate of temperature change with height with the dry adiabatic rate of temperature change will reveal whether the air is stable or unstable. Stability and instability in this situation refer to the action of a parcel of air initially set in motion. If this parcel is moved upward and arrives at the next level of the air warmer than its surroundings, it will be lighter and more buoyant than the surroundings and want to continue to

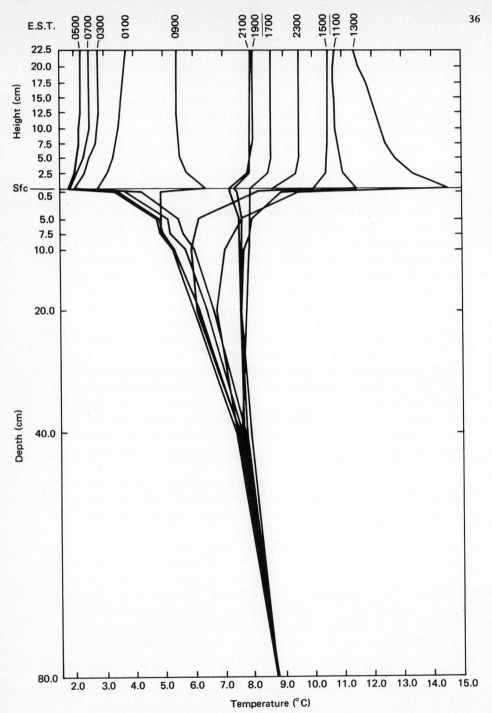

Profiles of air and soil temperatures at Seabrook, N.J., Dec. 5, 1956. FIG. 2-7
(*With permission of C. W. Thornthwaite Associates, from Davis*, 1957, *fig.* 24,
p. 50.)

move upward. The atmosphere is said to be unstable. If, on the other hand, the parcel moving upward arrives at the next layer and is colder than its surroundings, it will be heavier or more dense than its surroundings and tend to move back toward its starting place. The atmosphere is said to be stable since its temperature condition tends to resist the upward movement.

Figure 2-8 represents a portion of a standard temperature-pressure diagram (called an "adiabatic chart"), which illustrates these relations more clearly. Both height and pressure are indicated along the vertical scale. Temperature, on the horizontal scale, increases to the right. The solid slanting lines are the dry adiabatic lapse rates ($1°C/100$ m). Curve $ABCD$ is a profile of actual temperature vs. height through the atmosphere, measured by a temperature-measuring element sent aloft by a balloon. The layer of the atmosphere between AB has a greater decrease of temperature with height than the dry adiabatic rate, BC experiences a temperature inversion, while layer CD has a smaller temperature decrease with height than the dry adiabatic.

If a parcel of air at A is suddenly forced to rise to level B, it will cool by expansion at the dry adiabatic rate (assuming no condensation). Therefore, it will arrive at the level of B (at point B^1) warmer than the air around it which is denoted by point B. Since it is warmer and less dense than the surrounding air, it will want to continue rising. The layer AB therefore is called "unstable." Air layers through which the temperature change is greater than dry adiabatic are unstable; as a result of turbulent mixing of the air these air layers do not remain in existence very long.

If parcels of air at points B and C are suddenly set in motion upward to levels C and D, respectively (arriving at points C^1 and D^1), both parcels (cooling dry adiabatically) will be somewhat colder than the surrounding air (denoted by the temperatures of points C and D). Thus the parcels of air that started at B and C, respectively, will be more dense than the surroundings at levels C and D and will want to return to the level from which they started. Air layers BC and CD are called "stable" but they differ markedly in their degree of stability. Layer BC, being an inversion, is strongly stable. The parcel of air starting at B and cooling dry adiabatically as it moves toward C will be very much colder than its surroundings and therefore will tend to resist all movement away from B, while the parcel starting from C, being only slightly colder than its surroundings, will only weakly oppose any movement away from C.

In this discussion, we have considered only dry adiabatic changes. However, if cooling of the air due to expansion results in saturation of the air and subsequent condensation of moisture, some heat of condensation will be released into the air. Thus the rising air parcel will be subject to two effects—cooling by expansion at $1°C/100$ m of rise and heating by release of the latent heat of vaporization. This amount of heating can vary from essentially $0°C/100$ m (no condensation) to about $1°C/100$ m, depending on the amount of moisture being condensed. As a result, the combined rate of

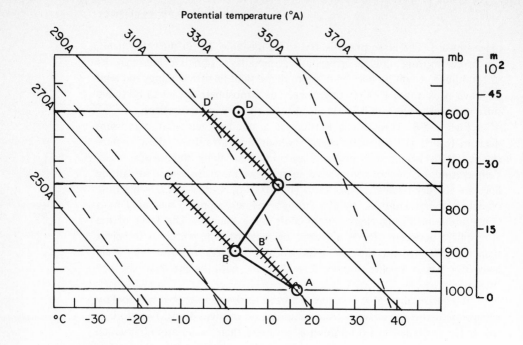

Potential temperature (°A)

┣┼┼┼┼┼┫ Path of parcel rising dry adiabatically
Solid sloping lines are dry adiabats
Broken sloping lines are wet adiabats

Temperature profile with height, illustrating stable and unstable air layers.   FIG. 2-8

cooling by adiabatic expansion and heating by condensation (the so-called "wet adiabatic lapse rate") varies from about 0.9°C/100 m to about 0.2°C/100 m, averaging around 0.4 to 0.6°C/100 m (shown by the dashed lines on Fig. 2-8).

Because temperature does change with height, comparison of temperatures at different elevations can be made by comparing the potential temperature of the air at the different levels. Potential temperature is defined as the temperature an air parcel will have if brought dry adiabatically to 100 mb pressure. Because of this relation the dry adiabats on Fig. 2-8 are also labeled in terms of potential temperature (°A or K) to simplify comparisons between levels.

2 Instruments to Measure Temperature

Temperature is most commonly determined by measuring the amount of expansion or contraction of different substances when subjected to heating or cooling. The standard instrument is a sealed glass tube with uniform bore and a bulb or reservoir at one end filled with mercury or with alcohol. A change in temperature will result in a greater expansion or contraction of the alcohol or mercury than the glass container itself. Graduations on the tube provide a direct measure of the temperature of the bulb and, in turn, of the environment of the bulb. If carefully constructed, calibrated, and exposed, mercury-in-glass or alcohol-in-glass thermometers provide reliable means of measuring the temperature of a particular place—which, or course, may or may not be "the" air temperature.

To use glass thermometers to read maximum or minimum air temperatures there must be some way for the instrument to record the desired temperature for later observations. This is accomplished in the maximum thermometer by having a constriction in the bore just above the mercury bulb. Heating of the bulb end will force mercury past the constriction up into the bore. However, cooling of the thermometer will not be accompanied by any downward movement of the mercury column into the bulb because it cannot pass the constriction. Thus the height of the column of mercury above the constriction is a measure of the maximum temperature reached since the last observation (at which time the thermometer had been reset to the existing air temperature by spinning it rapidly to force the mercury down past the constriction).

The minimum temperature is usually measured with a spirit thermometer (alcohol-in-glass) exposed horizontally. An elongated dumbbell-shaped index is contained within the bore in the column of alcohol. As the alcohol contracts with cooling, the concave surface of the alcohol column forces the index down the bore ahead of it. When the temperature rises, the alcohol will move around the index and up the bore, leaving the upper end of the index to mark the minimum temperature that has occurred in the period since the last reading. The instrument is reset by tipping it up and allowing the index to slide up to the concave top of the alcohol column marking the current air temperature.

A second type of expansion thermometer makes use of the varying effect of temperature on two unlike metals welded together as a bimetallic bar. An increase in temperature will cause the two metals to expand a different amount so that a slight bend results in the bar. This curvature is translated by means of a series of levers and gears to an indicator or pen arm. Thus it is possible to obtain a continuous record of temperature over any convenient time period. Such an instrument is called a "thermograph."

A Bourdon tube is another form of temperature-sensitive element used in a thermograph. The Bourdon element is an elongated oval tube, slightly curved, made of phosphor bronze filled with alcohol. The tube changes its shape as a result of the differential expansion of the bronze and alcohol. This change in shape is translated by means of levers and gears to a pen arm.

All these instruments have a relatively slow response speed to step changes in temperature. For more rapid response, a thermocouple is often employed. A thermocouple consists essentially of a closed loop of two unlike metallic wires such as copper and constantan joined at two junctions. If these junctions are at different temperatures, a small electromotive force is set up in the loop. This force can be measured by a galvanometer. The size of the force is proportional to the temperature difference between the two junctions. Knowing the temperature of one of the junctions, the temperature of the other junction can be determined from the galvanometer reading. Often, one of the thermocouple junctions is kept at 0°C by means of a melting ice bath in a thermos bottle equipped with a mixing device.

The thermocouple provides very rapid response to temperature changes, and if the wires used are fine enough, it is not seriously influenced by radiation conditions. It is usually better, however, to expose the thermocouple juncture within a well-ventilated, flat-plate radiation shield.

There are several other types of temperature-measuring instruments with limited application to applied climatology. Two are the electrical resistance thermometer (for example, thermistors) and the radiometer.

The electrical resistance of nearly all metals varies with temperature. Platinum and nickel are common metals used for resistance thermometers, although copper can also be used in special cases. The sensor itself consists of a series of coils of wire supported by the air or some inert medium such as mica and protected by a convenient sheath. Middleton and Spilhaus (1953) report that a meter of platinum wire, 0.1 mm in diameter, possesses a resistance of about 15 Ω at 0°C. For use as a resistance thermometer possibly 5 m of wire could be coiled around a mica frame.

More recently, ceramic elements having electrical resistances proportional to the temperature have been used either in the form of thin rods or as thermistor beads. The resistance of the thermistor beads changes about 4 to 5 percent/$^\circ$C change in temperature.

A radiometer is a device to measure the radiant energy given off by a surface. This will be a function of its temperature, emissivity, and area. Radiometers generally consist of some form of thermopile, with one set of thermoelectric elements exposed in the apex of a hollow cone that can be directed toward the radiating surface in question, the other set being maintained at air temperature. A small electric current will be set up between the two thermoelectric elements that will be a function of the radiation absorbed. Often the emissivity of the surface is assumed to be near unity; if it is not, it may significantly influence the actual determination of temperature. The composition of the layers of air between the surface and the sensor can also be important if the sensor is at some distance, such as on a satellite. The radiometer has many advantages: it does not disturb the surface under consideration in any way; it can obtain observations of temperatures remotely; and it can be used to measure the temperature of objects (such as furnaces) in which normal thermometers cannot be exposed.

3 Distribution

Maps of the distribution of mean monthly and annual temperatures are commonly available. Temperature is one of the most frequently mapped of the climatic elements, and we have a reasonable understanding of its variation both vertically and horizontally. The dependence of temperature on radiation from the sun is clearly demonstrated by any world map of temperature. If we lived on a completely homogeneous globe, the isotherms (lines of constant temperature) would exactly parallel the latitudes since radiation is everywhere the same along a particular latitude. The isotherms are very nearly parallel with the latitudes in the Southern Hemisphere where the surface is nearly homogeneous around the globe. In the Northern Hemisphere the isotherms bend as they cross from the ocean to the land areas since these surfaces will react differently to solar energy. In the summer the land is warmer than the water; any given isotherm bends poleward as it crosses from water to land. In the winter the land is colder than the water, and isotherms bend equatorward as they cross from the water to the land surfaces.

Study of long-period records of monthly or annual temperatures can sometimes reveal climatic trends or possibly even the influence of local environmental factors such as the growth of a town or the increasing pollution of the atmosphere. The distribution of monthly and annual temperatures at nearby stations in Delaware, Maryland and Virginia over the past 60 years reveals two types of variation (see Fig. 2-9). First, using mean January temperatures, grouped by 10-year running means to eliminate some of the year-to-year variation, there is a marked cyclic movement with about 22 years between successive peaks. Whether this cyclic pattern, found widely in the eastern United States, is really climatic or has been inadvertently introduced by the use of running means, as is often the case, needs to be studied. Second, at Wilmington and Onley, no long-term trend appeared in the 60 years of record while at Princess Anne, located between the other two stations, a slight warming trend exists (see dashed straight line). This trend is probably the result of exposure of the thermometer itself and the growth of the town. The lack of such a warming trend at Wilmington and Onley possibly reflects a better instrument exposure over the years.

These station records reveal much about the problems that can develop in any simple analysis of the distribution of temperature. Valid climatic trends are often revealed in such records, but care must be exercised to ensure that the results are not biased by cultural or man-made factors, by the exposure of the sensors, or by certain characteristics of the sensors themselves.

Records of the extreme temperatures that have occurred at a station are of value for planning purposes, for they provide some indication of the limiting conditions that can sometimes occur. While extremes can always be exceeded, they do give some suggestion of the severity of conditions for the purposes of design and specification. Table 2-7 provides values of a few of the record extreme temperatures that have been found.

4 Limitations

A thermometer openly exposed in the atmosphere will achieve a temperature which is a compromise between that of the air and that of all the surrounding objects including the sun with which it is mutually exchanging radiation. Thus, if only air temperature is to be measured, great care must be exercised

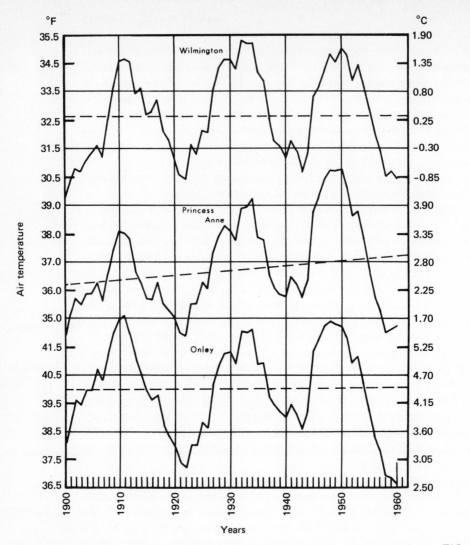

Ten-year running means of air temperature in January at locations on the Delmarva Peninsula.   FIG. 2-9

to ensure that any heat transfer by radiation is minimized. The atmosphere is fairly transparent and does not absorb significant amounts of the short-wave radiation. Normally the thermometer is exposed within a polished metal shield or a wooden-louvered box in order to eliminate the radiation effect. However, in fairly calm weather, the various radiation shields are themselves warmed or cooled more than the air around them. The enclosure within the shield can therefore be at a slightly different temperature from the ambient air. Artificial ventilation of the shield or screen is the only method that can eliminate this effect within the radiation shield. Errors of $1°C$ can result from inadequate ventilation within the shield or screen.

Weather shelters are supposed to be exposed with their floors about 1.2 m above a grassed surface in English-speaking countries and about 1.7 to 2.0 m above the ground in central Europe. Standardization of height is desirable because of the great variation in temperature that can exist near the ground surface. To provide observations that truly reflect regional differences in temperature rather than local differences in exposure, it is necessary to follow such established procedures for exposing and reading the thermometers. Stevenson screens exposed on post office roofs or on other buildings (where many early observations in climatology were taken) are of little value because of the unique conditions of their exposure.

Temperature varies appreciably from place to place within relatively short distances due to the influence of different surface features such as trees, buildings, lakes, roads, etc. (see changes across Toronto, Fig. 10-3). Observations with a thermocouple or other fast-response thermometer reveal rather large changes in air temperature as the small atmospheric eddies (see the discussion of wind later in this chapter) move by the sensor. Changes of several degrees within a matter of 2 or 3 min are possible. Considering these rather rapid variations in both space and time, Middleton and Spilhaus (1953) conclude that it gives a false sense of accuracy and reliability to record air temperatures to closer than whole degrees Fahrenheit. However, if humidity

TABLE 2-7

World extremes of temperature in °C*

AREA	HIGHEST TEMP., °C	PLACE	ELEVATION, M	DATE
Africa	57.8	Azizia, Libya	116	Sept. 13, 1922
North America	56.7	Death Valley, Calif.	−54	July 10, 1913
Asia	53.9	Tirat Tsvi, Israel	−220	June 21, 1942
Australia	53.3	Cloncurry, Queensland	190	Jan. 16, 1889
Europe	50.0	Seville, Spain	8	Aug. 4, 1881
South America	48.9	Rivadavia, Argentina	206	Dec. 11, 1905
Oceania	42.2	Tuguegarao, Philip.	22	Apr. 29, 1912
Antarctica	14.4	Esperanza, Palmer Pen.	8	Oct. 20, 1956
	LOWEST TEMP., °C			
Antarctic	−88.3	Vostok	3,420	Aug. 24, 1960
Asia	−67.8	Oymykon, USSR	800	Feb. 6, 1933
Greenland	−66.1	Northice	2,344	Jan. 9, 1954
North America	−62.8	Snag, Yukon	587	Feb. 3, 1947
Europe	−55.0	Ust' Shchugor, USSR	85	Jan. ?
South America	−32.8	Sarmiento, Argentina	268	June 1, 1907
Africa	−23.9	Ifrane, Morocco	1,635	Feb. 11, 1935
Australia	−22.2	Charlotte Pass, N.S.W.	. . .	July 22, 1947
Oceania	−10.0	Haleakala Summit, Maui	2,972	Jan. 2, 1961

* From ESSA, 1969, p. 27.

measurements are determined from the temperature readings, the thermometers should be read to the nearest tenth (p. 62). Yet, it is common practice among climatologists and laymen alike to read temperatures to tenths of a degree Fahrenheit (tenths of a degree Celsius are a bigger unit and should be used) and to publish such data implying an unwarranted degree of "precision."

Thermographs tend to lag behind the actual air temperature in reporting response to a change in temperature. The width of the ink line on the thermograph chart is such that it makes it impossible to read such a chart to closer than $\pm 0.2°$. Even this much detail is not possible if the thermograph has been allowed to operate for some time without cleaning. Dust in the levers and gears and deposits on the sensing element itself will make it even less responsive to temperature changes.

Many expressions developed for use in applied climatology (such as for heat flux from humans or other bodies, evaporation from a water surface, etc.) require knowledge of the surface temperature. This is difficult to obtain since any instrument exposed at the surface will disturb it sufficiently to alter the resulting temperature. What is meant by "the surface" is also subject to some question in the case of an undulating vegetation cover or even a surface of small waves on the water. Thus, in most cases, some modification of the expression is sought to allow use of a temperature taken near the surface either in the material or in the air above as a surrogate for the surface temperature. This usually results in some error in the final derived value.

ATMOSPHERIC MOISTURE

1 Basic Relations

The term "atmospheric moisture" covers a wide variety of climatic factors: moisture in vapor or gas form known as "humidity," water in small-droplet form called "clouds" or "fog," and water in large-solid or liquid-droplet form called "precipitation"—rain, snow, sleet, hail, or graupel.

Several terms are used to express moisture relations in the air. At any given temperature, saturated air contains a definite maximum mass of water vapor (Table 2-8). Relative humidity is the ratio of the actual amount of moisture in the air to the maximum amount in saturated air at that temperature. When air already saturated with vapor cools, some moisture must condense out of the air. The temperature at which condensation begins is known as the "dew point temperature."

If air at $5°C$ actually has 5.0 g of water/m^3, the relative humidity becomes (5.0/6.8) 100 or 74 percent. Relative humidity is a nonconservative factor in that its value may change either by the addition or removal of moisture or by a change in temperature, which will affect the maximum amount of water vapor in air at saturation—the denominator of the expression. Air saturated at $0°C$ (with 4.847 g/m^3), if now heated to $5°C$, will decrease from 100 to 71 percent relative humidity with no actual change in moisture content of the air.

Two other terms, less frequently used but often of more practical value, are absolute humidity and specific humidity (or mixing ratio). They are of value, however, for they are better indicators of the atmosphere's ability to give up moisture as precipitation. Absolute humidity is defined as the mass of moisture in a given volume of air, while specific humidity is the mass of moisture in a given mass of air. Values of absolute humidity are included in Table 2-8.

Specific humidity is usually not measured directly but rather obtained from the relation

$$q = \frac{0.622e}{p - 0.378e}$$

where q = specific humidity, g/kg
  e = actual vapor pressure
  p = actual air pressure

Specific humidity is the mass of the moisture in a given mass of air including the vapor, while the term "mixing ratio" is used to represent the ratio of the mass of the vapor to the mass of dry air only. Thus, since nearly all air has some moisture in it, the mixing ratio will be slightly larger than the specific humidity. In actual practice they differ by only a few percent and so they can be used fairly interchangeably.

Water vapor that has condensed into a visible small-droplet form is called "cloud" or "fog." As air is cooled and its moisture content approaches saturation, moisture will begin to collect around tiny hygroscopic particles of salt and dust in the atmosphere. These are known as "condensation nuclei" and are necessary for the initiation of the condensation process. Condensation must not be confused here with precipitation for they are quite different

TABLE 2-8

Maximum mass of moisture possible in a unit
volume of air at different temperatures*

AIR TEMP., °C	MASS OF MOISTURE IN AIR AT SATURATION, g/m³	AIR TEMP., °C	MASS OF MOISTURE IN AIR AT SATURATION, g/m³
−30	0.45	5	6.80
−25	0.70	10	9.40
−20	1.07	15	12.83
−15	1.60	20	17.30
−10	2.36	25	23.05
− 5	3.41	30	30.38
0	4.85	35	39.63

* With permission from List, "Smithsonian Meteorological Tables," 6th rev. ed., table 108, 1949.

TABLE 2-9

Characteristics and height ranges of common cloud types

HEIGHT RANGE	STRATIFORM CHARACTERISTICS	TYPES	CUMULOFORM CHARACTERISTICS	TYPES
High: > 20,000 ft (> 6,000 m)	Ice crystal clouds; thin, wispy, milky, sun usually shines through. Cs, continuous layer, gray; Cc, lumpy with breaks, white	Cirrus Cirrostratus, Cs Cirrocumulus, Cc	Extreme vertical development, 20,000–60,000 ft (6,000–18,000 m) thick. Anvil top with cirrus, strong vertical currents in cloud, turbulent	Cumulonimbus
Middle: 6,500–20,000 ft (≈ 2,000–6,000 m)	Water and ice crystal, slightly more vertical development. As, continuous layer, gray, foreboding; Ac, lumpy with breaks, whitish appearance	Altostratus, As Altocumulus, Ac	Base 1,500–3,000 ft (450–900 m), top 5–15,000 ft (1,500–4,500 m) higher; building. Cauliflower top, boiling upward, strong turbulence	Cumulus congestus
Low: 0–6,500 ft (0–2,000 m)	Water and ice crystal, slight vertical development. Stratus is continuous layer, gray foreboding; Sc, lumpy with breaks, whitish appearance, fog is stratus at ground; Ns is stratus with rain falling out	Stratus, St Stratocumulus, Sc Nimbostratus, Ns Fog	Flat base at 1,500–3,000 ft (450–900 m), top 2–5,000 ft (600–1,500 m) higher; building, lumpy top, some turbulence, very white in appearance, often as little puffs of cotton	Cumulus

processes. Condensation results in the slow build-up of moisture on the various hygroscopic particles in the atmosphere, finally becoming large enough to be visible to the naked eye. These cloud or fog drops are still not ready to fall as precipitation (it takes a million cloud drops to form one good-sized raindrop). The precipitation process involves the relatively rapid growth of the cloud droplets to a size at which they are able to fall freely out of the air.

We normally identify two major classes of clouds: those with vertical development, the cumuloform or pilelike, and those without vertical development, the stratiform or layerlike clouds. Clouds are further separated on the basis of height of occurrence, in the case of the stratiform, or amount of vertical development, in the case of the cumuloform. Table 2-9 identifies only a few of the more common types of clouds.

The precipitation process is still not too clearly understood. As air cools to saturation, condensation will occur, but how the various cloud droplets grow rapidly enough to produce, in some cases, heavy rain in just a few hours is still a matter of study. The saturation vapor pressure with respect to ice is less than it is with respect to water at the same temperature; if ice crystals and water droplets are both present in a cloud, the cloud (with an average saturation vapor pressure between that for ice and for water) will be supersaturated with respect to the ice crystals and unsaturated with respect to the water droplets. This will result in moisture evaporating from the cloud droplets and condensing more rapidly on the ice crystals. The ice drops will grow at the expense of the water drops. This is clearly one possible precipitation process.

There are cases, however, where clouds without ice crystals give heavy precipitation within short periods of time, so some other process must also be at work. There are different-sized condensation nuclei present in the atmosphere, large salt particles vs. small particles. It is felt that this size difference may influence the rate of growth by condensation, with the larger nuclei gaining moisture at the expense of the smaller. A third possibility involves the action of electrical charges on the droplets and electrical forces within the atmosphere. Some of the ideas are hardly more than educated suggestions at the present time.

Precipitation comes in many different forms. It is sometimes difficult to distinguish between quite similar types. Even the definitions may vary from place to place (e.g., Great Britain—sleet is melting snow; United States—sleet is freezing rain), resulting in further confusion. Table 2-10 however, gives the basic descriptions of some of the more commonly recognized hydrometeors based on definitions in the "Glossary of Meteorology" (Huschke, 1959).

Table 2-8 shows that the maximum amount of moisture in the air at any temperature increases at an increasing rate with temperature. If saturated air at 35°C is cooled 10°C, it will give up 16.5 g/m^3, while saturated air at 15°C, cooled the same amount, will lose only 6.0 g/m^3. Thus the cooling of warm air provides the opportunity for producing greater amounts of

TABLE 2-10

Descriptions of common hydrometeors*

HYDROMETEOR	DESCRIPTION
Drizzle	Droplets < 0.5 mm diameter; intensity usually less than 1 mm (0.04 in.)/hr; very small, numerous drops that seem to float in air following air currents. Mist is often interchanged with drizzle although it more specifically refers to a condition of reduced visibility due to microscopic water droplets in air
Rain	Droplets > 0.5 mm diameter, intensity generally more than 1.25 mm (0.05 in.)/hr. Larger but fewer drops than drizzle so that there is generally less reduction in visibility except for the heavier rainfalls
Snow	White or translucent ice crystals often agglomerated into the form of flakes. Size of flakes depends principally on water content and amount of moisture surrounding crystal
Snow pellets (soft hail, graupel)	White, opaque, round or conical ice particles with snowlike structure, 2 to 5 mm diameter. Pellets are crisp and easily crunched
Sleet	Transparent, solid grains of ice formed from freezing of rain or slight melting and refreezing of snow as it falls. Sometimes confused with small hail
Snow grains	Very small, white, opaque particles, the solid equivalent of drizzle. Will not shatter or bounce when they hit surface
Hail	Balls or lumps of ice, often with concentric layers of clear and milky ice, diameter > 5 mm. If diameter is less, it is called "ice pellets" or "snow pellets" (formerly small hail)
Virga	Water or ice particles falling from clouds but evaporating before reaching earth
Rime	White or milky granular deposit of ice formed by rapid freezing of supercooled water droplets hitting exposed object
Glaze	Coating of clear smooth ice on exposed objects by freezing of film of supercooled water deposited by rain, drizzle, fog. Denser than rime. (Glaze associated with large water-drop size, rapid collections, slight supercooling, slow dissipation of heat of fusion while opposite effects favor rime)
Dew	Water condensed on objects near the ground whose temperatures are above freezing but below dew point—the temperature at which the air becomes saturated with moisture. Frost (or hoarfrost, which is the same) occurs if temperatures are below freezing. White dew is frozen dew resulting from temperatures falling below freezing after dew has formed

* Reproduced with permission of the American Meteorological Society, from Huschke, 1959.

precipitation than the cooling of colder air. Precipitation amounts in the equatorial region are generally larger than in middle or polar latitudes.

The only way to produce cooling of large masses of air is by expansion of rising air currents. Three major processes resulting in rising air currents are (1) convective activity, (2) orographic uplift, and (3) frontal or cyclonic activity. A fourth, stirring or turbulent mixing of air, can result in small amounts of precipitation. Each method provides distinctive patterns of precipitation. Convective activity, developing from the heating of masses of air over some local source, results in the creation of vertical air currents, cumuloform clouds, and shower activity. Orographic precipitation, resulting from air moving upward over mountain barriers, produces a pattern of heavy precipitation on the upwind side of mountains (precipitation increases upward to some level above which the moisture content is so low that further cooling results in smaller amounts of moisture being squeezed out of the air). The lee slope of the mountains will suffer from a lack of precipitation (rainshadow effect). In cyclonic motion, air spirals inward in a counter-clockwise direction in the Northern Hemisphere. As it approaches the center, the air rises and cools. If unlike masses of air are brought together in one of these cyclonic areas, the warmer air will be forced to ride up over the colder air mass (frontal activity); the warmer air will cool and clouds and precipitation may develop.

Rainfall intensity is closely related to duration of the observing period. Table 2-11 gives a few world's record observed point rainfalls. When plotted on logarithmic paper these values fall on a fairly straight line. The data show that intensity decreases rapidly with time. That most of these points fall along a straight line would suggest that even with the limited sampling network available the data collected are still quite representative of actual maximum intensities (except in the case of the 1-min rainfall, which seems to fall well below the straight-line relation).

2 Instruments to Measure Humidity and Precipitation

Relative humidity is usually measured by means of a hair hygrometer or a psychrometer. The hair hygrometer makes use of the fact that the human hair will expand or contract with changing moisture content. If the hair is stretched taut and attached to suitable levers and indicating arms, changes in relative humidity will change the length of the hair and, thus, the position of the indicating arm.

The hair hygrometer does not measure accurately at the extreme ends of the scale, and its response to changes in humidity is slow. Certain substances such as lithium chloride give a more rapid and complete response to changes in humidity than the human hair. The resistance of lithium chloride to an electric current varies with moisture content.

A psychrometer consists of two mercury-in-glass thermometers, the bulb of one covered by a muslin wick. This wick is kept moist. Evaporation cools the surface and produces a lower "wet bulb temperature." To ensure adequate opportunity for evaporation, the psychrometer is usually aspirated or rotated rapidly (the sling psychrometer). Preferably, distilled water should be used on the wick. Tables or graphs are used to determine the moisture content of the air based on wet bulb depression and the air temperature. Other tables provide values of vapor pressure and dew point temperature from wet and dry bulb data.

Precipitation is measured by means of a raingage. The standard Weather Bureau model is an 8-in. (200-mm)-diameter metal tube with a funnel having a tapered knife edge on top. The funnel directs the collected water into an inner measuring tube just one-tenth the area of the funnel top. This permits a 10:1 magnification of the depth of water in the tube. This depth is measured by means of a dipstick to the nearest tenth of an inch (2.54 mm) (precipitation is therefore given to the nearest one-hundredth of an inch or 0.25 mm). The gage is 23 in. (580 mm) high.

If intensity measurements or time-duration studies are needed, a

TABLE 2-11

World's greatest observed point rainfalls*

DURATION	DEPTH IN.	MM	LOCATION	DATE
1 min	1.23	31.2	Unionville, Md	July 4, 1956
8 min	4.96	126.0	Füssen, Bavaria	May 25, 1920
15 min	7.80	198.1	Plumb Point, Jamaica	May 12, 1916
20 min	8.10	205.7	Curtea-de-Arges, Roumania	July 7, 1889
42 min	12.00	304.8	Holt, Mo	June 22, 1947
2 hr. 10 min	19.00	482.6	Rockport, W. Va	July 18, 1889
2 hr. 45 min	22.00	558.8	D'Hanis, Tex. (17 mi NNW)	May 31, 1935
4 hr. 30 min	30.8+	782.3	Smethport, Pa	July 18, 1942
9 hr	42.79	1087	Belouve, La Réunion	Feb. 28, 1964
12 hr	52.76	1340	Belouve, La Réunion	Feb. 28-29, 1964
18 hr. 30 min	66.49	1689	Belouve, La Réunion	Feb. 28-29, 1964
24 hr	73.62	1870	Cilaos, La Réunion	Mar. 15-16, 1952
2 days	98.42	2500	Cilaos, La Réunion	Mar. 15-17, 1952
3 days	127.56	3240	Cilaos, La Réunion	Mar. 15-18, 1952
4 days	137.95	3504	Cilaos, La Réunion	Mar. 14-18, 1952
5 days	151.73	3854	Cilaos, La Réunion	Mar. 13-18, 1952
6 days	159.65	4055	Cilaos, La Réunion	Mar. 13-19, 1952
7 days	161.81	4110	Cilaos, La Réunion	Mar. 12-19, 1952
8 days	162.59	4130	Cilaos, La Réunion	Mar. 11-19, 1952
15 days	188.88	4798	Cherrapunji, India	June 24-July 8, 1931
31 days	366.14	9300	Cherrapunji, India	July 1861
2 mo	502.63	12767	Cherrapunji, India	June-July 1861
3 mo	644.44	16369	Cherrapunji, India	May-July 1861
4 mo	737.70	18738	Cherrapunji, India	April-July 1861
5 mo	803.62	20412	Cherrapunji, India	April-Aug. 1861
6 mo	884.03	22454	Cherrapunji, India	April-Sept. 1861
11 mo	905.12	22990	Cherrapunji, India	Jan.-Nov. 1861
1 yr	1041.78	26461	Cherrapunji, India	Aug. 1860-July 1861
2 yr	1605.05	40768	Cherrapunji, India	1860-1861

* *With permission*, Monthly Weather Rev., *from Paulhus, 1965.*

recording gage is necessary. The two principal recording gages are the weighing gage and the tipping-bucket gage. The weighing gage has an 8-in. (200-mm) funnel directing the precipitation into a collecting bucket which rests on the platform of a spring scale. The movement of the platform is transmitted through a series of levers usually to a pen arm which records on a chart mounted on a clock-driven cylinder. Daily or weekly records can be obtained depending on the gear movement of the clock-driven cylinder.

The tipping-bucket gage substitutes a small tipping collector for the weighing bucket. The tipping collector has a capacity of just 0.01 in. (0.25 mm). When full, the collector tips rapidly and an empty collector becomes aligned under the funnel. The tipping collector is connected with a recording mechanism that counts the number of tips with time. With heavy rains a small amount of precipitation may be lost if it falls during the moment when the full collector is being replaced by the empty one.

The tipping-bucket gage will not measure snowfall. The weighing and standard gages can be used in winter provided they have been prepared for it by removing the 8-in. (200-mm) funnel. Snow will fall directly into the open tube or the weighing bucket and is not blocked by the funnel constriction. The snow must be melted before measurement by means of a dipstick.

All raingages, but especially clear plastic gages or those that are not read daily, may lose some precipitation due to evaporation. Small amounts of oil added to the gage reduce later evaporation of any water collected by the gage.

Snow-depth measurements are usually not made in gages but rather by means of measuring rods at a number of places where the effect of drifting has been eliminated as much as possible. Replicate observations will eliminate significant errors due to drifting and provide fairly representative values of total snow depth. It is the water content of the snow that is usually reported in the precipitation records.

3 Distribution

There are 12,000 to 13,000 raingages in the United States alone. Of these some 3,000 are recording raingages while the rest are read daily by cooperative observers. With an area (exclusive of Hawaii and Alaska) of just over 3 million mi^2 (7.8 million km^2) this provides an average of one gage every 250 mi^2 (650 km^2), not a particularly dense network although it is considerably better than in most parts of the globe.

The surface area of a raingage is only 0.0324 m^2 or 1/30,000,000 km^2. Collecting all the United States raingages together would result in a total catch area of about 1/800 km^2, or less than the infield area of one baseball diamond. When viewed in this light we realize that we have a very small sample of precipitation indeed. Our observations are only point values. We assume these point values are representative of average conditions over a much larger area, but really such an assumption can err significantly in the presence of topographic features and special local sources or sinks of precipitation.

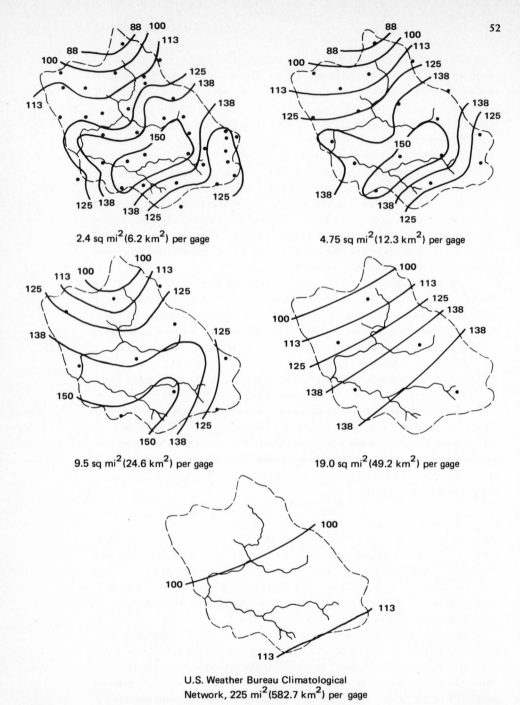

2.4 sq mi² (6.2 km²) per gage

4.75 sq mi² (12.3 km²) per gage

9.5 sq mi² (24.6 km²) per gage

19.0 sq mi² (49.2 km²) per gage

U.S. Weather Bureau Climatological
Network, 225 mi² (582.7 km²) per gage

Effect of raingage density on isohyetal pattern (isohyets in mm). Storm of   FIG. 2-10
July 16–17, 1950, in central Illinois on a 95-mi² basin. (*With permission
Illinois State Water Survey, from Hudson, Stout, and Huff, 1952, fig. 12, p.
10.*)

Such assumptions might seem to be fairly satisfactory for monthly or annual totals but not at all realistic for individual storms. Foster (1948) has given an instructive example of the storm of May 30, 1935, near Rush, Colorado, which caused severe flash flooding of the South Platte and Republican rivers. The maximum 1-day rain at Rush, the only official measuring station in the area, was 76 mm (3.0 in.), not sufficient to cause the flooding which occurred. Rush was located near the southern edge of the storm. However, other receptacles exposed in the storm area to the north of Rush measured quite different totals of rainfall (Table 2-12). While the reliability of some of these samples might be questioned, the general impression that they give is that the Rush value was highly unrealistic for this particular storm. Such so-called "bucket surveys" are extremely helpful in providing needed insight into the actual distribution of precipitation from individual storms.

Hudson, Stout, and Huff (1952) investigated possible errors in the analysis of rainfall distribution arising from changing the density of the raingage network. Using a 95-mi^2 (246-km^2) watershed in Illinois, they have provided a series of maps (Fig. 2-10) of the isohyetal pattern over the basin for the storm of July 16–17, 1950, based on a distribution of 40, 20, 10, 5, and 0 gages within the basin area. With five or fewer gages in the basin (19.0 mi^2 per gage or more) the area of intense precipitation (> 150 mm) in the southern part of the basin was missed entirely. Over a number of months of record (summer months for 3 years), and assuming the 20-gage distribution (4.75 mi^2 per gage) to give the true mean, they found that the average error in total basin rainfall ranged from 3 percent for the 10-gage network density to 9 percent for the 1-gage network. The largest percent errors in each case

TABLE 2-12

Rainfall totals collected by other receptacles during storm of May 30, 1935*

OTHER RECEPTACLES MEASURED	RAINFALL, MM
Water tank (2.4 X 0.6 m)	457 (est.)
Water tank (2.4 m diameter)	610
Standard raingage and bucket (0.3 m deep)	610
11-liter can in yard	152 or 203
Barrel in yard	102
Half-bushel measure (0.3 m diameter X 0.35 m deep)	305
Bucket (0.25 m diameter X 0.3 m deep)	229
Coffee can in yard	102
Oil can in yard	203
Pail	102
Can (0.2 m deep)	> 216 (overflowed)
Stock tank	381
Can in yard	212
Can beside road	178

* *With permission, from Foster, 1948, p. 47.*

were associated with the low rainfall amounts. The authors concluded that a network density of one gage per 100 mi² (one per 360 km²) in Illinois should be sufficient for the determination of monthly rainfall for most purposes.

From available recording raingages, rainfall intensities for various durations and for return periods of up to 100 years have been computed statistically. These are valuable for engineering works or hydroelectric developments. It must be remembered, however, that the so-called "100-year storm" may actually occur twice in the same year or in successive years. The Weather Bureau has compared values of the maximum observed and 100-year computed daily rainfall for 15 cities in the United States (Table 2-13). In all cases, the maximum observed 24-hr rainfalls over the period of observation (from 74 to 91 years) agree quite closely with the statistically computed 100-year rainfalls.

From the long years of record at the available network of stations, maps of the average monthly and annual distribution of precipitation over various continental and world areas have been prepared. Such maps have only limited value for most purposes since the values are long-term averages and the detail is quite general.

Much more useful for engineering-design studies are large-scale maps of precipitation over small areas for short time periods, or depth-area-duration curves similar to Fig. 2-11. Two general depth-area relationships may be considered. One is developed using only rainfall data from storm centers, while the other, involving a geographically fixed relation, uses rainfall data

TABLE 2-13

Maximum observed and 100-year, 24-hr rainfalls for 15 U.S. cities*

CITY	YEARS OF RECORD†	MAX. 24-HR RAINFALL, MM	DATE‡	100-YEAR, 24-HR RAINFALL, MM
Boston, Mass.	88	213	Aug. 18, 1955	183
New York, N.Y.	91	243	Oct. 8, 1903	201
Washington, D.C.	91	185	Aug. 11, 1928	190
Jacksonville, Fla.	90	258	Sept. 5, 1950	182
Cincinnati, Ohio	91	133	Mar. 12, 1907	142
Chicago, Ill.	91	158	July 12, 1957	155
St. Louis, Mo.	91	223	Aug. 15, 1945	226
New Orleans, La.	91	356	Apr. 15, 1927	368
Galveston, Tex.	91	364	July 13, 1900	343
Denver, Colo.	90	166	May 21, 1876	99
Albuquerque, N.Mex.	74	57	Sept. 27, 1893	66
Salt Lake City, Utah	88	69	May 2, 1901	56
Portland, Oreg.	91	195	Dec. 12, 1882	127
San Francisco, Calif.	91	119	Jan. 29, 1881	109
Los Angeles, Calif.	84	187	Dec. 31, 1933	193

* *From NOAA, National Weather Service.*
† *Prior to 1952.*
‡ *Of first day when 24-hr rainfall extends into two calendar days.*

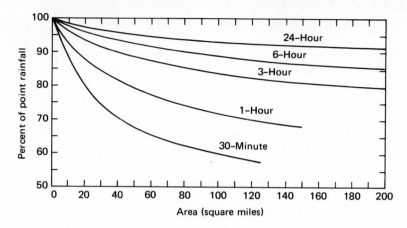

Generalized depth-area curves for rainfall-frequency data. (*From U.S.*   FIG. 2-11
Weather Bureau, 1957, *fig.* 1-3, *p.* 7.)

from different parts of the storm, some near the center and some not. By averaging the data the geographically fixed curves which are illustrated in Fig. 2-11 are flatter than curves developed from storm-centered data. The relation shown in Fig. 2-11 is based on data from 20 dense networks of raingages in different areas; no systematic regional variation of the relation has been found. Storm size also does not seem to be a factor in the relation. The ordinate scale is the ratio of the 2-year 24-hr rainfall over the area to the average of the 2-year 24-hr value for individual points in the area expressed as a percent. This diagram provides in convenient form all pertinent information needed for design of culverts or other water-carrying facilities, or for the design, construction, and operation of reservoirs and hydroelectric works.

There are some maps of relative humidity measured at specific times of the day. However, in view of the fact that relative humidity reflects both moisture and temperature parameters, such maps can be of no more than general informational value. Detailed data on specific or absolute humidity on an hourly basis would be desirable for any applied purpose.

4 Limitations

Many uncertainties exist in precipitation and humidity data, some of which have already been discussed. The lack of an adequate sampling network is obvious. The problem of extrapolating data from point observation sites is another. The effect of wind and exposure on the catch in the gage can be cited as a third.

The precipitation sampling network, one gage per 250 mi^2 (650 km^2) in the United States on the average, actually varies considerably nationwide. In populated areas, the network is denser than that [for example, 13.8 gages per 1,000 mi^2 (2,600 km^2) in New Jersey], while in sparsely settled areas, possibly critical from a precipitation point of view, the spacing of stations is considerably less [1.7 gages per 1,000 mi^2 (2,600 km^2) in western Colorado and 2.6 gages per 1,000 mi^2 (2,600 km^2) in Utah]. This density of stations is hardly adequate to give the information needed for water resources evaluation.

Any review of the recent debate concerning the anomaly in the precipitation record at La Porte, Indiana, only serves to emphasize the problem of uncertainties in precipitation data. Over the period 1927 to 1963, measured precipitation at La Porte, just downwind of extensive industrial developments around Chicago and Gary, was considerably higher than at nearby stations. Whether this resulted from increased hygroscopic nuclei in the air—thus a real, though man-made, phenomenon as some say (Changnon, 1968; Hidore, 1971)—or whether it resulted from observer error only as others claim (Holzman and Thom, 1970), cannot be resolved as of now. Precipitation figures in the past few years have failed to reveal the continuation of this anomaly at La Porte. There has also been a change of observers. After the fact, there are few ways to check the reliability of a record; even with a reasonably dense network there are still vast areas unsampled to compound the evaluation process.

Point rainfall observations, especially for record high amounts, cannot really be extrapolated for even short distances. They should be accepted as point values only and their contribution to area studies interpreted accordingly.

Precipitation is measured after being caught in a raingage. The exposure of the gage is quite important, for there must be no obstructions in the vicinity to affect wind flow or prevent free entry of precipitation into the gage. Since it is the total precipitation at the surface that is wanted, the gage should be exposed as near the surface as possible. Tests have shown that the amount of precipitation caught in a gage varies directly with the wind velocity. Wind velocity increases with height, so that the greater the height of the gage above the ground, the lower will be the precipitation catch. Koschmieder (1934) has provided figures on the deficit from true catch with different wind speeds (Table 2-14).

TABLE 2-14

Effect of wind speed on raingage catch*

| WIND SPEED | | DEFICIT FROM |
M/SEC	MPH	TRUE CATCH, %
0	0	0
2	4.47	4
4	8.94	10
6	13.41	19
8	17.88	29
10	22.4	40
12	26.8	51

* *With permission,* Monthly Weather Rev., *from Koschmieder, 1934.*

The limitations of humidity observations—the lack of reliability in the measurement of relative humidity, the slow speed of response of many sensors, as well as the dependence of relative humidity on both moisture and temperature—raise serious questions concerning the use of this quantity. Both humidity and precipitation values should not be accepted uncritically by those unfamiliar with the problems involved in their measurement; incorrect or misleading conclusions can easily result.

EVAPORATION

1 Basic Relations

Evaporation represents a significant mass and energy transfer from the ground to the atmosphere. Natural evaporation is much more than the reverse of rainfall or the return of moisture to the atmosphere; it is also a means by which the downward flow of energy from the sun can be balanced. Vast amounts of energy are required to bring about evaporation and are transferred to the air with the vapor as latent heat. Natural evaporation can proceed only when the vapor pressure of the ambient air is less than the vapor pressure at the evaporating surface, and it can continue only while there is an external source of energy.

There is a difference between total and net evaporation, for the former reflects the total movement of molecules of water away from the surface while the latter recognizes that there can be some downward movement of moisture to the surface, a recondensation of moisture. The difference between the evaporation and the condensation, the net evaporation, will be less than the total evaporation. It is what is referred to loosely as the evaporation. Evaporation is usually given as a depth of moisture so that it can be compared directly with the depth of precipitation.

Transpiration refers to the loss of water from living plants (as perspiration refers to water loss from humans and other animals). The amount of water lost by transpiration varies considerably from one plant species to another, largely in response to rooting depth and vegetative area. The "transpiration ratio," the quantity of water needed to produce a unit weight of dry matter, is often used to compare the transpiration by different species of vegetation.

The combined evaporation from the surface of the ground and the transpiration from the vegetation covering the ground has been called the "evapotranspiration." It represents the net water loss from the total surface. Mather (1949) has shown that when the ground is well covered with plants the principal loss is by transpiration. Evaporation actually constitutes only a small fraction of the total water loss from a well-vegetated forest or pasture area since little energy for evaporation can reach such a surface.

Thornthwaite (Wilm, Thornthwaite et al., 1944) has defined one further term, the "potential evapotranspiration," as the water loss from an extensive, closed homogeneous cover of vegetation that never suffers from a lack of water. He later (1954b) specified that the albedo of the vegetated area should be of the order of 20 to 25 percent. Potential evapotranspiration differs from actual evapotranspiration in that the plant has all the water it wants to use; water loss is never restricted by a lack of available moisture. Actual evapotranspiration or water loss from a surface can be shown to depend on (1) climatic factors of net radiation, wind velocity, and humidity; (2) type of soil; (3) soil moisture content; (4) type of vegetation and depth of rooting; and (5) land management practices; while potential evapotranspiration depends almost entirely on the climatic factor of energy from the sun. As long as the evaporating area (and the buffer zone around it) is sufficiently large, wind and humidity factors are of limited importance. With adequate moisture always available, type of soil, land management practices, and soil moisture content are unimportant. Even the type of vegetation or depth of rooting becomes less significant. Mather (1954a) found that with always adequate water many different species of vegetation all used essentially the same amount of water, an amount closely related to what would be evaporated by the energy available from the sun. This relation is found only if the strict definition of potential evapotranspiration is fulfilled: extensive vegetation coverage to eliminate advection effects, closed vegetation cover, homogeneous stand, and no deficit of water in the soil.

Tanner and Fuchs (1968) developed a more rigorous definition for potential evaporation as the water loss occurring from a surface having any particular configuration under a postulated set of meteorological factors, provided there is no saturation deficit at the surface. Under these conditions, in which the water supply cannot limit evaporation, potential evaporation is primarily determined by meteorologic conditions, although differences in both plant canopy and soil conditions can influence the evaporation by changing the soil heat flux, the albedo, and the transport coefficient. Wetting a surface may change both surface and meteorological conditions, resulting in a new value of potential evaporation. In the case of a wet surface of infinite extent (so that saturated air will exist in the surface layer, thus eliminating the need for a convective term expressing the potential heat supply from unsaturated air), Tanner and Fuchs express potential evaporation as

$$Ep = \frac{S}{S + \gamma} (R_n - G)$$

where S = slope of saturated water vapor curve vs. temperature, bar/K
γ = psychrometric constant, bar/K
R_n = net radiation flux density, W/m^2
G = soil-heat flux density, W/m^2

Although the air in the surface layer is saturated, gradients of temperature and vapor pressure still exist.

Various relations have been developed to express concepts similar to potential evapotranspiration. Dalton (1802), who generally receives credit for developing the first mass transport formula for evaporative flux, used the expression

$$E = C (e_0 - e_a)$$

where C = a constant, empirically determined, usually containing a wind speed term

      e_0, e_a = saturation vapor pressure at evaporation surface and in air above the surface, respectively

This vapor-pressure gradient has proved difficult to determine accurately because it needs the measurement of the saturation vapor pressure at the surface of the evaporating material. This is not possible to do at present in any simple fashion. Many variations of the Dalton expression substitute vapor pressures based on the more easily measured air and water or air and moist soil temperatures. The surface temperatures are taken somewhat within the medium rather than on the exact surface, with a resulting loss of accuracy.

Evaporation is proportional to the vapor-pressure gradient between the evaporating surface and the air above, but not necessarily to the vapor-pressure deficit or, as it is sometimes called, the "saturation deficit" (Thornthwaite, 1940; Leighly, 1937). Mather (1959a) discussed the rate of evaporation from two moist surfaces having exactly the same temperature into two samples of air both having relative humidities of 60 percent but with temperatures of 15.6°C (60°F) and 26.7°C (80°F), respectively. While it might at first appear that the rate of evaporation would be greater into the air at 26.7°C, such is not the case. Figure 2-12 depicts roughly the relation between air temperature, relative humidity, and vapor pressure of the air. The air at 26.7°C and 60 percent relative humidity has a vapor pressure of just over 21 mb. This is the same vapor pressure that would exist at a water surface at a temperature of about 18°C (64.5°F). If the surface-water temperature were below this temperature, the vapor pressure at the surface would be below the vapor pressure of the air and condensation onto the surface would be occurring instead of evaporation.

The air at 15.6°C and 60 percent relative humidity has a vapor pressure of about 10.6 mb, equal to the vapor pressure at a water surface at a temperature of about 7.5°C (45.5°F). Thus, if the moist surface in our example had a temperature between 7.5 and 18°C, there would be evaporation into the air sample at 15.6°C and condensation from the air sample at 26.7°C. Only if the temperature of the moist surface is above 18°C will evaporation into both samples occur. Even then the rate would be greater into the air at 15.6°C because the evaporation is proportional to the vapor-pressure gradient.

Terms such as "evaporability" or "evaporating power of the air" have also been used to express a concept similar to potential evapotranspiration. Thornthwaite (1954b, p. 209) writes:

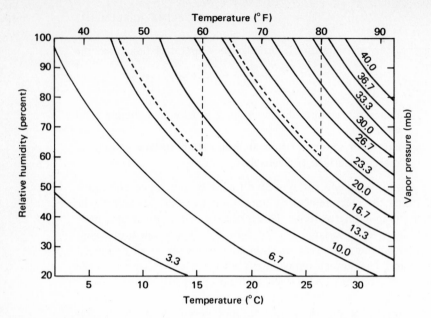

Relation between temperature and moisture content of the air. (*From Mather, 1959a, fig. 1, p. 35.*) FIG. 2-12

It should also be clear that there is no relation between potential evapotranspiration and expressions relating to the "evaporating power of the air" such as relative humidity or saturation deficit; they are not synonymous. The moisture content of the air is strongly influenced by the evaporation regime. Where the soil is dry and there is little evaporation, relative humidity is low and saturation deficit high. But if the conditions required for potential evapotranspiration are introduced, evaporation increases and more moisture is in the air. Thus, atmospheric moisture is not a conservative property of the air. Consequently, it is not possible to determine potential evapotranspiration by considering either relative humidity or saturation deficit.

In 1953, while participating in a field study in O'Neill, Nebraska, Halstead (1954) and his associates obtained a very extensive series of micrometeorological measurements from which it was possible to compute the various components of the net radiation for several different days. The computations, which are summed in Table 2-15, show that the percentage of the total net radiation used in evaporation $[E/(Q+S+E)]$ is almost directly proportional to the amount of moisture in the soil profile expressed as a percentage of field capacity.

From the results given in Table 2-15 it might be concluded that net radiation could be used to compute evapotranspiration. However, there is still the problem of advection or stored heat supplying some additional energy for evapotranspiration.

During 1955 and 1956 a net radiometer was in operation at the evapotranspirometer installation at Seabrook, New Jersey. Figure 2-13 shows

TABLE 2-15

Heat used for convection, evaporation, and soil storage, and soil moisture content, O'Neill, Nebr., 1953*

DATE	HEAT USED FOR CONVECTION, Q, CAL/CM²	HEAT STORED IN SOIL, S, CAL/CM²	HEAT USED FOR EVAPORATION, E, CAL/CM²	TOTAL Q+S+E CAL/CM²	$\frac{E}{Q+S+E}$, %	SOIL MOISTURE IN 0–460 MM PROFILE, † MM	SOIL MOISTURE AS % OF FIELD CAPACITY, %
Aug. 8–9	115.0	2.2	273.8	387.3	71	60	60
Aug. 18–19	59.1	–4.8	287.8	324.1	84	80	80
Aug. 22	98.4	19.0	216.2	333.6	65	64	64
Aug. 25	181.9	41.5	131.8	355.2	37	37	37
Aug. 31	242.3	28.3	44.5	315.1	14	20	20

* With permission C. W. Thornthwaite Associates, from Halstead, 1954.
† Moisture content at field capacity 22 percent by volume or 100 mm in a 460-mm profile.

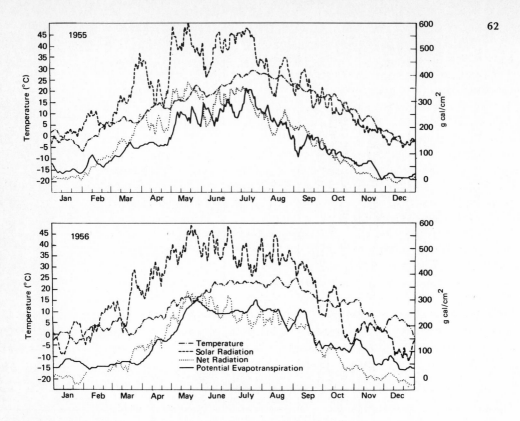

March of temperature, radiation, and potential evapotranspiration, Seabrook, N.J., 1955 and 1956.   FIG. 2-13

the march of temperature, radiation, and evapotranspiration at Seabrook. The data are presented as 10-day running means to smooth the daily fluctuations somewhat. The potential evapotranspiration is given in g cal/cm^2, the energy needed to evaporate the measured water loss; the curves of net radiation and potential evapotranspiration are, thus, directly comparable. The close correspondence between these two curves indicates that for Seabrook, at least, practically all the net radiation is utilized in evaporating water when the soil is moist. In March, April, and May not all the net radiation is used for evapotranspiration, for with cold soil and air in springtime slightly more net radiation goes into heating the air and soil than at other times of the year. In October, November, and December evapotranspiration is greater than one would expect from net radiation alone. At Seabrook, advection and stored soil heat play only minor roles in the energy budget over a moist area.

2 Instrumental or Theoretical Means to Determine Evaporation
Scientists have tried various ways to determine the amount of water lost from a vegetation-covered surface. One of the earliest attempts was to remove leaves or branches from a plant, weigh them, allow them to dry for a time,

and reweigh them to determine the water loss. Plants were placed in sealed containers and the amount of moisture accumulated in the confined air was also determined. Plants were grown in pots that were weighed periodically to determine the evapotranspiration losses. All these methods are highly artificial and the results cannot be used for anything but gross generalizations.

Evaporation pans, inflow and outflow measurements in lakes, and measurements of the water needed in large irrigation enterprises provide possibly better measures of evaporation or evapotranspiration, but (1) only a few fairly reliable results are available for study and (2) the results are so dependent on the size and exposure of the evaporating surface that great care must be exercised in evaluating them.

Middleton and Spilhaus (1953) separate the evaluation of evaporation into two general areas which they call relative evaporation gages or atmometers and absolute evaporation measurements. They point out that instruments providing actual measurements of evaporation are purely relative; for comparable results the evaporating surface would always have to have the same degree of wetness, the same roughness characteristics, and be exposed in exactly the same way to sun and wind. Even with similar instruments this is difficult. Absolute values of evaporation are possible from mathematical expressions based on factors of the moisture and heat exchange.

Instrumental methods to measure evaporation usually consist of various-sized tanks or pans filled either with water or with soil and vegetation (depending on whether evaporation or evapotranspiration is being measured), or atmometers having small water-filled porous porcelain bulbs or wetted pieces of paper as the evaporating surface. The small pans or the atmometers introduce many measurement problems as a result of exposure, size, surface configuration, color, and availability of water for evaporation and usually can provide only a rough estimate of the rate of evaporation under given meteorological conditions. Somewhat better are the larger evaporation tanks. They include, for example, the Weather Bureau class A pan and the Bureau of Plant Industry pan, 1.2 and 1.8 m in diameter, respectively. The class A pan is 0.25 m deep, and the pan is supported 0.15 m above the ground surface. Radiation can fall on the sides of the pan and air can move around and under the pan. Evaporation is obtained by measuring the change in water level in a stilling well. Rain is allowed to fall on the pan surface. Its amount is determined by a nearby raingage.

The Bureau of Plant Industry (BPI) pan is 0.6 m deep, and the pan is buried in the ground to a depth of 0.5 m. Thus it responds better to soil temperature and is better shielded from side-wall radiation than the class A pan. Another version of the BPI pan has the water surface covered with a wire screen. Evaporation for all these pans must be adjusted by coefficients that vary from season to season in order to provide estimates of lake evaporation. In the case of the class A pan the coefficient is estimated at 0.7 to 0.8, while it is near 0.9 for the BPI pan and approaches 1.0 for the screened sunken pan. These values are only approximate. Since in each case the coefficient is a function of certain sensor characteristics as well as the humidity and wind velocity, it can vary appreciably with time.

Soil-filled tanks or lysimeters have also been used to measure evapotranspiration. Operated under natural conditions, they measure the actual evapotranspiration; if they are supplied daily with water by sprinkling or by subirrigation, they can provide values of potential evapotranspiration. Thornthwaite developed an evapotranspirometer (anywhere from 4 to 8 m² in surface area and approximately 0.7 m deep to allow room for proper root development) to determine the potential evapotranspiration at a place. Various watering practices and crops have been tested in these tanks and in the surrounding buffer areas. If operated carefully, the evapotranspirometer can provide a good measured value of evapotranspiration, but it is extremely easy to obtain a biased reading if the vegetation or the soil moisture in the tank is unlike the surroundings in any respect.

In recent years, there has been a rapid development of indirect methods for the absolute determination of evapotranspiration. These techniques have considered evaporation or evapotranspiration to be a physical process responsive to the supplies of available energy reaching the surface or to the gradients of wind and vapor at a point. While it is beyond the scope of the present work to develop in detail each of the different approaches, it is proper to review briefly the form and nature of several of the more recognized expressions for determining evaporation or evapotranspiration.

Aerodynamic or profile techniques have been applied to the evaporation problem. They involve certain assumptions concerning the mechanisms by which heat and water vapor are diffused in the atmosphere by turbulence. It is assumed that the rate of vertical diffusion is proportional to the product of the height gradient of the temperature or moisture content and a turbulent-diffusion coefficient that reflects the intensity of turbulence in the air flow over the ground. The turbulent-diffusion coefficients are assumed to depend on the gradient of the wind velocity profile. These coefficients are found to be approximately equal only when the atmosphere near the ground is at or very near neutral stability (the heat flux is zero or very small). When this condition is not satisfied, the buoyancy of the atmosphere influences the diffusion coefficients. The coefficients for heat and water-vapor flux differ increasingly as the stratification departs from neutrality.

One of the earliest expressions for evaporation using the aerodynamic approach is due to Thornthwaite and Holzman (1942) and is of the form

$$E = \rho k^2 \frac{(q_2 - q_1)(u_2 - u_1)}{[\ln(Z_2/Z_1)]^2}$$

where ρ = air density
k = von Karman's constant (approximately 0.4)
q = specific humidity
u = mean wind velocity
Z = height

The subscripts 1 and 2 denote particular levels above the surface. Note that this expression involves the product of the logarithmic height gradients of specific humidity and the mean wind velocity. The product of the velocity gradient and von Karman's constant are, in effect, a diffusion coefficient.

Since the expression takes no account of the temperature structure of the atmosphere, its use is limited to neutral stratification. Many workers following Thornthwaite and Holzman have pointed out this limitation and suggested corrections for varying stability conditions (Deacon and Swinbank, 1958; Munn 1961; Monin and Obukhov, 1954; Crawford, 1965).

One current approach employs the so-called "eddy correlation technique." In a turbulent velocity field, a net transfer or diffusion of heat or water vapor in the upward direction can only occur if upward-moving turbulent eddies tend to be warmer or more moist than downward-moving eddies. The magnitude of the flux can be obtained from simultaneous measurements of vertical velocity and temperature (or moisture). The average product (not the product of the averages) of vertical velocity and temperature will, when multiplied by air density and specific heat, give the flux of heat due to turbulent transport plus any flux due to the average vertical air motion. This latter term, called the "mean flux," is calculated from the product of average vertical velocity and average temperature, density, and specific heat. The difference between these two terms is the flux due to turbulent eddies. The required measurements must be made with sensitive and fast-responding instruments and are difficult to accomplish. Appropriate instrumentation is being developed and tested by many groups (e.g., see Dyer and Maher, 1965; Mather (ed.), 1967).

Energy-budget techniques involve partitioning the available energy R_n (net radiation) into its various categories of use at the earth's surface and determining evaporation as a residual of the other measured terms. Ignoring the energy used in photosynthesis and other minor miscellaneous exchanges, the energy budget at the earth's surface can be written

$$R_n = S + H + LE$$

where S = soil heat flux
       H = atmosphere heat flux
      LE = energy going into evaporation of water

Bowen (1926) felt that S would be very small when adequate moisture was present in the soil, and so he developed a relation $B = H/LE$. Later Thornthwaite and Mather (1955c) suggested that under conditions of always-optimum soil moisture content almost all of R_n went into LE and that both S and H could be neglected as a first approximation. Energy advection was neglected in this suggestion.

If the exchange coefficients for heat and vapor are equal, it would be a simple matter to evaluate the Bowen ratio. Despite their probable lack of equality, Suomi and Tanner (1958) have solved the energy-budget equation for LE as follows:

$$LE = \frac{R_n - S}{1 + B}$$

Under humid conditions, errors in B due to the lack of equality of the appropriate turbulent diffusion coefficients result in only very small errors in LE.

Penman (1948, 1956a) has combined the energy-budget approach with the aerodynamic approach to achieve a relation that eliminates normally unmeasured terms. His expression for open-water evaporation (E_0) is

$$E_0 = \frac{\frac{\Delta R_n}{\gamma L} + f(u)\,(e_a - e_d)}{\Delta/\gamma + 1}$$

where Δ = slope of saturation vapor-pressure curve vs. temperature, mb/K

γ = psychrometric constant, mb/K

R_n = net radiation, erg/cm^2 sec

L = latent heat of vaporization for water, cal/g

$f(u)$ = wind function equal to 0.35 (0.5 + u_0/100) mm/day over open water

$e_a - e_d$ = saturation deficit (e_d is vapor pressure corresponding to dew-point at screen height)

This expression is converted into potential evapotranspiration over a grass surface by multiplying by an empirically derived coefficient that is 0.8 in the period from May to August, 0.7 in September, October, March, and April, and 0.6 in November through February for southeastern England. Other values must be determined for other areas.

Many workers have suggested more empirical techniques by which to compute potential and actual evapotranspiration. Possibly the best known and probably the most widely used is due to Thornthwaite (1931, 1948).

Thornthwaite's expression for evapotranspiration is simple to evaluate, requiring only information on temperature and day length. However, since it assumes average values of wind speed and humidity, it does not work as well as some other expressions in monsoon climates or in areas where there is a strong seasonal change in humidity or wind. To its advantage, however, is the fact that it can be evaluated almost everywhere so that some estimate of the areal and temporal variations in evapotranspiration can be obtained.

Thornthwaite's formula for unadjusted potential evapotranspiration (in cm/month) is

$$e = 1.6 \,(10\,\frac{T}{I})^a$$

where T = monthly temperature, $^\circ$C

I = annual heat index (sum of 12 monthly heat-index values (i)

determined from $i = (t/5)^{1.514}$

a = nonlinear function of heat index, approximated by expression $a =$
$6.75 \times 10^{-7} I^3 - 7.71 \times 10^{-5} I^2 + 1.79 \times 10^{-2} I + 0.49$

Without nomograms and tables, evaluation of evapotranspiration would be a time-consuming process. The unadjusted potential evapotranspiration, so obtained, is for an average 12-hr day and a 30-day month. The value must be adjusted by multiplying by a factor that expresses how the particular month varies in hours per day and days per month from the average. We shall use this expression in the next few chapters in evaluating the factors of the climatic water balance.

Many other expressions for evaporation or evapotranspiration could be outlined, but those already included provide an indication of the nature of the approaches being used with considerable success at the present time. Interested readers are referred to Budyko (1956), van Bavel (1966), and Baier et al. (1965, 1968; see also chapter 5) for three additional approaches.

3 Distribution

Several maps of the areal distribution of evaporation from class A pans or lakes or the evapotranspiration as computed by one of the various empirical formulas have appeared. Thornthwaite (1948) has provided a map of mean annual potential evapotranspiration (Fig. 2-14) which, while having the same general pattern as recent National Weather Service (ESSA, 1968) maps of lake evaporation, differs appreciably in absolute values in arid areas. Annual evaporation based on estimates from lakes exceeds 86 in. (218 cm)/year along the California-Arizona border, while Thornthwaite shows just 60 in. (152 cm) of evapotranspiration. Throughout the northern Great Plains up into Montana, annual lake evaporation varies from 30 to 40 in. (76 to 102 cm), while Thornthwaite indicates no values over 30 in. (76 cm) and most below 24 in. (61 cm). The lake evaporation map should, of course, indicate somewhat higher values since it reflects the influence of both solar and advective energy, while Thornthwaite's potential evapotranspiration assumes no influence of advection.

Geiger (1965a) has also prepared a world map of actual evaporation or evapotranspiration based on energy-budget considerations and estimates of the moisture storage available for evaporation. Discussion of Geiger's map will be saved until the next chapter.

Budyko (1956) has achieved a detailed world map of the annual expenditures of heat for evaporation (LE in kcal/cm^2), using information on net radiation and his expression for evaporability. Isolines on the Budyko map are not continuous but break sharply at the continental borders as would be expected because of the greatly different opportunities for evaporation over the land and water surfaces. The map shows very low values of LE in the major desert areas. In land regions of the moist tropics, energy for evaporation values reaches 60 kcal/m^2 year, equivalent to about 1 m depth of

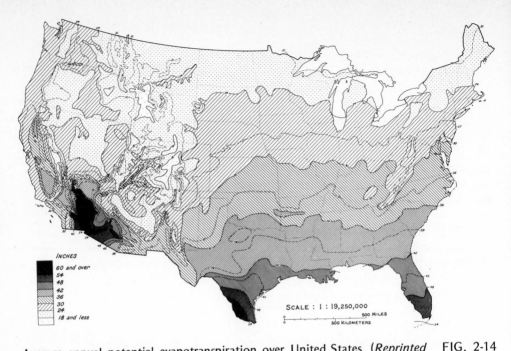

Average annual potential evapotranspiration over United States. (*Reprinted*   FIG. 2-14
from the Geogr. Rev., *vol.* 38, 1948, *copyrighted by the American
Geographical Society of New York.*)

evaporation. Over the subtropical ocean areas, maximum values reach 120 to
140 kcal/m² year, resulting in up to 2 m of evaporation/year.

4 Limitations

When a psychrometer is used to determine the humidity of the air, a
thermometer bulb is moistened and becomes an evaporating surface. The
water evaporating from the wet bulb thermometer cools the bulb. The surface
area of the bulb is small and the amount of water vaporized is very small.
Heat flows into the water film on the bulb from the warmer surrounding air,
and the evaporation process will reach equilibrium at a rate and at a wet bulb
temperature such that the energy appropriated from the air is just sufficient to
maintain the evaporation. Solar radiation contributes almost no energy to this
process. The water from the wet bulb moistens the air but the amount is so
minute that the effect on the moisture content of the air is completely
negligible.

The Weather Bureau class A evaporation pan on a summer day in a dry
situation may evaporate 8 liters of water. Solar radiation contributes an
important share of the energy for evaporation, the amount depending on the
turbidity of the water and on the albedo of the pan, which varies greatly with
type, age, and condition of the material used. Additional energy for
evaporation is available from the air. The amount of water evaporated from
the pan will do little to modify the moisture content of the air, but

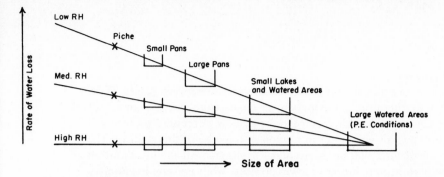

Schematic relation between rate of water loss under different humidity   FIG. 2-15
conditions and evaporation surface areas. (*From Thornthwaite and Mather,*
1955b, fig. 1.1, p. 16.)

immediately over the water surface the humidity is raised, the moisture
gradient reduced, and the evaporation impeded. The extent of this influence
depends on the rate at which fresh air passes across the evaporating surface
from outside.

  If the area of the evaporating surface is large, the influence of the
moisture condition of the air passing over the evaporating surface becomes
small and solar radiation is the primary source of energy for evaporation. The
influence of the evaporation on the atmospheric humidity is very important
because this influence cannot fail to have a reciprocal effect on the
evaporation. In moist air the temperature of the evaporating surface will rise
to a point above the dew point of the air such that the evaporation will just
use the energy that is available. Similarly, in dry air rapid evaporation will
lower the temperature of the evaporating surface until the evaporation is in
accord with the available energy. The size of the water body or moist land
area necessary to ensure potential evapotranspiration is difficult to determine.
The size of the area under the high moisture conditions has to be large
enough so that the evapotranspiration from the area is not affected by
external factors such as the advection of moist or dry air masses and their
modification by local conditions. Water pans which vary from one another
only in size are influenced differently by such conditions and, thus, result in
different values of evaporation. Most land evaporation pans are certainly too
small to be uninfluenced by external conditions and, in some cases, even soil
pans surrounded by swamps and marshes appear to be too small. Expressed
schematically the relation of the various pans to the larger evaporation
surfaces is shown in Fig. 2-15.

  This diagram shows that water loss is fairly independent of the size of
the measuring instrument when the humidity is high, but, as the air becomes
drier, the size of the evaporating surface greatly influences the rate of
evaporation or evapotranspiration from it. Because of this limitation, pans or

other instruments like the Piché evaporimeter do not give reproducible measures of evapotranspiration except possibly under conditions of high humidity.

WIND

1 Basic Relations

Wind results from differences in atmospheric pressure between places. These may develop as a result of differences in air temperature or in the amount of upper-level convergence or divergence in the air flow. A pressure gradient will be established and air will be set in motion in response to this gradient. The pressure-gradient force will be proportional to the gradient and in a direction perpendicular to it, from higher to lower pressure. While pressure is a primary meteorological factor in synoptic meteorology and weather forecasting, it is of less importance in applied climatology.

Direction, speed, gustiness, and frequency of calms are all important characteristics of wind. Wind is a very unstable parameter in most parts of the world, fluctuating markedly within a matter of minutes or hours, changing directions with passing weather systems. It is influenced by both local environmental as well as general atmospheric conditions.

The variability of the wind is revealed in both the wind direction and wind speed. The word "velocity" includes both the direction and speed properties of the vector quantity wind. Velocity, unfortunately, has been used too often to refer to speed alone. Wind direction always refers to the direction from which the wind is coming. A north wind blows from the north and toward the south. Wind speed is usually given in knots or nautical miles per hour although most of our earlier records give winds in miles per hour. For research purposes, speeds are often expressed in centimeters or meters per second.

Figure 2-16 provides a record of the horizontal wind speed (indicated by "cup") as well as the vertical motion of the wind (indicated by "vane") at 16 and 32 m above the surface on a June evening at Centerton, New Jersey. The response of these particular sensors is rapid and the whole record covers only 7 min. The horizontal wind at 32 m shows several changes in speed of over 800 cm/sec within 15 sec, while variations in the vertical wind exceed 300 cm/sec in the same interval.

Commonly, the term "wind" refers to the horizontal motion of the air—that motion which can be measured by means of convenient sensors such as the three-cup anemometer. This type of sensor responds well to the horizontal component of the wind but not as well to wind arriving at some more vertical angle. Wind is not just a horizontal motion for there are circular-moving eddies in the air that pass a particular sensor providing both upward- and downward-moving currents as well as forward- and backward-moving flows. Most anemometers are not sensitive enough to measure motions within the smaller eddies; they give only some averaged value of the horizontal wind movement over a period of time. However, the action of the individual eddies and the sum total of the up and down movement of the

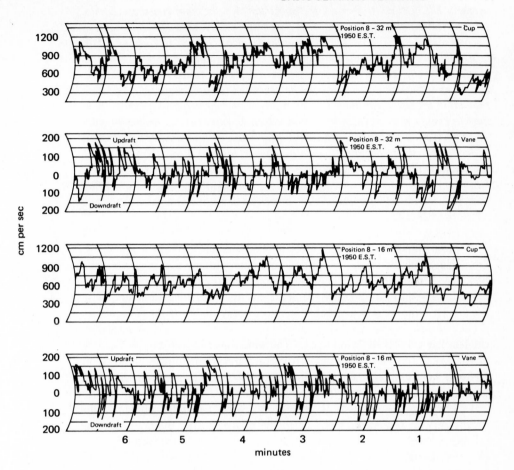

Vertical and horizontal wind speeds at selected levels, Centerton, N.J.,   FIG. 2-16
June 14, 1961. (*With permission, from Thornthwaite, Superior, and Mather,*
1961, fig. 16.)

wind at a place over a period of time are both important aspects that need to
be understood.

Turbulence consists of traveling or stationary eddies superimposed on the
smooth mean flow of the atmosphere. Much of it arises from frictional forces
within the air itself, or along the earth-air interface. Some, however, arises
from buoyancy, i.e., from thermal inequalities that lead to free convection. A
wind record taken at any level within the bottom 3,000 ft shows that all
three velocity components, u, v, and w, of the wind vector V are subject to
eddy fluctuations with a wide range of frequency and period (which are
reciprocals of each other). The largest eddies have periods of some hours, and

belong to the field of mesometeorology; thunderstorms and tornados belong in this class. Below this scale the spectrum is unbroken; there are eddies taking many minutes to pass the observer, others that last only a few microseconds—measurable by the so-called microscale of turbulence. It is probable that most of the energy involved in turbulence comes from the mean flow, and enters the spectrum at the longer periods from the large eddies, and "flows" down through the spectrum towards the smallest eddies; the latter are so small that the viscosity of the air rapidly converts their kinetic energy into molecular motion which is expressed as ordinary heat. The large eddies, in other words, constantly break up into smaller and smaller eddies until their energy is wholly converted into heat. The so-called spectrum of turbulence, *referred to above, is a mathematical device actually a frequency distribution, whereby one specifies what fraction of the total energy is associated with each "size" or period of eddy.* (Thornthwaite et al., 1959, pp. 122–123)

Most published wind data involve a considerable amount of averaging. Prevailing wind direction, that most commonly found over a given time period, is often published along with average wind speeds. Such data are of limited use; a single direction to represent the prevailing direction of a fluctuating wind is not justifiable in most areas of the globe. Frequency studies provide better data; often wind statistics are presented by means of circular diagrams called "wind roses." The wind rose provides some indication of the amount of time the wind is from each direction (usually on an 8- or 16-point compass) and a rough indication of the frequency of different ranges of wind speeds with different directions (Fig. 2-17).

The term "gustiness" in relation to wind can refer either to its variability with respect to speed or to direction. Weather Bureau observing practice is to report gusts when the peak wind speed reaches at least 16 knots (8 m/sec) and the variation in speed between peaks and lulls exceeds 9 knots (4.6 m/sec). The duration of a gust is usually less than 20 sec. The ratio of the total range of wind speed between peaks and lulls to the mean wind speed averaged over both peaks and lulls is called the "gustiness factor."

Singer and Smith (1953) have used the term gustiness in relation to wind direction and have developed a fivefold classification of gustiness based on the character of the chart trace of wind direction. Sample sections of five wind-direction traces at 355 ft (108 m) above the ground are shown in Fig. 2-18 to indicate the characteristics of each type.

Air flowing along the surface of the ground encounters a frictional resistance and is slowed down. The effect of this resistance is felt for hundreds of feet aloft. Within this zone of frictional influence, each successively higher layer of air moves faster than the one immediately beneath so that shearing stresses are set up between them, resulting in upward and downward displacement of small masses of air. Each mass tends to preserve the horizontal velocity that it had previously. If a slower-moving mass of air moves upward to a faster-moving layer, the process of mixing will tend to retard the air in the higher layer. Similarly, a mass of air moving

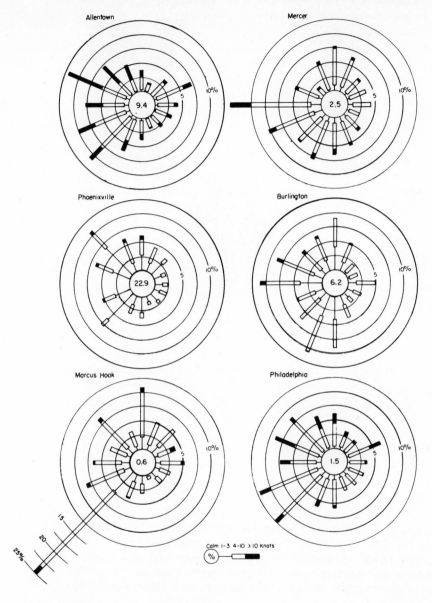

Annual wind roses at selected stations in the Delaware Valley for 1965.   FIG. 2-17

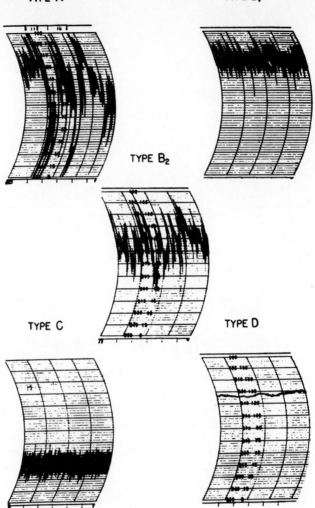

Selected traces of wind direction, illustrating different classes of gustiness. *A*, FIG. 2-18
fluctuations in wind direction $> 90°$. B_2, fluctuations 45 to 90°; B_1,
fluctuations 15 to 45°. *C*, fluctuations $> 15°$; unbroken solid core of trace.
D, fluctuations are short term, $< 15°$; trace approximates a line. (*With
permission of the American Meteorological Society, from Singer and Smith,
1953, fig. 1.*)

downward will tend to speed up the air in the lower layer. Thus the
difference in velocity between two adjacent levels tends to be equalized as
horizontal momentum is transferred downward and ultimately dissipated at
the ground. The tendency toward equalization of velocity depends on the
rate of vertical mass interchange or the intensity of turbulent mixing.

The greater the differences in velocity between adjacent layers, or the
stronger the shear, the greater is the tendency for turbulence to develop and

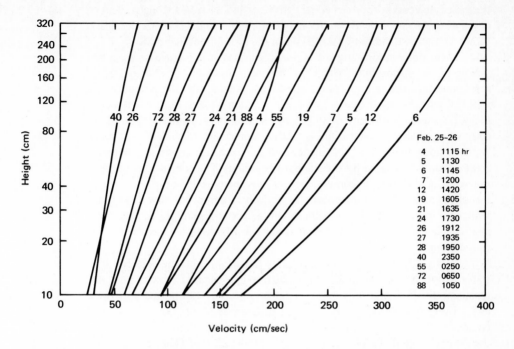

Wind profiles near the ground during the period Feb. 25–26, 1957,   FIG. 2-19
Centerton, N.J. (*With permission of C. W. Thornthwaite Associates, from
Davis, 1957, fig. 22, p. 42.*)

reduce the shear. If there were no external influences helping to maintain the
differences in wind velocity at different levels, turbulent mixing would
eventually equalize the velocities in all levels and the air would move
uniformly. However, since the supply of momentum in the upper air is
practically inexhaustible, and since momentum is constantly being converted
into heat and lost at the ground surface, wind velocities at different levels are
never completely equalized. Thus, while turbulence tends to destroy shear,
still the strongest mixing at a fixed level near the ground is associated with
the strongest shear (Thornthwaite, 1943a; Thornthwaite and Holzman,
1942).

The influence of atmospheric stability on the wind speed near the
ground may be seen in a series of 10-min profile observations obtained at
Centerton, New Jersey, during several hours on February 25 and 26, 1957
(Fig. 2-19). The weather was mild and sunny during the first day, changing to
cloudy with showers in the evening and with heavy rain beginning the next
day. The data are plotted on a semilogarithmic scale; if the observations fall
along straight lines, it indicates that wind velocity varies as the logarithm of
height. Actually only one single series (line 55) follows a straight line. The
daytime observations usually plot along lines having concave curvature (note
curves 4, 5, 6, 7, and 12). The curvature shifts from concave to convex in the

early evening between 1735 and 1907 hr. During the night, as the clouds thickened, the curves approached straight lines and remained straight the next day during the rain. The difference between a sunny and a rainy day is clearly shown by the first and last observations in the series (lines 4 and 88).

These observations suggest a reasonable physical picture of turbulent mixing in the atmosphere. When the thermal lapse rate is adiabatic, as for brief periods in the morning and again in the evening, the logarithmic law is fulfilled, and the coefficient of turbulent mixing is directly proportional to the wind speed. Turbulent mixing is mechanical only. In the middle of the day, when thermal convection is active, convective turbulence is super-imposed upon mechanical turbulence and the exchange coefficient becomes greater than is indicated by the standard equations. The concave curvature of the wind gradients is a consequence of this increase in the coefficient of turbulent mixing. As momentum is actively transported downward, the tendency for differences in velocity in adjacent layers to be equalized increases, and the wind profile becomes more nearly vertical except very close to the ground surface.

At night when the ground, cooled by radiation, cools the adjacent air, a stable lapse rate is created. The exchange coefficient becomes less, and turbulent diffusion is strongly inhibited. Momentum is transported downward less actively, the tendency for velocity differences in adjacent layers to be equalized diminishes, and the velocity profile departs more than ever from the vertical except within a foot or two of the ground (Thornthwaite and others, 1943).

While the pressure gradient causes the air to start in motion and establishes its initial direction of movement, this relationship does not hold for long. As soon as motion over the surface occurs, not only does friction take its toll, slowing the wind speed near the earth, but because the earth is rotating on its axis, the wind direction appears (to an observer on the earth's surface) to curve to the right in the Northern Hemisphere and to the left in the Southern Hemisphere. We describe this effect by saying that a Coriolis force has been applied to the moving air parcel, and it is this force that causes the deflection to the right or left of the parcel's original path. The Coriolis force is a maximum at the pole and zero at the equator.

The air parcel will continue to curve to the right (Northern Hemi-sphere) of its initial path of motion until it is finally moving at right angles to the pressure-gradient force or along the lines of constant pressure. The Coriolis and the pressure-gradient forces are then exactly in balance and in opposite directions, resulting in no further turning of the wind. If further turning to the right did occur, the wind would blow against the pressure gradient, which is not possible. If the final wind (assuming no effect of friction) blows parallel to straight lines of equal pressure, it is called the "geostrophic wind," while if these lines are curved, it is called the "gradient wind." The geostrophic wind actually is a special case of the gradient-wind relationship when the turning acceleration is zero.

Under average conditions there is a fairly regular diurnal pattern of variation of wind velocity. Maximum speed is usually found early in the

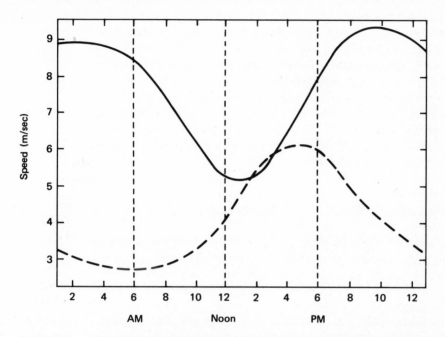

Comparison of diurnal course of wind speed at two elevations (data averaged   FIG. 2-20
1899–1911). (*With permission*, Monthly Weather Rev., *from Wright*, 1916.)

afternoon (between about 1 to 4 P.M.) and the minimum speed will be found
in the early morning hours just before sunrise. The reverse of this pattern is
found at some elevation above the surface (Fig. 2-20) (Wright, 1916). Since
the wind speed increases with height, convectional activity during the day
results in a good exchange of air between the surface layers and those above,
increasing wind speed in the surface layers and slowing upper level winds. At
night, slower-moving air near the surface does not mix with the air aloft, and
so the upper air layers reach their maximum speed during the nighttime
hours.

Friction will reduce the wind speed near the surface. It will also prevent
the wind from turning to the right (in the Northern Hemisphere) as much as
it would if only under the action of pressure-gradient and Coriolis forces.
Thus, with friction, pressure-gradient, and Coriolis forces all acting, the actual
wind at the earth's surface will move across the lines of constant pressure at
some angle (to the right of the pressure gradient direction in the Northern
Hemisphere). The angle depends on the amount of friction; over a very
smooth surface the wind direction will essentially parallel the lines of
constant pressure, while over a very rough surface, wind will cross the

constant-pressure lines almost perpendicularly. Thus one would expect a shift in wind direction with height as friction decreases. In the Northern Hemisphere, the wind direction spirals clockwise from its surface direction as the wind vane is moved upward in the lower atmosphere.

2 Instruments to Measure Wind

Early wind observations, made long before instruments were available, consisted of information on wind direction and some estimate of the effect of the wind speed. Admiral Beaufort, in 1806, developed a wind scale based on the effect of the wind on a nineteenth-century British man-of-war under sail which has been extended, modified, and somewhat corrupted over the years. Actual velocities cannot be identified with the original Beaufort scale since that type of ship went out of service before anemometers came into use. More recent versions of the Beaufort scale have been developed for land use and refer to the influence of wind on leaves, tree branches, and buildings. The Beaufort force numbers are now identified with velocity ranges measured by means of anemometers. The persistence of the Beaufort scale is one of those peculiar anachronisms of history that are most difficult to explain in the light of modern developments in anemometry.

Wind direction is usually measured by means of a wind vane which can be anything from a sock or streamer free to rotate, to a moveable arrow with suitable fins so that the vane will orient itself into the wind. The vane thus points toward the approaching wind even though wind direction is always given as the direction from which the wind is coming. The wind direction is recorded on the basis of an 8- or 16-point compass usually, although recent observations are given by $10°$ increments (a 36-point compass) starting with north as 0. The change in wind data from the 16- to the 36-point compass sometimes creates some data-summary problems in trying to assign certain wind directions. (For example, with a 16-point compass, each direction can be assigned a range of $20°$. But 4 of the compass directions will have to cover $30°$ intervals to divide the 360 into 16 directions. Bias will exist if 12 directions represent wind from $20°$ arcs and 4 directions represent wind from $30°$ arcs, so various compromise procedures have been suggested to permit interchange of data between the two scales. None of the compromises is entirely satisfactory.)

To measure wind speed, several types of anemometers have been developed and used over the years. The most straightforward is the pressure-plate anemometer, a flat metal pendulum hinged at the upper edge and mounted on a swivel so that the plate will always face directly into the wind. As the wind hits the plate, it causes it to swing upward; the amplitude of this swing is directly related to wind speed.

The most commonly used anemometers are the three- or four-cup rotating type. These sensors consist of three or four light metal or plastic hemispheres, mounted on arms free to rotate about a vertical axis. The number of rotations can be counted by means of various gear assemblies or photoelectrically. Several problems can develop from the operation of most rotating anemometers. First, the convex sides of the cups themselves will provide some wind resistance so that the instrument does not respond immediately to sudden wind changes. The weight of the cup assembly also contributes to the difficulty of getting the whole assembly rotating (high starting speed), and, of course, once the assembly is rotating, it may continue to rotate due to inertial overshoot even though the wind has stopped. Only if the rotating anemometer is designed out of lightweight materials with a minimum of internal frictional resistance will it be responsive to small-scale wind eddies. C. W. Thornthwaite Associates has developed a Portable Wind Profile Recording System to measure the microstructure of the air with minimum interference to the natural flow of air past the sensor. The anemometer cups are lightweight plastic reinforced with aluminum rings. Small stainless steel tubes form the spokes and the hub is of aluminum. The entire cup assembly weighs only 7 g. Miniature ball bearings are used to guide the rotation of the shaft. A shutter mounted on the shaft interrupts a light beam from a 3-V grain-of-wheat lamp housed in the base of the anemometer mount. This interruption of the light beam is recorded by a photocell which in turn activates an electromechanical counter. Weight and friction are thus minimized; the starting speed is less than 0.2 mph (8.94 cm/sec).

A bridled anemometer differs from a rotating anemometer in that it has more cups (usually 32) and is prevented from rotating by means of a spring. The force required to keep the anemometer from rotating can be translated into wind speed.

Pressure-tube anemometers such as a Pitot tube on an airplane are also used to measure wind speed. The tube opening must be directed into the wind. Differences in pressure due to the force of the wind will activate a pressure gage which has previously been calibrated to give wind speed.

Another rotation anemometer in fairly common use at present is the Aerovane, manufactured by the Friez Instrument Division of Bendix Aviation. It consists of a three-bladed propeller-type plastic windmill mounted on a wind vane so that it is always directed into the wind. Thus direction and speed are obtained from a single sensor mounting.

Upper air winds are usually measured by following a free balloon either by theodolite or radar as it rises through the atmosphere at a predetermined rate. Some errors result because the balloon will not always move as rapidly as the wind, and as it rises it has some self-induced vertical motions which make it difficult to identify its exact height. These errors are of little consequence since the data are usually not needed to that degree of precision.

3 Distribution

Maps of prevailing monthly or annual wind directions or even average wind roses for selected time periods are of limited significance for applied climatology because of the generalization. Winds are so variable in direction and speed that frequency studies or wind roses under particularly significant events or for specific time intervals are much more useful for applied studies.

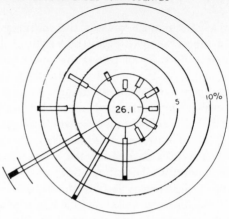

LOW SUSPENDED PARTICULATES

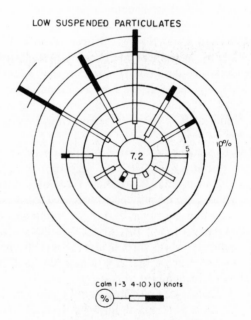

Composite wind roses, Delaware Valley area, during periods of high and low   FIG. 2-21
concentrations of suspended particulates, 1965 (data based on three wind
stations in vicinity of Philadelphia).

For example, Mather (1968*b*) prepared a series of wind roses for the
Philadelphia area showing prevailing wind directions and speeds at times of
low and high atmospheric pollution (Fig. 2-21). These particular wind
relationships are much more important to the particular subject under
investigation (meteorology and air pollution in the Delaware Valley) than
average wind roses or prevailing wind directions would have been. From these
wind roses one obtains specific information not only on the type of weather

situation which must be associated with various pollution conditions but also on the significant directions and speeds of movement of pollutants at particularly limiting times.

Various other wind summaries are available and useful for applied climatology purposes. For example, Table 2-16 gives the annual percent frequency of wind speed by various speed groups for different cities around the United States. Even without direction these data provide some indication of wind force to be considered for various engineering activities. Table 2-17 provides the detailed information on the fastest mile of wind ever recorded for each month of the year along with its direction for a number of United States stations. Again such extreme data have great usefulness in different engineering applications. Mean-value climatologies provide little help in such situations.

Many local circulations of wind flows exist because of local topographic and temperature differences. This is especially true when the general circulation is weak or ill defined. At those times, the influences of local environmental differences can result in significant local wind circulations. Examples of these locally important winds are the land-sea and the mountain-valley breezes, as well as the more restricted downslope or katabatic winds resulting either from the presence of a pressure gradient or because of gravity effects.

A warm, dry wind moving down the lee side of a mountain under a fairly strong pressure gradient is known as a "foehn" or "chinook" wind. Heating of the downward-moving air results from adiabatic compression. Foehn winds are found in most mountain areas, but the exact strength of the foehn or the amount of heating depends on a number of factors such as the strength of the wind flow over the mountain, the topography, and its orientation to the wind flow. Foehn or chinook winds can result in significant rises in temperature in lee valleys.

Very cold, descending winds are found in some mountainous areas. In these cases, the temperature of the original air (often a high-lying plateau region interior to the mountain range) is so cold that adiabatic heating as the air flows down the mountainside is not sufficient to raise its temperature even to normal for the region in question. The air, thus, appears in the lowland area as a cold, dry flow of wind, often in the form of rapidly moving gusts.

Land-sea breezes or mountain-valley breezes form when the general circulation is light and local topographic and heating effects become more influential. In each case, the air movement is from the area of cooler temperatures to the area of warmer temperatures; the unequal heating has resulted in the creation of weak high- and low-pressure systems respectively over the cool and warm regions.

Water surfaces remain relatively cool in the daytime while adjacent land areas are heated. Valley areas, likewise, being more shaded, will remain cooler in daytimes than the exposed mountain surfaces. Thus by afternoon sea breezes or up-mountain breezes may develop, moving air from the cooler to

TABLE 2-16

Annual percentage frequency of wind by speed groups and the mean speed at selected U.S. stations*

STATION	0–3 MPH	4–7 MPH	8–12 MPH	13–18 MPH	19–24 MPH	25–31 MPH	32 MPH AND OVER	MEAN WIND SPEED, MPH
Atlanta, Ga.	13	24	36	21	6	1		9.7
Boise, Idaho	15	30	32	18	4	1		8.9
Boston, Mass.	3	12	33	35	12	4	1	13.3
Dallas, Tex.	9	21	32	28	9	1		11.0
Denver, Colo.	11	27	34	22	5	2		10.0
Detroit, Mich.	8	23	37	26	5	1		10.3
Duluth, Minn.	6	15	33	31	11	4		12.6
Great Falls, Mont.	7	19	24	24	15	9	1	13.9
Huron, S.Dak.	10	18	29	29	10	3	1	11.9
Miami, Fla.	14	30	34	20	2			8.8
Phoenix, Ariz.	38	36	20	5	1			5.4
St. Louis, Mo.	10	29	36	21	3	1		9.3
San Francisco, Calif.	16	21	26	22	11	3		10.6
Seattle, Wash.	13	16	35	26	8	2		10.7
Washington, D.C.	11	26	35	22	5	1		9.7

* From U.S. Weather Bureau, Climatography of the United States 82; Decennial Census of the United States Climate Summary of Hourly Observations, 1951–60 (Table B).

the warmer areas. Sea breezes may move inland 30 to 60 km although they are most noticeable in the vicinity of the coast where the temperature differences are greatest. Over the hotter land the air will rise and flow back aloft to the ocean area as a small convective circulation cell.

At night, the land or the mountain area cools more than the water or the valley region. This leads to a reversal of the circulation and land breezes (to the ocean) or down-valley winds develop. The land breeze and the up-valley breeze are not as well developed as the sea breeze and the down-valley breeze. This is due to stronger temperature differences existing in the daytime across the land-water interface and the influence of gravity reinforcing the temperature differences at night in the mountainous region.

There are areas aloft in which the air is flowing much more rapidly than it is in neighboring areas. These fast-moving atmospheric currents are called "jet streams." The jet stream acts much like a high-speed river in the air that meanders from north to south high in the atmosphere. Even within the jet stream the air speeds up and slows down in a pulselike movement. The jet may be as much as 1,000 m thick and 500 to 600 km wide. Velocities in the stream may reach 90 to 130 m/sec (200 to 300 mph).

Several jet streams have been identified in each hemisphere, the subtropical jet and the middle-latitude jet, both in the troposphere (of the order of 9,000 to 15,000 m above the surface), and the polar jet (in winter only) in the lower part of the stratosphere above. Jet streams influence the movement of weather systems at the earth surface; their immediate influence in applied climatology lies in their effect on aircraft operation.

4 Limitations

Any instrument introduced into the atmosphere will disturb the atmosphere and thus make measurement of its properties suspect. Anemometers act in this manner by creating their own eddies and currents and so disturbing the natural flow of the wind. Probably more important, however, is the problem of exposure of the anemometer.

Since wind speed varies markedly with height, the level at which the anemometer is exposed becomes critical. No standards have been established, so that wind observations are usually not comparable from one place to another. Efforts have been made to place sensors on towers or high buildings but exposures vary appreciably; each wind observation can be accepted as only representative of that particular observation site. However, in the free air, wind and pressure will not change too rapidly over a horizontal area, so that some extrapolation between adjacent areas is possible. Near the surface, the influence of surface conditions is such that extrapolation does create uncertainties.

Mather (1968b) analyzed wind data from six nearby stations within the Delaware Valley area to show some of the problems of exposure and sensor reliability (Fig. 2-17). Two of the stations, Allentown and Philadelphia, are regular Weather Bureau observing stations with observations taken by meteorologically trained personnel. The exposures are excellent, and the observations can be considered quite reliable. The other four stations, Mercer, Phoenixville, Burlington, and Marcus Hook, are all industrial observation sites.

The annual wind rose at each of the stations for 1965 (Fig. 2-17) indicates, as expected, a predominance of winds from the westerly direction in the Delaware Valley area. At Allentown, the winds are almost equally from the northwest and southwest quadrants while, at Philadelphia, the southwest quadrant prevails. Phoenixville is much like Allentown with the northwest

STATION	YRS.	JAN.	FEB.	MAR.	APR.	MAY
Atlanta, Ga.	50	54 NW	59 W	66 NW	68 SE	59 NW
Boise, Idaho	22	50 SE	56 W	52 W	50 W	50 W
Boston, Mass.	43	66 SW	58 NE	73 E	63 NE	55 W
Dallas, Tex.	48	66 SW	61 SW	59 W	58 N	65 S
Denver, Colo.	89	51 SW	50 W	53 NW	56 NW	65 W
Detroit, Mich.	28	57 SW	49 W	68 SW	56 N	61 NW
Duluth, Minn.	35	65 NW	67 E	75 NW	75 NE	61 W
Great Falls, Mont.	18	65 SW	72 W	73 W	70 W	65 SW
Huron, S.Dak.	29	57 NW	56 NW	68 NW	73 SE	70 NW
Miami, Fla.	51	50 SW	68 SE	53 S	70 N	48 SW
Phoenix, Ariz.	24	41 WNW	49 SSE	50 WNW	45 NW	59 SSE
St. Louis, Mo.	45	59 SW	57 SW	82 SW	61 SW	61 SW
San Francisco, Calif.	24	60 SW	62 SW	52 NW	48 W	56 W
Seattle, Wash.	28	63 SW	64 S	60 SW	65 S	45 SW
Washington, D.C.	13	56 NW	57 SW	60 E	56 N	48 S

* *From ESSA, Environmental Data Service, "Climatic Atlas of the United States," 1968, p. 74.*

and southwest quadrants both equally favored, but the large number of light winds and calms, 22.9 percent, as compared to 1.5 at Philadelphia and 9.4 at Allentown, would suggest that the instrument was not as responsive to winds at Phoenixville as at Allentown or Philadelphia.

The Marcus Hook observations show a high predominance of southwesterly winds. Winds are from the southwest almost 25 percent of the time. The next most frequent direction is north with about 12 percent. While

TABLE 2-17

Fastest mile and direction of wind at selected U.S. stations*

JUNE	JULY	AUG.	SEP.	OCT.	NOV.	DEC.	YEAR
70	56	49	49	47	46	63	70
NE	SE	NW	N	NW	NE	W	NE
50	61	50	50	56	57	56	61
SW	W	SE	SE	SE	NW	NW	W
46	52	52	87	63	80	73	87
NW	NW	NE	S	NW	SE	W	S
65	77	56	48	61	56	47	77
N	N	NW	W	NW	W	S	N
56	57	47	50	47	58	51	65
SE	NE	N	NW	S	NE	NW	W
56	77	50	52	56	66	59	77
SW	NW	NW	SW	S	W	NW	NW
59	72	68	60	61	68	72	75
NW	W	N	NW	S	E	NW	NE
70	73	71	73	73	73	82	82
W	W	SW	NW	W	SW	SW	SW
65	77	70	64	72	73	56	77
SE	NW	S	NW	W	NW	NW	NW
48	53	62	132	122	94	52	132
NE	SE	SW	E	S	SE	S	E
46	71	60	75	48	45	68	75
NE	N	SSW	SW	SSW	SSW	W	SW
60	56	60	73	62	65	56	82
SW	SW	SW	W	SW	SE	SW	SW
48	50	45	48	51	55	53	62
W	W	WNW	S	SW	SSW	S	SW
54	38	35	55	63	57	60	65
SW	SW	SW	SW	S	SW	SW	S
57	54	49	56	78	60	62	78
NW	E	NE	SE	SE	E	SW	SE

southwest should be a prevailing direction, the extreme percentage from southwest, coupled with the high percentage from north, casts some doubt on the representativeness of the exposure of the instrument.

The Burlington wind rose seems to be quite similar to that at Allentown or Philadelphia. At Mercer, however, there is a very low percentage of winds from the northwest, a direction quite favored at the other observing sites. This low value of wind from the northwest shows up in every month of the year and must result from a local exposure problem.

Before any wind observation is accepted, the user should visit the site, study the exposure of the instrument, and compare its operating characteristics with other nearby observations. It is a relatively easy matter to obtain erroneous and misleading wind data.

THE WATER BUDGET AND ITS ROLE IN CLIMATOLOGY

3

The term water budget can have several meanings. To some, it is literally the long-term balance of water over the whole world, synonymous with the hydrologic cycle. The sum total of all available water remains essentially unchanged over time although the form in which the water appears—liquid, solid, or gaseous—or its particular composition—salt or fresh, polluted or unpolluted—may change. The term water balance can, therefore, mean the accounting of all the water over the globe on an annual basis in all its various forms and states.

To others, the term water budget recognizes the fact that a strict balance is not always possible because inflows do not equal outflows at a particular place and time. The term has more of a local connotation, referring to the actual short-period accounting of the moisture condition at a given spot on the globe. Of course, summation of such water budget information on an annual basis and from all areas of the globe results in the water balance for the world.

In the present volume we are more concerned with the second meaning of the term water budget—its use in expressing the moisture relationships, both instantaneously and longer-term, at individual places or over local watersheds, although we will deal briefly with the world water balance as well. The moisture factor in climate is of such fundamental importance in

applied climatology that we must be fully aware of both the global balance
and the local budget aspects of the term.

Precipitation, evaporation (or evapotranspiration as the combined water
loss from soil and vegetation is more properly called), and runoff are the
principal elements of the water budget. Since precipitation and evapotranspiration result from different processes, they are often not the same
either in amount or distribution through the year at a place. In some regions,
more precipitation occurs month after month than is needed by the
vegetation or is lost by evapotranspiration. The excess water (called
"surplus") moves over the soil surface or through the upper soil layers to
form streams and rivers and ultimately runs off back to the oceans. In other
areas, there is less precipitation month after month than is needed by the
vegetation, the soil dries out, and a moisture deficit occurs in every month.
There is no excess of precipitation and no runoff except locally, where the
soil cannot absorb the falling rain fast enough to prevent its runoff.
Consequently there are no permanent rivers, and there is no drainage to the
ocean. In most areas of the world, however, the precipitation is deficient in
one season and excessive in another so that a period of drought is followed by
one with a surplus and with runoff. The march of precipitation through the
year never coincides exactly with the changing demands for water in any part
of the world.

We need to be familiar with the occurrence and distribution of such
factors as water surplus and deficit, both locally and worldwide, as well as
understand the role the water budget plays in such things as the distribution
of vegetation, the scheduling of irrigation, and the determination of quality
and quantity of stream runoff. We must also recognize the relation of the
water budget factors to other environmental factors since both willful and
inadvertent changes in the physical environment have profound effects on the
factors of the water budget. In such a context, knowledge of the water
budget becomes an important area of human influence and, thus, an area of
great concern in any study of the application of climatic information to the
practical problems of everyday life.

THE WORLD WATER BALANCE

The dramatic dust bowl conditions of the 1930s probably did more to
stimulate the study of the hydrologic cycle than any other event for many
decades. President Roosevelt called for the creation of a belt of forest trees
stretching through the Great Plains from North Dakota to Texas. The idea,
supported by newspaper publicity, was built up by popular imagination into a
grandiose plan to change the climate of the entire Plains region and to
eliminate droughts and dust storms merely through the planting of trees.

A most significant contribution to our understanding of the hydrologic cycle followed in 1937 in articles by Holzman and Thornthwaite
in which they showed conclusively that the local precipitation could not be
greatly changed by local cultural practices. Holzman clearly emphasized the
mobility of water vapor in the atmosphere and discussed the modification of
air masses as they moved from land areas to ocean areas or vice versa. He

concluded that the moisture for precipitation over large land masses came primarily from maritime air masses, with only a small part being contributed by continental evaporation. Dry continental air masses are not capable of releasing their moisture immediately in most cases and may move off the continental areas with large increases in moisture content.

Information on the hydrologic cycle is usually expressed by means of a simple diagram showing annual quantities of water evaporated and precipitated, with arrows to indicate the direction of the water or vapor movement. Both Thornthwaite and Holzman clearly pointed out the errors in previous diagrams of the complete hydrologic cycle. In their modified version (similar to Fig. 3-1) they emphasized the important aspect of the two separate air masses and stressed the important role of maritime air masses in supplying moisture for land precipitation.

Benton and his associates (Benton and Blackburn, 1950; Benton et al., 1950) carried Holzman's study one step further and attempted to determine the actual amount of precipitation that originated in maritime and continental air at Huntington, West Virginia, and to estimate the factors of the water budget for the Mississippi River Valley. They determined that in 1946 at Huntington at least 85 to 90 percent of the precipitation came from maritime air masses, while probably no more than 10 percent of the annual precipitation on the watershed of the Mississippi River came from land evaporation within the watershed. No more than 10 to 14 percent of the precipitation was probably derived from land evaporation either in or beyond the watershed (Benton et al., 1950, p. 72).

Budyko (1956) came to almost the same conclusions in a study of the components of the water balance over European Russia. His values of the percentage of total precipitation that resulted from water vapor of local origin (land-based) varied from a low of 4.1 percent in October to a maximum of 18.4 percent in May. For the year as a whole he estimated that 10.9 percent of the precipitation came from land evaporation.

The actual quantities of water involved in the different steps of the annual world water balance have been variously estimated over the years. In most cases, the latitudinal estimates have been based on the analysis of detailed world maps of annual precipitation and evaporation. As these maps change with the inclusion of new or more recent data, the totals of precipitation and evaporation change.

Recently Rudolf Geiger has prepared excellent new world maps of average annual precipitation and effective evapotranspiration.[1] Making use of the most recently available data sources, Geiger's maps extend the analysis to the ocean areas, where great care has been exercised to maintain the internal consistency of the presentation.

---

[1] The whole series of wall maps published by Justus Perthes, Darmstadt, Germany, on a scale of 1:30,000,000 includes maps of temperature, radiation, air currents, ocean currents, sunshine, and atmospheric perils as well as a brief text to describe the maps themselves (see Geiger, 1965a).

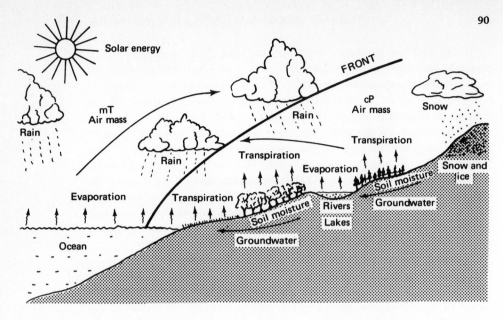

The hydrologic cycle. (*van Hylckama, 1956, with suggested modifications after Holzman, 1937, and Thornthwaite, 1937.*)   FIG. 3-1

To use the maps to provide data on the distribution of precipitation and effective evapotranspiration, the world was divided into 10° squares of latitude and longitude. The percentage of each individual square in each different precipitation or evapotranspiration class was determined by dividing the square into 100 smaller squares and counting the number in each precipitation or evapotranspiration zone. Multiplying the area in each zone by the average depth of evaporation or precipitation gives the volume of water contributed or removed by each area. Global values of precipitation and evaporation by 10° latitude belts are included on Table 3-1 along with similar values obtained by others over the past 50 years.

The present results (Mather, 1969*a*) are larger than those given by Meinardus (1934) and earlier by Brooks and Hunt (1930). The present study shows a maximum amount of precipitation (over 87,000 km^3) in the 0 to 10°N belt. Brooks and Hunt also have the highest precipitation in this belt, but their maximum figure is only 69,000 km^3. While the Meinardus figures agree better in the lower latitude regions, agreement is somewhat poorer in the middle latitudes. Wüst (1922) and Budyko, et al. (1962) have given latitudinal values of evaporation from the whole earth. Wüst's latitudinal values are 8,000 to 12,000 km^3 below those found here in each latitude belt from 30°N to 40°S, while Budyko's are slightly greater in every latitude belt except the equatorial zone 10°N to 10°S. Budyko's total for the whole earth would be about 5 percent higher than the figures found here if all latitude belts were included.

Several authors have also given latitudinal values of precipitation and evaporation by land and water areas. Table 3-2 lists some of these values along with values obtained by evaluating Geiger's maps. While Brooks and Hunt have found a larger total of precipitation on the ocean areas than has Mather (1969a), their value is considerably smaller in the 0 to 10°N belt. In fact, the most significant difference in the two studies is the much larger variation in precipitation within the tropical region in the Mather study. Land precipitation does not vary appreciably between the two studies.

Wüst's (1954) figures for the ocean areas are considerably smaller than those found in the more recent study. While his values of precipitation and evapotranspiration are less in every latitude belt, the most significant differences are still in the tropics.

The recent latitudinal values of precipitation and evaporation from Table 3-2 have also been plotted graphically in Fig. 3-2. Note that the latitude scale has been modified to take into consideration the actual area between each 10° of latitude. The upper two curves represent the latitudinal distribution of

TABLE 3-1

Summary of estimates of global annual precipitation (*P*) and evapotranspiration (*E*) by latitude from various authors (all values X 10 km³)*

LATITUDE	WÜST (1922)		BROOKS & HUNT (1930)	MEINARDUS (1934)	BUDYKO (1962)	MATHER (1969a)	
	P	*E*	*P*	*P*	*E*	*P*	*E*
80-90°N	60	20	48	44			
70-80	330	100	216	225		250	258
60-70	730	230	796	789	641	799	493
50-60	1780	970	2290	1947	1215	1877	1022
40-50	2600	1590	2797	2874	2029	2489	1844
30-40	1880	2590	3255	3223	3640	2724	3417
20-30	1740	3640	3482	3273	4974	2790	4383
10-20	3050	4670	4756	4867	5873	4638	5638
0-10	6490	4560	6888	8188	5380	8743	5882
10-0°S	5100	5120	6240	6352	5679	6428	6165
20-10	3250	4850	4622	4967	6526	4235	6465
30-20	2190	3840	3495	3456	5656	2828	5089
40-30	3090	3080	3793	3399	4566	3252	3973
50-40	2890	1820	3790	3798	2830	3783	2302
60-50	1790	590	2460	2729	1346	2766	1189
70-60	520	160	683	868		917	486
80-70	310	60	75	98		248	160
90-80	120	20	20	7			
Total	37920†	37910†	49706	51104	50355	48767	48767

* *With permission of American Water Resources Association, from Mather, 1969a.*
† *The fact that total precipitation and evaporation do not balance results in part from differences in rounding off multiplications.*

TABLE 3-2

Estimates of annual precipitation and evapotranspiration from land and water areas of the globe from various authors (all values $\times 10$ km^3)*

LATITUDE	BROOKS & HUNT (1930)		WÜST (1954)		MATHER (1969a)			
	P(LAND)	P(WATER)	P(OCEAN)	E(OCEAN)	P(OCEAN)	E(OCEAN)	P(LAND)	E(LAND)
80–90°N	2	45	40	20			62	40
70–80	50	167	130	80	188	218	482	221
60–70	408	388	280	90	316	271	759	418
50–60	713	1577	1000	490	1118	604	852	585
40–50	844	1953	1600	1150	1637	1259	851	608
30–40	914	2341	1660	2430	1873	2809	892	535
20–30	1018	2464	1500	3380	1898	3848	937	712
10–20	916	3840	2750	4220	3700	4926	1597	1085
0–10	1397	5491	5650	4070	7146	4797		
10–0°S	1588	4652	3600	4420	4408	5020	2020	1145
20–10	1021	3600	2850	4600	3095	5682	1140	784
30–20	615	2880	2050	3960	2307	4694	521	396
40–30	233	3560	2580	3320	3016	3802	236	172
50–40	77	3713	3460	2050	3725	2262	58	39
60–50	20	2440	2470	660	2748	1183	18	6
70–60	33	650	750	160	882	466	35	20
80–70	68	7	40	20	95	67	153	93
90–80	20							
Total	9937	39768	32410	35120	38152	41908	10613	6859

With permission of American Water Resources Association, from Mather, 1969a.

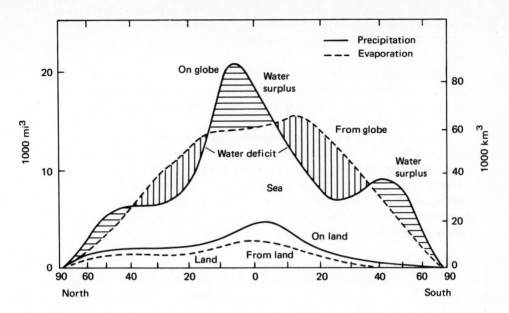

Latitudinal variation of evaporation and precipitation on the globe as well as   FIG. 3-2
on the land and sea areas. (*From Mather, 1969a, diagram prepared by Prof.
Dr. Rudolf Geiger and used with kind permission.*)

precipitation and evaporation on the whole globe, while the lower two curves
represent precipitation and evaporation on the land areas alone. The area
between the upper and lower precipitation curves thus represents graphically
the precipitation on the ocean area, while the area between the two
evaporation curves represents the evaporation from the oceans of the world.

Figure 3-2 clearly shows that a water surplus, especially marked in the
belt 0 to 10°N, exists in the equatorial zone of convergence. In both
hemispheres, evaporation exceeds precipitation in the belt from 10 to 40°
latitude corresponding generally to the subtropical high-pressure areas of
reduced precipitation. The deficit of moisture is more strongly marked in the
Southern Hemisphere than in the Northern. From 40 to 90° precipitation
exceeds evaporation and a water surplus exists. Again, the water surplus is
greater in the Southern Hemisphere. Precipitation and evaporation are quite
small and nearly in balance in the polar regions.

Availability of data on precipitation and evapotranspiration by 10°
squares of latitude and longitude provides the opportunity for quantitative
evaluation of many aspects of the water balance. For example, since positive
values of $P - E$ represent a surplus of water and since storage change for the
year as a whole can be considered zero, total $P - E$ from the land areas must
represent the runoff or return flow of water from the land to the ocean.
Summing the values of $P - E$ by continents provides a measure of the runoff
from each continent. These values are given in Table 3-3 along with similar

values obtained by Budyko from a study of the heat and water balance of the earth and by the Geological Survey from an analysis of available stream gage records for comparison purposes. The Geological Survey figures are somewhat less than the others since, in part, they do not include subsurface flow to the oceans.

The largest discrepancy occurs in the value of runoff for Asia, where there is a difference of 5,000 km³ between the Geological Survey estimate and the results of the $P - E$ analysis. Agreement is reasonably good in Africa, Europe, and South America.

Dividing by the area of each continent gives the depth of runoff by continents. Table 3-4 shows that South America has a much greater depth of runoff than the other continents, evidently because of the much greater percentage of land area in the equatorial region of high precipitation. Asia, Europe, and North America follow in that order in terms of depth of runoff according to the present author, although the Geological Survey figures put Europe second while Budyko's results place North America second.

TABLE 3-3

Comparison of runoff values by continents from three different analyses (all values $\times 10$ km³)*

CONTINENT	(MATHER (1969a)			BUDYKO (1956)			U.S. GEOL. SURVEY†
	P	E	RUNOFF	P	E	RUNOFF	RUNOFF
Africa	2149	1762	387	2019	1537	482	333
Asia	3038	1835	1203	2692	1721	971	705
Australia	354	283	71	362	316	46	41
Europe	641	374	267	598	359	239	236
North America	1546	944	602	1637	978	659	516
South America	2700	1549	1151	2399	1528	871	1113
Antarctica	188	113	75				
Total			3756			3268	2944

* With permission of the American Water Resources Association, from Mather, 1969a.
† Based on computations of the U.S. Geological Survey, Water Resources Division, 1966, kindly furnished by W. H. Durum.

TABLE 3-4

Depth of runoff (in mm/year) by continents from various authors*

CONTINENT	MATHER (1969a)	BUDYKO (1956)	U.S. GEOL. SURVEY
Africa	127	160	112
Asia	272	221	160
Australia	91	61	53
Europe	267	241	236
North America	246	269	211
South America	648	490	627
All continents	274	244	218

* With permission of American Water Resources Association, from Mather, 1969a.

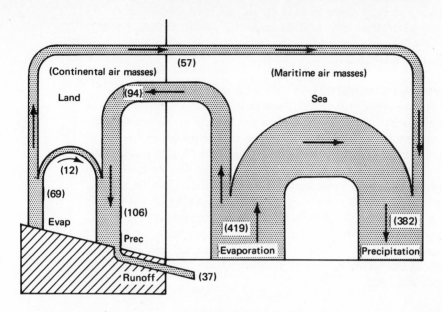

Schematic representation of the hydrologic cycle. Width of each tube   FIG. 3-3
proportional to volume of water involved in that phase of hydrologic cycle.
(Values in parentheses are water volumes in 1000 km³ .) (*Diagram prepared by
Prof. Dr. Rudolf Geiger and used with kind permission.*)

    If we accept the foregoing figures of annual precipitation and
evapotranspiration over the land and water areas of the globe as well as the
figure of 11 percent of the land precipitation being derived from land
evaporation, we are in a position to identify all the terms in the world water
balance (Table 3-5).

    Expressed schematically, the annual hydrologic cycle can be repre-
sented by a series of tubes whose widths are proportional to the volume of
water involved (Fig. 3-3). The volumes of water represented emphasize the

TABLE 3-5

Factors of the annual world water balance

FACTOR	AMOUNT, KM³
Precipitation on oceans	382,000
Evaporation from oceans	419,000
Precipitation on land	106,000
Evaporation from land	69,000
Runoff to oceans	37,000
Land precipitation from land evaporation	12,000
Land precipitation from ocean evaporation	94,000

great mobility of water since both atmospheric transports of moisture from land to water and from water to land exceed the total amount of surface transport of water from the land to ocean by runoff. Holzman (1937) described this mobility of water; the present quantitative values fully substantiate his understanding.

THE ATMOSPHERIC MOISTURE FLUX OVER A HEMISPHERE

Starr and Peixoto (1958, 1964, 1965) and their associates undertook a most ambitious evaluation of the hemispheric water balance by determining the vertically integrated water vapor flux and its horizontal divergence or convergence over the Northern Hemisphere. One of their studies involved the evaluation of data from some 321 upper air stations at selected levels up to 500 mb for the year 1958 while another used fewer data points for the year 1950. The data needed for the study include only values of the specific humidity q, the eastward and northward wind components u and v, respectively, and the acceleration of gravity g. If λ = longitude, ϕ = latitude, p = atmospheric pressure, and t = time, then the zonal and meridional components of the horizontal flux of moisture, Q, can be written as (Starr, Peixoto, and Crisi, 1965, p. 463)

$$Q_\lambda = \frac{1}{g} \int_0^{p_0} qu \, dp$$

$$Q_\phi = \frac{1}{g} \int_0^{p_0} qv \, dp$$

The expressions are evaluated from the surface pressure (p_0) to the top of the atmosphere $(p = 0)$. Averaging the expressions over a time interval τ results in mean values of the zonal and meridional fluxes. The time interval is 1 year (either 1950 or 1958) in the studies of Starr and Peixoto. These scientists show that the divergence of the mean vector field over a year $\nabla \bar{Q}(\lambda\phi)$ can be given by

$$\nabla \bar{Q}(\lambda\phi) = \frac{1}{R \cos \phi} \left[\frac{\partial \bar{Q}_\lambda}{\partial \lambda} + \frac{\partial}{\partial \phi} (\bar{Q}_\phi \cos \phi) \right]$$

where R is the radius of the earth, assumed to be constant. For a column of air from the surface to the top of the atmosphere, the net sources of all water in the column (Σ) can be expressed as

$$\frac{\partial W}{\partial t} + \nabla Q = \Sigma$$

where W is the precipitable water in the atmosphere. Over a time period of 1 year it is possible to assume that $\partial W/\partial t$ is zero.

Evaporation and precipitation constitute the major sources and sinks of water vapor in the air, while only small quantities of liquid or solid water are

Distribution of the horizontal divergence of the vertically integrated total   FIG. 3-4
annual flux of water vapor for 1958 (g/cm² year). Solid lines for divergence
(evaporation), dashed lines for convergence (precipitation), intervals 100
cm/year. (*With permission of Svenska Geofysiska Foereningen, from Starr,
Peixoto, and Crisi, 1965, fig. 3.*)

carried in the air in comparison with water vapor. As a first approximation, Σ
can be equated to $\overline{E - P}$. The expression for the divergence of the yearly
mean vapor flux becomes equivalent to the difference between evaporation
and precipitation. Positive values of the divergence expression indicate that
evaporation exceeds precipitation, while negative values (convergence) indicate
that precipitation exceeds evaporation.

Starr, Peixoto, and Crisi (1965) determined the mean $E - P$ distribution
over the Northern Hemisphere by using a 5° latitude-longitude grid. The
pattern of mean total horizontal divergence found for 1958 is given in Fig.
3-4. The resulting pattern shows greater precipitation than evaporation over
the equatorial Atlantic and Pacific ocean areas. Areas of excess precipitation
over evaporation are also found in east Africa and over central India and

Burma in source regions for the Nile, Indus, Ganges, and Brahmaputra rivers. Extensive areas of divergence (evaporation greater than precipitation) exist over the subtropical areas of the Atlantic and Pacific oceans.

More interesting perhaps is the strong area of positive divergence (excess evaporation) over the Iranian desert stretching westward through the Middle East and across the Sahara Desert. In an earlier paper, these authors (1958) had suggested that considerable moisture was evaporated from such desert areas not only as a result of the often interior-drainage systems from wetter surrounding areas but also from influent streams crossing the areas and from groundwater flowing through aquifers into the areas. More detailed study of these surprising conclusions is needed.

In midlatitudes, there are many smaller areas of both convergence and divergence of the water vapor flux. One extensive area of convergence extends across the favored region of storm tracks from eastern North America to northwestern Europe, while another crosses the North Pacific from northern Japan to the Gulf of Alaska. Weak divergence or excess evaporation over precipitation is found in central and northern Siberia.

While the distribution of Fig. 3-4 represents conditions only for 1958, the pattern is sufficiently like the pattern these authors found for 1950 to suggest that certain of these areas of convergence and divergence are fairly permanent features of the atmosphere. The usefulness of the information for water resources studies is so great that the approach needs to be extended to the Southern Hemisphere as well as to additional years in the Northern Hemisphere.

THE ANNUAL WATER BALANCE OF NORTH AMERICA

Hare (1972) and Hare and Hay (1971) have recently investigated the annual water balance over Canada and North America south of 60°N. Problems of precipitation measurement make it very questionable to carry the study further north. Hare starts with the standard expressions for the disposition of the precipitation and the energy balance on an annual basis per unit area

$$P* = N* + E$$

where P = mean annual precipitation
N = mean annual surplus (runoff plus percolation)
E = mean annual evapotranspiration loss
$*$ = directly measured observations

$$R_n = H + LE$$

where R_n = net radiation
H = convective heat (enthalpy) flux
E = convective latent heat flux
L = latent heat of vaporization

Substituting in the Bowen ratio $B = H/LE$, he obtains the familiar evaporation equation

$$E = \frac{R_n}{L(B+1)} = P* - N*$$

Direct measurements of P and N are available, but R_n must be estimated by manipulation of the fairly readily available data on solar radiation, cloudiness, and atmospheric humidity. Considering net radiation as an input variable (double asterisk) and utilizing the previous relation for E, Hare (1972) obtains the expression for H as follows:

$$H = R_n** - L(P* - N*)$$

This and the previous expression for E permit determination of the evapotranspiration and convective heat fluxes by using only observed or easily computed variables. The needed data are global solar radiation, cloudiness, surface relative humidity, precipitation, and runoff plus percolation, all on an annual basis. Continent maps are available and can be used in the preparation of continent maps of the various derived fluxes.

The equations for E and H (as well as for R_n) were solved on a monthly basis for each 5° square between 30 and 60°N for the land area of North America by Hare and his associates. The resulting map of annual latent heat transport due to evapotranspiration, reflecting a compromise between the availability of water and energy for vaporization, shows values below 20 kly/year in the arid western regions and at high latitudes. Maximum values over 60 kly/year are found in the warm moist regions of the southeastern United States. The map of convective heat flux shows very high values as expected in the hot, arid Southwest with much lower values in the cooler and more moist areas. The results of these distributions were expected, but the computations provide quantitative values.

Hare also provides maps of the Bowen, runoff, and dryness ratios which Lettau (1969) defined as

$$B = \frac{H}{LE} \qquad C = \frac{N}{P} \qquad D = \frac{R_n}{LP}$$

as well as maps of the short-wave, long-wave, and net radiation fluxes.

While the various maps are no better than the original data used in their derivation, Hare has been able to achieve them by using readily available information without the need for complicated assumptions or many empirical relations. The work contributes not only to our understanding of the water balance but also to our knowledge of the radiation and convective heat fluxes as well.

THE WATER BUDGET OF A PLACE

Over the years, many scientists have worked on the problem of how to express the daily or seasonal water budget of a place or area. These studies were advanced significantly in the 1940s by the work of C. W. Thornthwaite in the United States, H. L. Penman in England, and M. I. Budyko in the USSR. The approaches of these three major contributors to water balance climatology are all quite different; the myriad of modifications of each of the schemes that these and other scientists produced later have resulted in innumerable variations on the basic techniques. Each of the many water-budgeting schemes and models that have appeared in the last quarter-century has something to contribute to the understanding of our environmental water relations.

Any water budget is limited in part by the actual data that are entered into it. Many of the scientists who worked in this field have produced their own methods to compute the evaporation or evapotranspiration of moisture from the surface to the air. Arguments abound as to which expression is less empirical or more physically sound or which provides values of evaporation under particular meteorological or time conditions with more accuracy.

It would appear that the emphasis has been somewhat misplaced. While precipitation is measured at a large number of places around the world (Chap. 2), we still have only a rough idea of the actual totals of precipitation except at the measuring gage itself. (Even here, wind and other exposure problems result in certain errors.) Extrapolation of precipitation values even a short distance can result in appreciable errors (see Fig. 2-10). Yet precipitation along with evapotranspiration are the major inputs to any water-balance bookkeeping procedure. Why should we accept precipitation in this context uncritically, yet produce scores of different techniques to estimate evapo-transpiration and argue in detail as to which should be used? Errors in the precipitation values entered into any water budget may be as large as errors in evapotranspiration, and, of course, they will produce as significant errors in the results.

The present volume will utilize the Thornthwaite expression for evapotranspiration, although it must be emphasized that other expressions are available and possibly just as useful. While others may produce more accurate values of evapotranspiration, they may not lend themselves as easily to worldwide evaluation of evapotranspiration as Thornthwaite's does. The important thing is the use of *some* evapotranspiration value to produce a *water budget*. It is the daily or monthly comparison of precipitation with evapotranspiration that is of ultimate importance in applied climatology rather than the particular method by which evapotranspiration is determined.

Realizing that one could not determine whether a climate was moist or arid without comparing the climatic moisture supply with the climatic moisture needs, Thornthwaite developed a climatic water budget utilizing a bookkeeping system of accounting for increments of water supply and loss on either a daily or a monthly basis. The climatic water budget developed during the early 1940s was modified by Thornthwaite and Mather (1955*b*) to make it more useful under a wide range of soil and vegetation conditions.

To illustrate the water-budget bookkeeping procedure, let us consider the average monthly march of water supply (precipitation) and climatic water

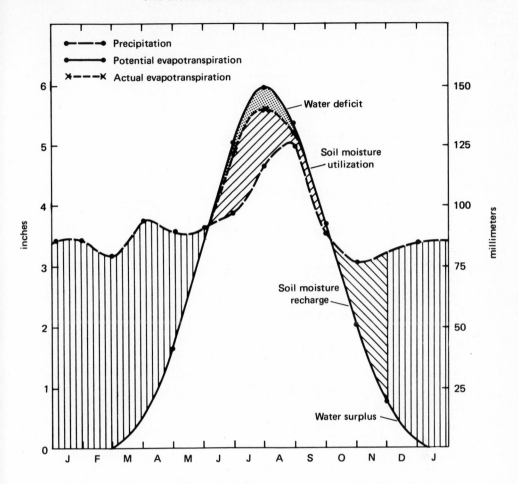

Average climatic water budget, Wilmington, Del.   FIG. 3-5

need (potential evapotranspiration[1]) at Wilmington, Delaware. The average monthly values are given in Table 3-6 along with the other steps in the climatic water-budget bookkeeping procedure. Figure 3-5 shows the monthly march of the factors of the climatic water budget graphically.

At Wilmington, average monthly temperatures vary from a low of 0.5°C in January to a high of 24.2°C in July. Potential evapotranspiration, which is closely related to temperature,[2] varies regularly through the year from low

---

[1] Potential evapotranspiration (*PE*) has been defined in Chap. 2, in the section on evaporation, as the water loss from an extensive, closed cover of homogeneous vegetation that never suffers from a lack of water. A fuller discussion of the concept is included in the earlier section.

[2] The complete relationship is provided in Chap. 2. Thornthwaite and Mather (1957) have prepared tables and nomograms to allow easy evaluation of the rather complex relations developed by Thornthwaite.

TABLE 3-6

Average climatic water budget at Wilmington, Del.*

	J	F	M	A	M	J	J	A	S	O	N	D	YEAR
Temp., °C	0.5	0.7	5.5	11.3	17.1	21.8	24.2	23.4	20.0	13.8	7.5	1.7	12.3
Pot. evap.	0	0	15	43	89	128	150	135	94	52	20	2	728
Precipitation	87	80	96	91	92	98	119	128	93	78	82	86	1130
P–PE	87	80	81	48	3	–30	–31	–7	–1	26	62	84	
Storage	150	150	150	150	150	122	99	94	93	119	150	150	
Change st.	0	0	0	0	0	–28	–23	–5	–1	+26	31	0	
Act. evap.	0	0	15	43	89	126	142	133	94	52	20	2	716
Deficit	0	0	0	0	0	2	8	2	0	0	0	0	12
Surplus	87	80	81	48	3	0	0	0	0	0	31	84	414
Runoff	42	50	56	54	44	35	28	23	18	15	18	31	414
Measured runoff Shellpot Creek	47	50	64	51	39	22	24	20	15	11	33	39	415

* Storage capacity in soil is 150 mm; all values except temperature are given in millimeters.

values of 0 mm in January and February when the mean temperature is near freezing to peak values of 150 and 135 mm, respectively, in July and August. Potential evapotranspiration increases rapidly in spring from less than 15 mm in March to 128 mm in June and decreases just as rapidly in fall from 94 mm in September to 2 mm in December. Total climatic water need for the year equals 728 mm.

Precipitation, or water supply, is much less variable through the year. The average annual total of 1,130 mm is evenly distributed, with each month receiving on the average more than 75 mm. Low values are found in fall and winter with 78 mm in October and 80 mm in February; highest values come in the summer period with 119 mm in July and 128 mm in August. Thus the wettest period from the standpoint of precipitation is also the period of greatest demand for water in terms of potential evapotranspiration.

Comparing precipitation with potential evapotranspiration on a monthly basis, one finds that they never coincide. There is too much precipitation in fall, winter, and spring and too little in summer. In November, the precipitation over and above that needed for evapotranspiration is stored in the soil and results in the upper layers being brought to field capacity.[1] (In the example of Wilmington, it is assumed that 150 mm depth of water can be stored in the root zone of the soil, equal to 25 mm/100 mm in a 0.6-m root zone.) After the soil is at field capacity, any precipitation not needed for evapotranspiration is considered to be surplus and ultimately lost as runoff. Surplus thus increases from 31 mm in November to 80 mm or over in each of the months from December to March.

Precipitation is greater than the potential climatic water need from October through May on the average. The soil remains at field capacity from November, and some water is added each month to surplus. June is the first month in which the rapidly rising climatic water needs finally exceed the supply of water from precipitation. Precipitation fails to supply the water needs by 30 mm in June ($P - PE = -30$). Some of this need is supplied by the water stored in the upper layer of the soil (change in storage = -28 mm), but ultimately 2 mm (30 -28 mm) is not supplied by either precipitation or stored soil water. This is, therefore, the water deficit. In July, 31 mm of water need are not supplied by precipitation. As the soil dries, water is less available from the upper soil layers and only 23 mm is removed from the soil to help supply the need. Deficit in July reaches 8 mm.

Storage, which has been at field capacity (150 mm) all winter and spring, decreases during the summer as more water is removed to supply at least some of the water needs. The lowest value of storage is found in September (94 mm, about 63 percent of field capacity). Storage is quickly brought back to field capacity in October and November as the precipitation in excess of the water need is retained in the upper soil layers.

[1] Field capacity is defined as the quantity of water held in the capillaries of an originally saturated soil against the pull of gravity.

The actual evapotranspiration (*AE*) or actual water loss equals the potential water loss during those periods when precipitation is greater than potential evapotranspiration. Sufficient water for evapotranspiration exists in those months. When precipitation is less than potential evapotranspiration, actual water loss equals the sum of precipitation and water removed from the soil (change in storage without regard for sign). While average potential evapotranspiration in Wilmington is 728 mm/year, actual water loss or evapotranspiration equals only 716 mm, resulting in a deficit of 12 mm. Since average annual precipitation is 1,130 mm and only 716 mm are actually lost by evapotranspiration, the annual surplus of water must equal the difference or 414 mm.

The surplus water, that moisture over and above what is needed for evapotranspiration or storage in the soil, is ultimately lost as runoff. This is not immediate for it takes some time for the water to move down to the groundwater table and through the soil to emerge again in surface streams or rivers. The rate of removal by runoff depends on the size of the watershed, its slope and ground cover, and on the type of soil through which the water must move. In the example given in Table 3-6, it is assumed that 20 percent of the surplus water available for runoff in any month actually is lost by runoff, while the remaining 80 percent is held over and added to the available surplus in the next month. Surplus and runoff must total the same amount on an average annual basis but the monthly distribution of each parameter can be quite different. Computed runoff reaches a maximum of 56 mm in March, using the 20 percent retention factor, and decreases to a minimum of 15 mm in October.

Agreement between measured and computed runoff is generally quite good on an annual basis, although local watershed factors which cannot be considered in the gross water budget computations often lead to a lack of agreement on a monthly basis. For comparison purposes, the measured runoff in Shellpot Creek at Wilmington, based on 19 years of record, is also included in Table 3-6.

The general agreement between measured and computed runoff values serves as a good verification of the whole computational system, although other comparisons (e.g., computed vs. measured soil moisture storage) are possible. The validity of the water budget, especially in middle-latitude areas without marked seasonal contrasts in moisture conditions, has been repeatedly substantiated in earlier studies by the author and others.

Other approaches (e.g., Penman, 1956*a*, see Chap. 2; Budyko, 1956) are also available to provide water budgets that permit derivation of actual evapotranspiration as a function of the potential evapotranspiration and the soil moisture content. In each, it is assumed that evapotranspiration is drawn from water in the soil between field capacity and the permanent wilting point since gravitational water is retained in the soil too short a time to contribute significantly to this water loss.

The moisture-holding capacity of a soil depends on the depth of the soil layer considered and the type and structure of the soil. It can vary from just a few millimeters on a shallow sand to well over 400 mm in a deep, well-aerated silt loam. The roots of plants compensate somewhat for the variable nature of soil; in sandy soils plants tend to be more deep-rooted, while in silts and clays they tend to be more shallow-rooted. Of course, young plants or mature trees will have root systems that ramify through markedly different depths of soil, and so they will have available to them quite different amounts of moisture.

As the soil dries, it becomes increasingly difficult for additional water to be lost by evaporation and transpiration. Thus, as the soil moisture content decreases, so also does the rate of evapotranspiration. Different investigators have suggested that the shape of the curve of decreasing evapotranspiration with soil moisture storage can be either concave or convex (i.e., a large decrease in evapotranspiration as soil dries, followed by a slowing rate of decrease the drier the soil, or a slow rate of decrease of evapotranspiration when the soil is moist, followed by an increasing rate of decrease of evapotranspiration as the soil becomes drier). Lowry (1959) investigated these suggestions and attempted to reconcile certain of them. He felt that unwarranted extrapolation of drying rate data had led to some of the discrepancies. Baier and Robertson (1966) and Baier et al. (1972) have combined the various proposals for the relation between AE/PE under different values of soil moisture content (Fig. 3-6). Curve A assumes transpiration is independent of available soil moisture, while curve B assumes very little influence of soil moisture content on evapotranspiration until the available soil moisture content is reduced by about 60 to 70 percent. Curve C assumes a linear relation between water loss and soil moisture content, while curves D, E, and F assume that water is equally available to plant roots up to some point at which demand exceeds supply to the roots. At that time there is a sharp decrease in the ratio of AE to PE since the plants have a great difficulty in obtaining needed water. Curves G and H are similar to D, E, and F except that they assume a linear relation after 30 and 50 percent of the soil moisture is removed at the potential rate first. Penman assumed that E equals PE until the field capacity deficit reached some figure such as 7.5 cm (the actual figure depends on the dryness of the spring, since a dry spring would encourage deeper rooting and permit a greater field capacity deficit to develop before E ceased to equal PE). Below this moisture content, he applied an empirically derived drying curve so that his relation might approximate curve D (Fig. 3-6). Budyko assumed E equals PE above a moisture content W_k. When W, the actual moisture content, drops below the critical value W_k, then $E = PE(W/W_k)$. This, of course, provides for a linear decrease in the actual evaporation as the moisture content drops below the critical value. This approaches the Thornthwaite and Mather suggestion (curve C, Fig. 3-6), which can be expressed by

$$AE = PE \frac{SM_a}{SM_{fc}}$$

where SM_a = actual soil moisture content
  SM_{fc} = soil moisture content at field capacity

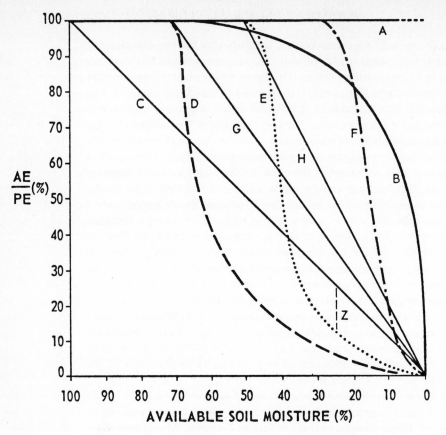

Various proposals for the relationship between the *AE/PE* ratio and available soil moisture. (*With permission Canada Department of Agriculture, from Baier, Chaput, Russelo, and Sharp,* 1972, *fig.* 1, *p.* 8, *after Baier and Robertson,* 1966.)

FIG. 3-6

APPLICATION OF THE FACTORS OF THE WATER BUDGET

When the seasonal course of precipitation is compared with the course of water need, in some form of water balance such as Fig. 3-5 or Table 3-6, one obtains information on many aspects of the water relations at a place. First, it is possible to determine the actual evapotranspiration or the actual loss of water from the plant and soil surfaces which, in almost all cases, is different from the potential water need. It is extremely difficult to measure actual evapotranspiration in practice because of the dependence of this quantity on such factors as soil type and method of land cultivation, type of plant cover, and moisture condition of the soil profile. Second, the difference between the potential and the actual evapotranspiration provides a measure of the moisture deficit of a place, the amount by which the available moisture fails to satisfy the demand for water. Knowledge of the moisture deficit is basic to any understanding of the economic feasibility of irrigation, for it provides information on the total volume of water needed at any time and gives a definitive measure of drought. When compared with the moisture surplus in other seasons it makes clear whether there is sufficient water accumulating during the year to permit large-scale irrigation during the dry period. At the same time, determination of the changes in soil moisture storage on a daily basis gives information on the state of the moisture in the ground at any time for use in the actual scheduling of the time and amount of supplemental irrigation (see Chap. 5).

Third, information on the water surplus, the amount by which the precipitation exceeds the water needs when the soil is at field capacity, is fundamental in any hydrologic studies which deal with the recharge of the groundwater table or with the runoff of water in streams and rivers. By definition, the water surplus is the water which does not remain in the surface soil layers but is available for deep percolation to the water table and overland or subsurface flow to the water courses. Thus information on water surplus, climatically determined from the water budget, provides a knowledge of stream flow which can only otherwise be obtained from extensive stream-gaging installations and data on flow to the groundwater table, which requires detailed well records (discussed in more detail in Chap. 5).

In oceanographic work, the water surplus is important in determining the amount of freshwater flow from the land to the ocean. This is particularly significant in the case of bays, estuaries, and seas which are almost entirely landlocked, for the volume of freshwater runoff to the water body will be important in determining the salinity, density, and other characteristics of the water (see, for example, the study of flow in the Delaware Bay, Chap. 5).

Fourth, the water budget provides information on the detention of moisture on and within the land areas of the earth through the year. The seasonal fluctuations in the amount of land storage are reflected in the worldwide change in ocean levels through the year, a subject of vital importance not only to oceanographers but to others who are concerned with problems of large-scale moisture flux between land and ocean (discussed earlier in this chapter as well as in Chap. 5).

Moisture storage in the land is also basic to any understanding of the movement of men and vehicles over off-road surfaces. On all but the most sandy soils, an increase in soil moisture content to or above the field capacity will result in a loss of bearing capacity and shearing strength and an increase in stickiness. This results in a lowered ability of men and vehicles to move over off-road surfaces. On sands, of course, the reverse is true; as the moisture content increases up to the point where quicksand conditions exist, the tractionability, or the ability of men and vehicles to traverse off-road surfaces, improves (discussed in more detail in Chap. 11).

Fifth, the water budget will result in information not only on the periods of moisture surplus and deficit but will permit the magnitude of these quantities to be compared with one another and with the water need in order to provide climatic indices that can be used in classification and correlation studies (see Chap. 4). Precipitation and the climatically determined water need are truly active factors in climate, and, as such, they can serve as the basis for the classification of the climates of the earth. At the same time, a comparison of moisture surplus or deficit with the water need provides indices of humidity and aridity which can be correlated with the distribution of vegetation (see Chap. 4).

Sixth, probably more important than the use of the water budget in any one of these many lines of research is the fact that with the ability to secure information on all phases of the moisture relationships of an area from readily available climatic data comes the ability to determine values of these moisture parameters for many years. This is significant for it not only makes data available for periods when actual measurements of these parameters were not made but it also permits the accumulation of a record which can be utilized in statistical studies. Many problems, such as the determination of probabilities of the amount of surplus water available for stream flow, of the occurrence of soil moisture conditions affecting the movement of men or vehicles over off-road surfaces, or of drought, can only be solved in this manner at present since measured values of the necessary parameters are not available for a long enough period of time. Such statistical studies provide the basis for evaluating the economic feasibility of any program of action and aid in forecasting the outcome of the program.

ASSUMPTIONS AND LIMITATIONS IN THE BOOKKEEPING PROCESS

Many investigators have pointed out certain shortcomings in the Thornthwaite water-budget bookkeeping procedure especially if it is used in daily computations. The most significant objection is that the value of computed potential evapotranspiration is not reliable on a daily basis. Thornthwaite's formula does not directly include humidity, radiation, or wind; it provides reasonable monthly values based only on monthly temperature by assuming that the other factors that also affect evapotranspiration do not deviate greatly from normal over a period as long as a month. In areas in which the humidity or wind changes markedly from month to month or from one season to another, the formula may not work well. Pelton, King, and Tanner (1960) and Jensen and Haise (1963) have both pointed out the lack of relation between daily temperature (the key to the Thornthwaite expression for potential evapotranspiration) and daily water loss. Penman (1956a) feels that Thornthwaite's formula for evaporation should be used only for periods longer than 5 days. Pelton, King, and Tanner report a correlation of 0.92 between monthly estimates of potential evapotranspiration and net radiation. However, they find a correlation coefficient of only 0.47 between the daily Thornthwaite potential evapotranspiration value and estimates of daily water loss based on energy budget calculations.

If daily values of potential evapotranspiration are needed, as they are for the determination of daily soil moisture content or irrigation scheduling, one of the many combination methods (see Tanner and Fuchs' method in Evaporation section, Chap. 2) should be used. Use of these daily values of evapotranspiration with daily values of precipitation will permit the day-to-day changes in soil moisture content, of water surplus, and deficit to be evaluated for use with various scheduling operations.

A second question that many investigators have raised is the fact that total precipitation is entered directly into the budget although, for some purposes, the overland storm runoff should be subtracted and only the so-called "effective" precipitation used in the computation. This is especially true when values of water storage, or moisture surplus or deficit, are sought.

The Soil Conservation Service (SCS) has published information on the infiltration characteristics for most soil series and has provided procedures for estimating storm runoff from values of daily precipitation, ground cover, and hydrologic soil characteristics (Soil Conservation Service, 1964). Albrecht (1971), using the values of effective precipitation derived from the SCS procedure, has obtained values of actual evapotranspiration from the water budget. He then evaluated the ratio AE/PE and related it to crop yield in the Missouri Valley area. He found that the ratio AE/PE computed by using the values of "effective" precipitation provided correlation coefficients with corn, sorghum, and wheat yield that were 0.09 to 0.11 higher than those found by using the AE/PE ratio computed on the basis of total, uncorrected precipitation.

In computing the depth of water in a soil column, the gravitational water or moisture above field capacity and the capillary water or moisture below field capacity should be considered separately. At field capacity the soil contains no surplus of gravitational water and no deficit of capillary water. Thus this value becomes an important and useful point in the computational process. When the moisture content of the soil is at field capacity or above, the surplus water or water that is added by precipitation is lost slowly by downward percolation regardless of whether there is evapotranspiration or not. This gravitational water is only detained briefly, the period depending on the depth and permeability of the soil and the amount of gravitational water. For example, it has been empirically found that, in a 1-m thickness of loam, about 90 percent of the gravitational water in the soil on any given day is held over in the soil until the succeeding day. This percentage becomes smaller as the thickness of the soil layer decreases. Also, the greater the amount of sand in the soil, the smaller is the percentage of gravitational water held over from one day to the next. Thus, in computing the drying of a soil from an initial value of moisture above field capacity, it is necessary to determine separately both the loss of water by evapotranspiration and by gravitational flow.

The use of various empirical formulas to provide the starting value of potential evapotranspiration for comparison with precipitation, the use of somewhat different assumptions concerning the way water is stored in the soil or how it is made available to the plant roots during a drying period, or the use of other methods of accomplishing soil moisture recharge and initiating runoff may all be accepted. These modifications represent only variations in technique rather than fundamental differences in the overall approach. While we will utilize the Thornthwaite approach in the present instance, our major concern is for an understanding of the overall value of a water budget at a particular place, its usefulness and applicability, and its function in clarifying the water resources relationships of a place or an area.

CONCLUSIONS

As man increases his demands for water on this planet, through population growth, intensified industrialization, and increased application of water-using devices, and as usable supplies of water decrease through pollution, we are faced with a water crisis of growing seriousness. We will not run out of water as some predict although we may run out of free or inexpensive water. Technology will be able to supply us with enough to satisfy most of our needs or will provide us with alternate ways of using water more effectively.

Man, at present, is working diligently at the task of providing additional supplies of water or using water with greater efficiencies. This work is directed toward (1) increasing precipitation through cloud seeding, (2) reducing evaporation through the use of films or mulches, (3) desalination of ocean or brackish water, (4) storage and interbasin transfers of excess water, and (5) tapping new underground sources of water. But consideration of any of these sources of water—how they occur, how they can be increased or decreased, what controls them, what their increased use will do to other aspects of the water economy over a large region—must ultimately focus back on the water budget or water balance.

Just the effects of vegetation changes, by grazing, cropping, substitution of species, or clearing, on the amount of water which enters or is retained in the soil, and hence on the water economy of the area, are illustrative of the interrelationships of water in all parts of the budget. The elimination of transpiration by stopping plant growth will result in additional water for soil moisture storage and for runoff. Uncontrolled runoff might result in harmful erosion and, hence, further change soil structure and the soil moisture relationships. Similarly, burning of vegetation destroys the above-ground parts of the plants as well as much of the surface organic material. Such action can also lead to increased runoff and decreased soil moisture storage. Grazing or overgrazing will result in two changes in environmental conditions: (1) the removal of vegetation will reduce transpiration and make more moisture available for soil storage or runoff, and (2) the compaction of soil by animals' hoofs will reduce the capacity of the soil to absorb water and hence make it less able to store water. Thus the additional water made available through reduced transpiration will probably run off with the possibility of erosion damage. Replacing a grassland or cropped vegetation surface with forest (allowing farmland to be reforested) will result in the retention of more moisture in the upper soil layers and increase actual evapotranspiration. This, in turn, will make less water available for stream runoff.

The water budget is an important area of human influence. Because knowledge of the water budget provides quantitative information on the periods of moisture surplus and deficit, and permits the determination of the amounts of moisture stored in the soil or lost through runoff at any time, it becomes a basic tool in many aspects of applied climatology.

CLIMATIC CLASSIFICATION, CLIMATIC INDICES, AND VEGETATION DISTRIBUTION

4

THE WATER BUDGET AND CLIMATIC CLASSIFICATION

The climatic water budget became the basis of Thornthwaite's 1948 climatic classification, a basic effort in applied climatology. The Thornthwaite scheme drew, in part, on nineteenth-century attempts at climatic classification undertaken primarily by biologists or botanists who recognized the existence of a relation between vegetation distribution and certain factors of climate. First de Candolle (1855) published an important monograph on the factors influencing the distribution of plant species; this was shortly followed in 1866 by Grisebach's world map of the distribution of vegetation. Those contributions encouraged other investigators to try to relate temperature and precipitation distributions with those of vegetation. Linsser (1867, 1869) undertook basic studies on the response of plants to heat and the effect of precipitation on vegetation growth and became possibly the first to produce a true climatic classification based on the distribution of vegetation.

By 1875, the idea that climates might be classified on the basis of vegetation or physiological response was accepted, but it was not until years later that the distinguished meteorologist and climatologist Wladimir Köppen was able to develop this idea sufficiently to produce a useful and practical climatic classification employing readily available climatic variables. The first Köppen climatic classification appeared in 1900, and now, some 70 years later, there are still many who know of no other scheme of climatic classification but that of Köppen. It is simple and easy to teach, and wall maps of the distributions are readily available for display and study. The Köppen system still leaves much to be desired for it attempts to relate vegetation development to certain monthly values of temperature and precipitation without adequate concern for such vegetation-limiting variables as water need in relation to supply, soil moisture storage, or actual evapotranspiration. At the same time, it only seeks to define fairly large climatic regions that are, in many cases, only poorly related to the major vegetation regions of the world. Temperature and precipitation are not the most active factors resulting in the growth and development of vegetation, so any close relation between distributions of temperature or precipitation and vegetation is largely fortuitous and based on the fact that these two climatic factors are related to other more active climatic parameters.

The development of the concept of a moisture index, usually expressed as a ratio of the amount of precipitation to the evaporation or need for water, introduced the so-called "modern" era of climatic classification, now some 100 years old. A moisture index of unity provides an absolute means for separating dry from moist climates. Other preselected values of the moisture index provide a rational scale of moisture adequacy based entirely on climatic data.

The moisture index concept emphasizes the important role of evapotranspiration in climate. But since actual evapotranspiration can never exceed precipitation (unless additional water is supplied by runoff from other areas), an index relating actual evapotranspiration with precipitation will not truly reflect the aridity of a climate. Only when some expression for the climatic demand for water is used with precipitation is it possible to determine quantitatively how adequate the available water supply is and whether a climate is really moist or arid. Köppen's expression for the moisture index involved the ratio of mean annual precipitation to mean annual temperature plus a constant. He changed the formulation of the moisture index several times, never being fully satisfied with its expression.

Carter (Carter and Mather, 1966) suggested that a common mistake in classification work is to consider that climates merely identify the regions between the various climatic isolines that coincide with the major vegetation boundaries on distribution maps. It is the boundaries of the climatic types rather than the core areas between isolines that should be the real focus of a classifier, and it is in these regions that recent major emphasis in classification has produced most significant results. Many climatic classifications are judged on the basis of how well the climatic mapping accords with the vegetation mapping. Carter believes there is not adequate recognition of the fact that a climatic classification has a special intellectual significance by itself, based on how well it expresses those climatic factors considered to be "active" within the particular group of natural physical processes under investigation. While recognizing that climate in all its complexity is never quite the same from one

place to another, it is of little value for classification purposes if we are
continually concerned with the minutiae of differences between places.
Rather we need to work with those climatic factors that are capable of some
degree of broad generalization or categorization.

Specification of the type of climatic classification or the use to which it
is to be put is important for it is clear that no simple classification will ever
serve all uses equally well. A climatic classification for human comfort or
health (see Chap. 8) or for air conditioning need will be far different from
one for irrigation needs or water resources. Similarly, energy and moisture
factors are not the only active factors of climate, and certain classifications
need to consider other factors. These two factors do seem to be of prime
significance for agricultural or vegetation purposes, however, the subject of
present concern.

ELEMENTS OF A CLASSIFICATION[1]

Primary emphasis in Thornthwaite's climate classification is assigned to the
moisture index, one of four factors that he uses to describe the climatic
situation of a particular place. The other factors include an index of thermal
efficiency, the summer concentration of thermal efficiency, and an index
expressing the seasonal variation of effective moisture. As is the case with
Köppen's classification, each of the four factors is expressed in terms of a
letter, so that the complete classification at a particular station involves four
letters, two of which are capitals, and two of which are lowercase with
various superscripts and subscripts.

1 The Moisture Index

The Thornthwaite expression for the moisture index[2] is $I_m = I_h - I_a$. The
moisture index is used to delimit nine separate moisture regions (Table 4-1).
Negative values of the moisture index are found in dry climates, positive
values in moist climates. A moisture index value of zero, therefore, separates
the dry and moist climates. It occurs in the subhumid climatic region between
the so-called "dry" subhumid and "moist" subhumid climates.

Mapping of the moisture regions In the mapping of climatic variables, the
numerical values apply strictly to the station locations themselves. Between
stations, other factors have to be employed to determine the placing of
isolines. For most climatic parameters, significant environmental factors are
oceanic and topographic influences.

---

[1] The following discussion is based on a 1955 revision of the original 1948
Thornthwaite classification carried out by Thornthwaite and Mather. As a result of this
revision, certain of the original limits and definitions have been modified and hopefully
improved.

[2] $I_h = 100 \ (S/PE)$; $I_a = 100 \ (D/PE)$, where S, D, and PE equal annual water surplus,
deficit, and potential evapotranspiration, respectively. Substituting for I_h and I_a, we get
$I_m = 100 \ [(S - D)/PE]$. It will be shown in a later section that this can also be written
$I_m = 100 \ [(P/PE) - 1]$.

Since many climatic factors vary with height in some rational fashion, the topographic map becomes an important guide to the interpolation of climatic values between stations. Reduced potential evapotranspiration, smaller values of moisture deficit, greater values of moisture surplus, and increased precipitation are all usually related to increased elevation. Potential evapotranspiration is much more closely related to elevation than is precipitation since the latter factor also is influenced by location on the mountain with respect to the prevailing wind. Since the moisture index depends on both precipitation and potential evapotranspiration, topography is only moderately well correlated with values of this derived parameter.

Coastal locations are generally cooler than locations in the immediate interior. This is especially true if cold upwelling of ocean water occurs along the coast. Oceanic influences are not as marked along wet coasts because temperature differences between wet land surfaces and ocean surfaces are less pronounced (Carter, 1954).

Values of the moisture index at more than 2,200 stations in North America have been plotted on a large-scale map for analysis purposes, using the foregoing general suggestions for interpolation. A reduced version of this map has been included here as Fig. 4-1. It shows that large, fairly homogeneous climatic regions exist in eastern North America and through the Great Plains region. In the mountainous areas of western North America, and especially in the southwestern United States, the topographic diversity results in a great variety of climatic regions within short distances. All the Thornthwaite climatic types are found in North America although the distributions of A and E climates are somewhat restricted. C and D climates, which cover most of eastern and central North America, are most extensive.

TABLE 4-1

The moisture regions (I_m), according to Thornthwaite's 1955 climatic classification*

CLIMATIC TYPE		MOISTURE INDEX
A	Perhumid	100 and above
B_4	Humid	80–100
B_3	Humid	60–80
B_2	Humid	40–60
B_1	Humid	20–40
C_2	Moist subhumid	0–20
C_1	Dry subhumid	−33.3–0
D	Semiarid	−66.7− −33.3
E	Arid	−100 − −66.7

* From Carter and Mather, 1966, table 5; reprinted in modified form from the Geogr. Rev., vol. 38, 1948, copyrighted by the American Geographical Society.

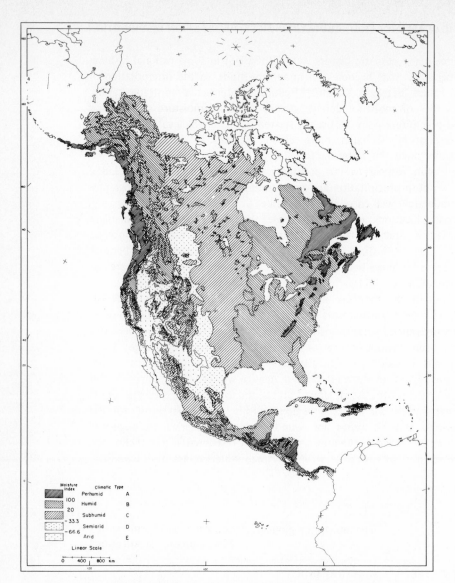

Moisture regions of North America. (*Analysis by Dr. D. B. Carter, artwork by K. Nishimoto, reproduced courtesy of C. W. Thornthwaite Associates, Elmer, N.J.*)   FIG. 4-1

The data coverage is quite complete so that the analysis can be viewed with considerable reliability.

2 Seasonal Variation of Effective Moisture

Köppen, in his classification, indicated seasonality of precipitation by the addition of lowercase letters *f*, *s*, and *w* to specify that the precipitation was well distributed through the year, that there was a dry summer period, or that there was a dry winter period, respectively. Thornthwaite considered the seasonality of *adequate* and *exceptional* moisture conditions. Exceptional moisture conditions are of two kinds: a surplus is exceptional in a dry climate and a deficit is exceptional in a moist climate. Since Thornthwaite was able to determine whether the climate is moist or arid on a basis of a moisture index, and since this factor takes prime consideration in his climatic classification, he also based his factor of the seasonality of exceptional moisture on information from the moisture index. Naturally, a dry climate must have some season of inadequate moisture. What Thornthwaite tried to make clear was not the seasonality of precipitation but whether a dry period existed in a moist climate or whether a moist period existed in a dry climate. In identifying the seasonality of precipitation, one is concerned whether, in a moist climate, the water deficiency is large, moderate, or small. Likewise in a dry climate, one is concerned whether there is a large, moderate, or small water surplus.

While Köppen used *s* for summer dry and *w* for winter dry, Thornthwaite applied the letters *s* and *w* to both moist and dry climates. He defined *s* as a seasonal variation in moisture with the drier season in summer; *w* as a seasonal variation in moisture with the drier season in winter. With no subscript, these letters indicate a moderate deficiency or surplus, while with a subscript 2 following these letters a large deficiency or a large surplus is indicated. Thornthwaite used the lowercase letter *r* to indicate little or no water deficiency in a moist climate, the lowercase letter *d* to indicate little or no water surplus in a dry climate.

The lowercase letters used to indicate the seasonal variation of effective moisture depend upon the first capital letter used in the classification. If the climate is moist $(A, B,$ or $C_2)$ followed by an *s*, there is a moderate summer water deficiency. In this notation the important consideration is the existence of a water deficiency in a moist climate or a water surplus in a dry climate. The *s* and *w* refer merely to winter or summer as the time for the period of water deficiency to occur (Table 4-2).

3 An Index of Thermal Efficiency

Thornthwaite used potential evapotranspiration as an index of thermal efficiency also. Since it expresses the amount of energy available in the climate at a particular place in terms of the water that could be evaporated by this energy if water was readily available, potential evapotranspiration combines both thermal and moisture aspects of the climate.

In his classification, Thornthwaite used nine capital letters, each expressed with a prime superscript, to provide a thermal classification including megathermal, mesothermal, microthermal, tundra, and frost types. He selected a potential evapotranspiration of 114 cm, a value that would result at a station having a mean temperature every month of $23°C$ $(73.4°F)$ (and no variation in day length), to separate megathermal from mesothermal climates. The megathermal he called A', the mesothermal, B'. The other breakpoints between the B'-C', C'-D', and D'-E' climates occur in a descending geometric progression of intervals. Table 4-3 gives the actual limits for each of the thermal regions, the letters that are used in the classification system, and the names of the climatic subtypes.

4 Summer Concentration of Thermal Efficiency

The fourth aspect of the classification is an index expressing how much of the thermal energy is received during the three summer months. In an equatorial climate in which the temperature never varies, one-quarter of the energy or thermal efficiency will be received during any 3-month period. In an arctic climate in which the temperature is above freezing for only 3 months of the year, it is clear that 100 percent of the thermal efficiency is received during the three summer months. Thus the scale of summer concentration of thermal efficiency must vary between 25 and 100 percent for any climate.

While astronomical considerations tend to result in a somewhat fixed latitudinal relationship between the summer concentration of thermal efficiency and the annual total, there are a number of different situations in which this relationship becomes modified. One can think of maritime locations in which summer temperatures are lower and winter temperatures are higher than would normally be expected for the latitude. Here the summer concentration of thermal efficiency is less than the latitudinal average. On the other hand, in continental locations far from any maritime influence, the summer concentration of thermal efficiency should be larger than the latitudinal average. In this particular usage, the summer concentration of thermal efficiency becomes an index of continentality or oceanity.

Table 4-3 shows the limits of the summer concentration of thermal efficiency for each of the nine climatic types that Thornthwaite has defined.

TABLE 4-2

Seasonal variation of effective moisture*

MOIST CLIMATES (A, B, C_2)		ARIDITY INDEX
r	Little or no water deficiency	0–10
s	Moderate summer water deficiency	10–20
w	Moderate winter water deficiency	10–20
s_2	Large summer water deficiency	20+
w_2	Large winter water deficiency	20+
DRY CLIMATES (C_1, D, E)		HUMIDITY INDEX
d	Little or no water surplus	0 –16.7
s	Moderate winter water surplus	16.7–33.3
w	Moderate summer water surplus	16.7–33.3
s_2	Large winter water surplus	33.3+
w_2	Large summer water surplus	33.3+

From Carter and Mather, 1966, table 6; reprinted in modified form from the Geogr. Rev., vol. 38, 1948, copyrighted by the American Geographical Society.

He uses lowercase letters ranging from a to d with prime superscripts to differentiate the classes as in the case of the thermal efficiency scale.

APPLICATION OF THE CLASSIFICATION TO SELECTED CLIMATIC DATA

The first step in determining the complete classification is to evaluate the climatic water budget at the particular station by using average monthly data (see Table 3-6). Average climatic water budgets have been published for nearly 9,000 stations for almost all land areas of the earth [Mather (ed.), 1962, 1963, 1964, 1965]. The seasonal course of potential evapotranspiration, actual evapotranspiration, and precipitation can be plotted to provide a graphical representation of the various factors of the water budget. Such plotted water budgets show immediately the periods of moisture surplus and moisture deficit. Figures 4-2 and 4-3 show the seasonal march of the factors of the climatic water budget at representative stations on the North American continent.

To evaluate the four letter symbols, the following information is needed from the water budget: the annual water need or potential evapotranspiration, the percent of the potential evapotranspiration (in this case the thermal efficiency) that comes during the three summer months, the annual precipitation, the water surplus, the water deficit, the surplus

TABLE 4-3

Thermal efficiency and its summer concentration*

THERMAL EFFICIENCY		CLIMATIC TYPE		SUMMER CONCENTRATION	
IN.	CM			%	TYPE
		E'	Frost		
5.59	14.2				
		D'	Tundra		d'
11.22	28.5			88.0	
		C_1'			c_1'
16.81	42.7		Microthermal	76.3	
		C_2'			c_2'
22.44	57.0			68.0	
		B_1'			b_1'
28.03	71.2			61.6	
		B_2'			b_2'
33.66	85.5		Mesothermal	56.3	
		B_3'			b_3'
39.25	99.7			51.9	
		B_4'			b_4'
44.88	114.0			48.0	
		A'	Megathermal		a'

* *Reprinted from the* Geogr. Rev., *vol. 38, 1948, copyrighted by the American Geographical Society.*

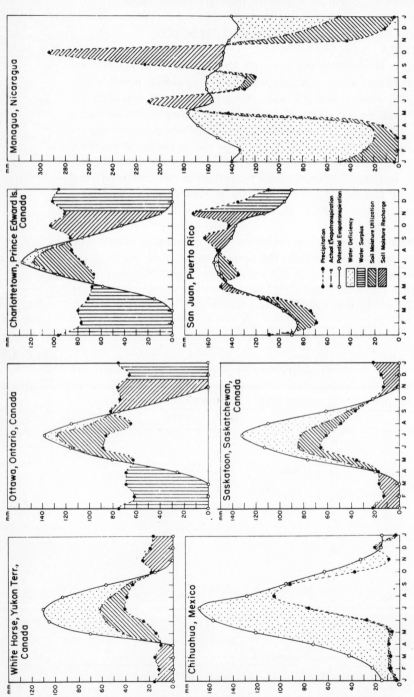

Average monthly water budgets at selected stations in North America.   FIG. 4-2

121

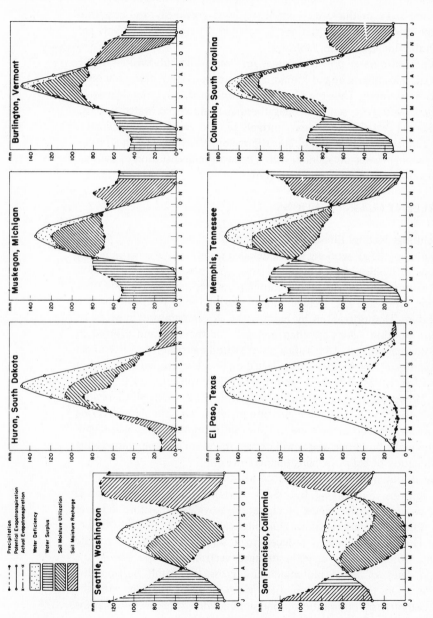

Average monthly water budgets at selected stations in the United States.   FIG. 4-3

expressed as a percent of need, the deficit expressed as a percent of need (these latter two factors express the humidity index and the aridity index, respectively), and the moisture index. The necessary data for classifying the stations, as illustrated in Figs. 4-2 and 4-3, have been obtained from the water budget computations and are included in Table 4-4. Using these data and the limits defining each of the letter symbols outlined previously, the complete classification for each of the stations is possible. These letter symbols are included in the final column in Table 4-4.

Thornthwaite never provided names for the various subdivisions of the mesothermal, microthermal, and humid climatic types. They have only been referred to by symbols. Thus the climatic description of the station Managua, in Nicaragua, would be semiarid, megathermal, little or no water surplus, with a thermal efficiency regime normal to the megathermal climate. Memphis, Tennessee, is in a humid climatic region, third mesothermal, has a moderate summer water deficiency, with a summer concentration of thermal efficiency equal to a third mesothermal climate.

OTHER RECENT CLASSIFICATIONS

1 The Holdridge Natural Life-Zone Classification

L. R. Holdridge (1959), working with a detailed knowledge of environmental relationships in tropical and subtropical areas, has proposed a classification scheme utilizing the mean annual biotemperature, the potential evapotranspiration ratio, and the average annual precipitation. To determine the biotemperature, Holdridge sums by months the air temperatures in excess of $0°C$. These monthly values are then averaged for the year and utilized to provide a thermal scale that identifies seven latitudinal regions or altitudinal belts. The limits of these belts are established by selecting geometric decreases from 24 (that is, 24, 12, 6, 3, and $1.5°C$). The value of $24°C$ for the boundary between tropical and warm temperate (low subtropical) regions corresponds to a mean annual temperature of $24°C$ since no below-freezing temperatures will occur in these areas. However, in middle-latitude or higher regions, the mean biotemperatures will exceed the mean air temperatures due to the elimination of below-freezing temperatures.

Annual biotemperature is converted to annual potential evapotranspiration (in mm) by multiplying by 58.93. Dividing the value of potential evapotranspiration by the annual precipitation provides a moisture index or potential evapotranspiration ratio as Holdridge calls it.

The Holdridge scheme is reduced to a simple triangular coordinate graph. The biotemperature scale is arranged vertically at one side of the graph, and lines of equal biotemperature cross the graph horizontally. Superimposed on the temperature field is an isosceles triangle, with the potential evapotranspiration ratio on one sloping side (from low values at the top to large values at the bottom) and the annual precipitation on the other sloping side (again with low values at the top). Diagonals at preselected values are drawn from these two sides to the base of the triangle, where the scale is labeled "humidity provinces." These diagonals, making angles of $60°$ with the

TABLE 4-4

Comparative moisture data at selected stations in North America

STATION	PE, MM	SUMMER NEED %	SUMMER NEED MM	SURP., MM	DEF., MM	I_h	I_a	I_m	CLIMATIC TYPE
North America (excluding U.S.):									
White Horse, Yukon, Canada	450	68.7	268	0	182	0.0	40.4	−40.4	$DC'_2\ dc'_2$
Ottawa, Ont., Canada	585	63.1	876	321	30	54.9	5.1	49.8	$B_2 B'_1\ rb'_1$
Charlottetown, Prince Edward Is., Canada	542	62.5	1000	476	18	87.8	3.3	84.5	$B_4 C'_2\ rb'_1$
Managua, Nicaragua	1821	27.5	1177	0	644	0.0	35.4	−35.4	$DA'\ da'$
Chihuahua, Mexico	934	48.5	406	0	528	0.0	56.5	−56.5	$DB'_3\ db'_4$
Saskatoon, Sask., Canada	534	55.4	368	0	166	0.0	31.1	−31.1	$C_1 C'_2\ db'_1$
San Juan, P.R.	1472	31.0	1546	83	9	5.6	0.6	5.0	$C_2 A'\ ra'_1$
United States:									
Seattle, Wash.	683	47.4	840	283	126	41.4	18.4	23.0	$B_1 B'_1\ sa'$
Huron, S.D.	633	62.2	487	0	146	0.0	23.1	−23.1	$C_1 B'_1\ db'_1$
Muskegon, Mich.	591	63.1	810	259	40	43.8	6.8	37.0	$B_1 B'_1\ rb'_1$
Burlington, Vt.	621	61.7	825	225	21	36.2	3.4	32.8	$B_1 B'_1\ rb'_1$
San Francisco, Calif.	702	33.3	551	36	187	5.1	26.6	−21.5	$C_1 B'_1\ da'$
El Paso, Tex.	960	52.7	218	0	742	0.0	77.3	−77.3	$EB'_3\ db'_3$
Memphis, Tenn.	923	52.5	1232	404	95	43.8	10.3	33.5	$B_1 B'_3\ sb'_3$
Columbia, S.C.	952	51.2	1108	193	37	20.3	3.9	16.4	$C_2 B'_3\ rb'_4$

biotemperature lines and with each other, form a series of hexagons. Because of the particular choice of limiting values, there are, in addition, certain triangular regions spaced among the hexagons specified by half the ranges of the limits for the hexagons. To eliminate these areas, Holdridge superimposes another set of hexagons over the earlier grid by drawing lines from the centers of each of the small triangles through the intersections of each of the limiting biotemperature, precipitation, and potential evapotranspiration ratios included on the diagram. This results in 30 hexagons, all representing major vegetation types. The hexagons are interlocked with no unaccounted triangles or vegetation types.

Holdridge considers that the life zones of "first-order ecosystems" (1959) so defined can actually be easily recognized in the field without the need to refer to the climatic data. Other factors such as day length and seasonal variations in radiation that may help to differentiate zones within the broader life zones are not included quantitatively on the life-zone chart because they are correlated directly with mean annual biotemperature and elevation.

To determine the life zone for a particular site only the mean annual biotemperature and precipitation are actually needed (along with elevation). The potential evapotranspiration ratio can be determined from the available data, but it is not necessary in order to locate oneself within one of the life-zone hexagons. The scheme provides a direct and simple method for determining the life zone at a place; it does not, however, take into account the seasonal patterns of temperature or moisture that are so necessary in order to differentiate between zones having the same annual temperature and precipitation amounts.

2 Budyko's Radiation Index of Dryness

In developing a scheme for the classification of climates, Budyko (1956) first produced a moisture index that he called the "radiational index of dryness" based on the ratio of the net radiation to the amount of energy needed to evaporate the available precipitation. If we assume that the value of net radiation is equivalent to the potential evapotranspiration, then Budyko's ratio becomes merely PE/P (all expressed in energy terms). Even though net radiation does not equal potential evapotranspiration exactly, the Budyko ratio approximates the inverse of the standard moisture index.

Budyko has prepared a world map of the distribution of the radiational index with scale intervals, 3, 2, 1, 2/3, and 1/3. He further suggests certain relations among these regions and various soils and vegetation zones; values of 1/3 or less are associated with tundra, 1/3 to 1 with forest of various types, 1 to 2 with steppe, 2 to 3 with semidesert, and greater than 3 with desert.

The radiational index is quite useful in evaluating the efficiency of the moisture supply. But since the net radiation term in the numerator includes energy for convection, the index value of 1, for example, is really equal to values less than unity obtained from the PE/P ratio. This has a tendency to expand the more humid categories at the expense of those on the drier side of the boundary value.

3 Bagnouls-Gaussen Classification

Gaussen (1955) proposed a simple means of determining a moisture index by merely comparing the monthly precipitation (in mm) with the value of the mean monthly temperature (in °C) doubled. If the value of temperature doubled represents the potential evaporation of water possible with the available monthly energy, then the ratio again becomes the simple *P/PE* index of earlier investigators. This moisture index was adapted by Bagnouls and Gaussen (1957) as the basis for a more formalized climatic classification. Walter and Lieth (1964) have also used it in modified form to produce climatic diagrams of the world, themselves useful as part of a climatic classification. Meher-Homji (1963) used the same basic index in a bioclimatic classification. Since it involves monthly data and classifies the moisture condition of each month, it represents a significant break from the older approaches more concerned with annual data.

As employed in a climatic classification by Bagnouls and Gaussen (1957), the number of dry months is used as a significant characteristic of climate. Twelve major climatic provinces are identified on the basis of (1) the number of dry months, and (2) three types of thermal regimes (no temperatures below freezing, temperatures above and below freezing, and no temperature above freezing).

Carter (Carter and Mather, 1966) has pointed out serious deficiencies in the Bagnouls-Gaussen scheme: (1) the lack of validity of the estimate of potential evapotranspiration; (2) the failure to consider how stored soil moisture might modify the dryness of a month; and (3) the lack of functional and rational criteria by which to define climatic types.

Walter (1955) modified Gaussen's scheme by plotting the temperature curve during warm months at three times the precipitation curve. In spite of this attempt to rectify Gaussen's underestimation of evapotranspiration Walter's scheme provides for a monthly water loss of only 90 mm if the air temperature is constantly 30°C (or an annual evapotranspiration of 1,100 mm), still well below the expected value on the basis of observations. Walter and Lieth (1964), using this modification, have identified 10 major climatic types among their many climate diagrams.

THE WATER BUDGET AND THE DISTRIBUTION OF VEGETATION IN NORTH AMERICA

Climate, soil, fire, relief, drainage, and man are among the many factors that influence the distribution of natural vegetation. The term "natural vegetation" in the present context refers to the predominant vegetation association existing under the normal pattern of climatic, environmental (including natural fires), and edaphic conditions, generally uninfluenced by the directed or willful action of man. It does not necessarily refer to a so-called climatic climax vegetation association.

Few investigators would really question that climate has an influence on the distribution of vegetation, although the exact nature of this influence under different environmental conditions may be quite variable. S. R. Eyre (1963) expresses the opinion that chemical composition and physical characteristics of the soil and parent material can be as important as climatic factors. He cites several examples of different vegetation covers on different soil areas under exactly the same climatic conditions to support this point. He does not eliminate the basic control of vegetation by climate, however. Hare (1954) and later Thornthwaite and Hare (1955) attempted to correlate potential evapotranspiration and the moisture index (I_m) with the distribution of vegetation. They found that in a temperate, humid climate, I_m can roughly distinguish different forest types, while the subarctic, boreal forest subdivisions are better correlated with PE. Major (1963), in an alternative approach, has considered a third factor, actual evapotranspiration (AE), and has suggested that vascular plant activity and growth might be related to the actual water loss of the vegetation since this factor expresses the real water activity of the plant rather than some unrealized potential value.

Mather and Yoshioka (1968) studied the relation among natural vegetation, the climatic moisture index, and potential evapotranspiration in all major vegetation regions across the conterminous United States. To broaden the scope, they also included data from tundra and tropical rainforest areas around the world as well as from coniferous and hardwood areas in Canada. The results of the study are plotted in Fig. 4-4. Information concerning the characteristic natural vegetation at each station has been indicated by means of a symbol on the diagram.

Vegetation distributions were obtained from a map by Shantz and Zon (1924) in the "Atlas of American Agriculture," supplemented by a more recent map by Küchler (1964). Sixteen principal vegetation areas were identified. Climatic stations were located in each of these 16 vegetation areas (in the case of discontinuous regions some stations were sought in all segments of the regions), and the data of the Thornthwaite moisture index (I_m) and potential evapotranspiration (PE) were obtained from the computations of the average water budgets for each station.

The moisture index $I_m = 100[(S - D)/PE]$. Using average annual data, $S = P - AE$ if there is no change in storage, and $D = PE - AE$. Substituting these expressions for S and D, the moisture index can be written as $100[(P - PE)/PE]$ or $100[(P/PE) - 1]$. Potential evapotranspiration occurs in both the abscissa and ordinate scales of Fig. 4-4, but since the precipitation P is entirely independent of PE, the ordinate variable is independent of the abscissa variable. Plotting the climatic and vegetation data in Fig. 4-4 reveals that many different climatic stations within the same vegetation association have somewhat similar thermal (PE) and moisture (I_m) factors. These two indices demonstrate an ability to distinguish among different vegetation associations.

The range of annual PE for stations within the conterminous United States is from about 500 to 1,250 mm. Below 500 mm, the vegetation is the generally coniferous (boreal) forest and tundra of Canada and Alaska. Above 1,300 mm tropical rainforests and semiarid or arid vegetation associations are found. The moisture index across North America ranges from well over +100

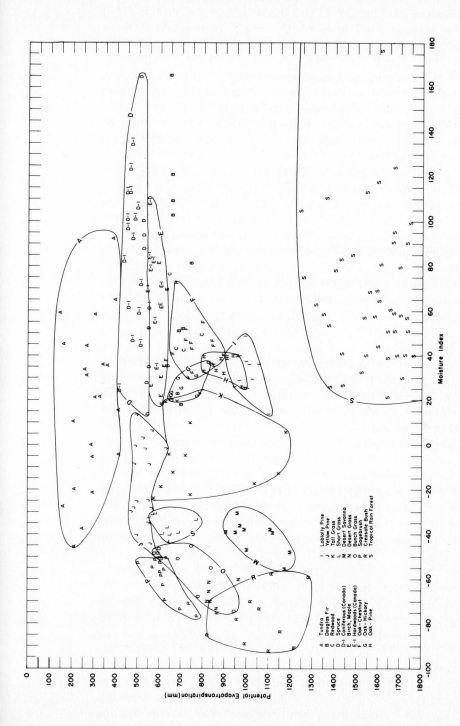

Relation between climatic moisture index, potential evapotranspiration, and natural vegetation at representative stations in the United States, Canada, and the Tropics. (*From Mather and Yoshioka, 1968, with permission of the Association of American Geographers.*)

FIG. 4-4

in the mountainous areas of both the northeast and the northwest to values close to -100 in the arid areas of the southwestern United States and northern Mexico. The range of stations used is sufficient to include all but one portion of these distributions of PE and I_m. There appear to be no stations with PE values over 800 mm and with I_m values over +60. These values represent a fairly warm, moist climate and should be found near latitude 30 to 35°N, close to the coast where sufficient moisture is present to give the high values of I_m. However, such climates do not generally exist anywhere in the world since these latitude belts fall in the so-called "subtropical high-pressure regions" where precipitation is low to moderate. I_m values above 50 seldom occur in these areas except possibly where orographic influences prevail.

The pattern of distribution shown in Fig. 4-4 supports the idea that it is possible to locate discrete and, in many cases, nonoverlapping vegetation areas that can be defined by the annual potential evapotranspiration and the climatic moisture index. If active factors of climate are chosen, it is possible to obtain a good relation between certain climatic and botanic distributions in midlatitudes. From the distributions shown in Fig. 4-4, one is clearly able to define not only a forest climate as opposed to a grassland or desert climate but also to identify a birch-maple as opposed to a spruce or oak-chestnut forest climate in a region such as North America. Where the areas on Fig. 4-4 are fairly distinct as in the case of the tundra, spruce, birch-maple, oak-chestnut, oak-hickory, tall grass, yellow pine, short grass, desert savanna, creosote bush, sagebrush, bunch grass–desert grass, and tropical rainforest, one feels fairly confident in emphasizing the dominant role that climate plays in vegetation zonation. It does not, of course, eliminate the possibility of edaphically influenced vegetation associations within large homogeneous climatic regions or the possibility of more than one vegetation association being able to survive and compete successfully within a given climatic range. Thus an overlap of Douglas fir and redwood points or an overlap of bunch grass and desert grass stations points is found in Fig. 4-4. In part, this overlap may be due to poor definition of the vegetation association itself or to the choice of poorly descriptive climatic parameters to delimit the particular vegetation association. Edaphic and other environmental factors must also contribute to these areas of overlap, but the general lack of overlap bears witness to the responsiveness of the botanic environment to significant climatic stimuli.

Relation between Vegetation and the Humidity and Aridity Indices

Both Thornthwaite (1952a) and Mather (1959b) have shown that the vegetation associations are also related to the two factors that make up the moisture index, namely, the aridity index $I_a = 100 \, (D/PE)$ and humidity index $I_h = 100 \, (S/PE)$. The average values of the aridity and humidity indices have been obtained for the United States stations used in Fig. 4-4 and the results are plotted in bar form in Fig. 4-5. The vegetation classes have been ranked by decreasing humidity index and increasing aridity index as much as possible. Figure 4-5 clearly shows the unique position of the redwoods and

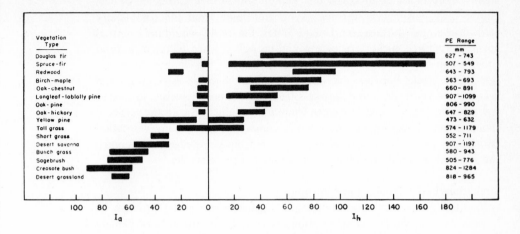

Ranges of humidity (I_h) and aridity (I_a) indices and annual potential evapotranspiration for representative vegetation types in the United States. (*From Mather and Yoshioka, 1968, with permission of the Association of American Geographers.*)

FIG. 4-5

Douglas fir among the forest associations. They can exist in a climatic region that results in as much aridity as grasslands experience, but at the same time they are also found in areas of very high rainfall with some of the highest humidity indices in the United States. All the other forest types except the yellow pine and the pinõn-juniper (not included) association exist in areas with significant humidity indices and insignificant aridity indices.

As expected, the tall-grass vegetation areas occur in those climatic regions with moisture indices very close to 0. Because of the lack of accord between the distribution of precipitation and evapotranspiration, periods of both moisture surplus and deficit exist, but since no long periods of deficit occur, the vegetation is luxurious and the soil is rich and productive.

The short grass, bunch grass, and other semiarid and arid vegetation associations occur in areas with markedly negative moisture indices. Essentially no moisture surplus occurs at stations where these vegetation associations are found, while increasing aridity indices accompany a shift in the vegetation from short grass, to bunch grass, to sagebrush, and creosote bush.

Figure 4-5 indicates clearly that it is possible to speak about a forest climate as opposed to a grassland climate and a desert shrub climate in midlatitudes. The occurrence of certain forest associations, such as the yellow pines between grassland associations, on Fig. 4-5 points up the need for caution in making all-inclusive statements but it does not vitiate the conclusion that one can speak of forest or grassland climates. Within the different vegetation-formation classes there would appear to be distinct breaks between certain types of forest associations and other forest groups, as well as between certain semiarid vegetation associations and others which are slightly more moist or dry. The distribution shown in Fig. 4-5, of course, cannot really distinguish the loblolly-longleaf pine from the oak-hickory hardwoods in the southeastern United States. Figure 4-5 would indicate that possibly either oak-hickory or pine would be well adapted to the region and either group might exist in the region as a result of the climatic conditons alone. The breaks between the spruce-fir and birch-maple associations and between the sagebrush and desert grassland associations, however, seem very real and significant. Especially when viewed in conjunction with the values of annual potential evapotranspiration it becomes clear that many vegetation associations can be identified in large measure from knowledge of the annual and seasonal ranges of just a few significant climatic parameters.

Limitations in the Tropics and Subtropics

Within the Tropics and subtropics and especially when the problem of tropical savanna vegetation associations is considered, the role of climate seems to become of less import. Correlations found in midlatitudes between climate and vegetation do not necessarily exist in the Tropics. For example, Parsons (1955) has described in some detail the Miskito pine savanna of the Nicaragua and Honduras coasts. Not only is this pine savanna significant in that it marks the most tropical penetration of pines in North America, but it is possibly the rainiest area of savanna vegetation anywhere in the Western Hemisphere. Average annual rainfall varies from 2,500 to 3,700 mm, and there is only a short dry season. Parsons feels that the limits of the pine savanna are well marked by the termination of the gravel shelf on which this vegetation association is found. Savanna develops as a result of the lowered water-holding capacities of the soils, for in the same area on soils with better moisture-holding characteristics, high forest occurs. Parsons also considers fire, hurricanes, and impeded subsurface drainage as possible contributing factors. There is a suggestion that hardwoods might be able to take over the area if fire and other cultural destructive agents were eliminated.

The Miskito pine savanna represents a challenge to all who seek neat causal relationships between distributions of active factors, whether they be climatic, edaphic, or cultural. Puerto Cabezas is located on the coast of Nicaragua near the middle of the area of pine savanna. It experiences 3,293 mm of precipitation on the average through the year, ranging from 458 mm in June to 50 mm in March. The moisture index for Puerto Cabezas is +106 with a humidity index of 109 and an aridity index of 3. With a potential evapotranspiration of 1,545 mm, this area, by all rights, should experience a tropical rainforest type of vegetation. Climate in this case may be a "permissive" factor, but it certainly is not a determinant one since savanna rather than rainforest exists there.

Denevan (1961), in a study of the upland pine vegetation of Nicaragua, also emphasizes the role of fire in the development of the pine savanna. He considers the frequency, severity, time, and extent of burning as determining factors; with frequent fires only grass can regenerate and survive. If the area can be protected long enough against fire, he feels pine forests will take over to be replaced in time by hardwoods.

The existence of a pine savanna in this region poses two separate problems: first, the problem of conifers in tropical and subtropical climatic regions, and second, the problem of the savanna vegetation and its relation, if any, to climate. Denevan feels that pines are not able to compete successfully against broadleaf vegetation in tropical and subtropical areas without the aid of fires and man. Other vegetation associations in tropical areas should be restudied with this possibility in mind.

The degree to which the distribution of vegetation can be explained on the basis of climatic conditions depends *in part* on the proper selection of climatic factors. Temperature and precipitation by themselves are poor descriptors of climate. Precipitation does not indicate whether a climate is moist or dry unless one is able to compare it with the water need of a place. Temperature does not really reveal the energy that is useful for plant development unless one knows also the moisture condition of the soil at the time. Thus the active factors of climate from a vegetation viewpoint are water surplus or water deficit (water supply in relation to water need), actual evapotranspiration (actual plant water use), and soil moisture storage. Knowledge of these factors is available from computations of the climatic water budget, and combinations of these factors are expressed in shorthand form in the moisture index or the humidity and aridity indices of the Thornthwaite climatic water budget.

The moisture index along with evapotranspiration is useful in differentiating among the three land biochores: grassland, forest, and desert. Even within each biochore, these active climatic parameters are able to reveal differences which seem to be related to the different formation classes. For example, careful analysis of the vegetation and climatic records reveals the oak-hickory vegetation association as developing under a different climate from birch-maple, or bunch grass in a different climatic environment from sagebrush.

Use of more expressive climatic parameters or, finally, use of the frequencies of some of these elements to describe distributions in environmental biology should yield increased appreciation of the role of climate. It should still not depreciate the role of edaphic, cultural, and other factors in influencing vegetation distribution. There are many examples of nonclimatic controls on vegetation such as the savanna and the low-latitude pine associations. It is, of course, only through a complete interpretation of all contributing influences that the most understandable picture of the geographic distributions can be obtained.

CLIMATE, HYDROLOGY, AND WATER RESOURCES

5

Ideally, any study of the water resources of an area should be based on long records of water supply (precipitation, water inflow, soil moisture storage) and water loss (evaporation, transpiration, water runoff, subsurface outflow). Actually, water resource scientists hardly ever encounter such complete records so they must begin with what is available, improvise and extend them in whatever way they can, construct new records from other available data, and theorize from data available from other areas under similar environmental situations. Climatic water budget approaches can contribute significantly to the study of water resources by making available values of such needed parameters as soil moisture storage, water deficit, water surplus, and water runoff.

SOIL MOISTURE STORAGE

Since precipitation adds to the soil moisture content and evapotranspiration and runoff subtract from it, it is possible to obtain the actual moisture content of the soil at any time from knowledge of the balance between these quantities. This can be done on either a daily or longer-term basis, depending on the climatic records employed.

The method most commonly used to determine soil moisture is to collect soil samples in cans that can be tightly sealed, to weigh them, dry them in an oven for a given period of time, and then to reweigh them. The change in weight of the sample indicates the soil moisture content, which is usually expressed as a percentage of the dry weight of the sample. It would be more practical to express moisture content on a volume basis, but since it is difficult to obtain soil samples of known volume, moisture determinations are seldom given volumetrically. Collection of samples to determine soil moisture content disturbs the soil structure and plant roots in the sampled area. Thus it is often not feasible to make repeated samples in an area in order to follow the seasonal course of soil moisture. Although the method is laborious and subject to a few limitations, it is reasonably accurate and is the basic technique against which the moisture values determined by instrumental methods are compared.

Some of the more important general categories of instrumental techniques that do not require disturbing the soil or vegetation in the sampling area once the instruments are in place are measurements of electrical conductivity, resistivity, thermal conductivity, or capacitance of the soil; determination of the weight of gravimetric plugs or absorption blocks placed in contact with the soil; measurement of the tension with which moisture is held in the capillaries of the soil; and determination of the scattering of high-energy neutrons in the soil mass as a result of collisions with hydrogen atoms.

In spite of the improvements that have been made in recent years in instrument design and operation, soil moisture instruments still do not give results that are too useful for either long- or short-term planning. The soil moisture content as measured by the different instruments or by the weighing and drying of soil samples is representative of only the area from which the sample or measurement is taken. A short distance away, changes in topography or soil type or structure can be such that a far different soil moisture content exists. Thus, to determine the moisture content of a field of just a few acres, it may be necessary to make a large number of measurements. Even if it were feasible to make the necessary observations, and in a few localities they are now being made, the length of these records is not great enough to permit their use in studies of the probability of occurrence of different levels of soil moisture at different times of the year. For long-range planning, such probability studies are essential.

A climatic approach to determination of soil moisture content is extremely simple and straightforward. Of all the avenues of water gain and loss by the soil, precipitation, evapotranspiration, and water runoff (including percolation) are by far the most important. As a first approximation, the soil moisture may be determined simply as a result of a bookkeeping procedure in which all precipitation occurring when the soil moisture content is below field capacity is treated as a moisture increment and moisture is lost from the soil only by evapotranspiration. Precipitation occurring when the soil is above field capacity is treated as surplus; it is lost by percolation and runoff and so

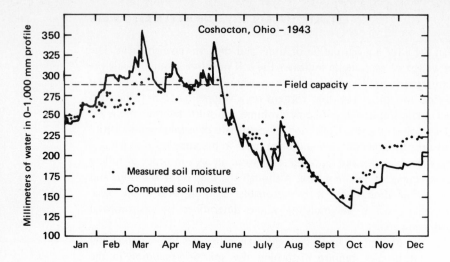

Comparison of measured and computed soil moisture content at Coshocton, Ohio, 1943. Soil moisture in 0–1 m profile on watershed Y102; measured values obtained by Soil Conservation Service from soil samples and weighing lysimeter. Computed values from climatic data using water budget method. (*From Thornthwaite and Mather, 1955b.*)

FIG. 5-1

does not enter into the soil moisture storage balance. Water movement by surface and subsurface flow into or out of the area is neglected.

Excellent results can be obtained through such a computational procedure. Day-to-day variations in soil moisture have been computed by using a climatic budgeting technique for a number of years at Coshocton, Ohio, and the results for 2 years have been compared in Figs. 5-1 and 5-2 with the actual soil moisture measurements (by weighing soil samples) as published by the Soil Conservation Service (Harrold and Dreibelbis, 1951). For the type of soil at Coshocton, a 1-m soil profile contained a total of about 300 mm of water at field capacity. It was assumed that 93 percent of the available gravitational water was detained in the soil each day.

At Coshocton during the winter there are a number of periods when the air temperature is well below freezing and the ground is frozen to some depth. During these periods percolation of gravitational water is greatly reduced or in some cases prevented entirely until thawing of the soil occurs. In making these computations, it was assumed that percolation did not occur on days when the mean temperature of the air was below $-1.0°C$.

Considering the assumptions and approximations made in computing soil moisture on the one hand, and the methods of soil moisture determination employed on the other, agreement between the measured and computed values of soil moisture at Coshocton is very satisfactory. Certain problems relating to the possible surface runoff of intense precipitation on sloping land (not considered in the present example) and the proper treatment of melting from a snow cover still need to be resolved, but in spite

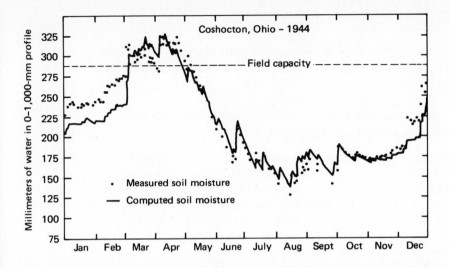

Comparison of measured and computed soil moisture content at Coshocton, Ohio, 1944. (*From Thornthwaite and Mather, 1955b.*) FIG. 5-2

of these limitations, it is possible to determine with fair reliability the day-to-day variation in soil moisture content over an area.

Versatile Budget Approach

A significant advance in the procedure to compute soil moisture content has been developed by Baier and Robertson (1966) and Baier (1967) in Canada. The method, called the "versatile budget," has developed from an earlier "modulated budget" approach formulated by Holmes and Robertson (1959), which accepted potential evapotranspiration as a possible maximum of actual water loss and subdivided the total available soil moisture storage into various layers in the soil with the same or different moisture capacities.

The modulated budget method assumed that moisture from the upper zone was lost by evapotranspiration only at the potential rate and all moisture had to be lost from this layer before removal from the next lower layer could begin. Exponential decay drying rate curves were used to express the removal of water from lower layers in the soil. The newer "versatile budget" method permits the simultaneous removal of moisture from the various layers in the soil by roots on the basis of the data on potential evapotranspiration and available soil moisture content. The budget also permits modifications for the surface runoff of heavy precipitation, and use of different types of soil-drying curves (Fig. 3-6).

A simplified version of the general Baier and Robertson expression for the versatile budget, giving values of actual evapotranspiration (in inches) for any day i (AE_i), from the sum of changes in the soil moisture content in each zone (j) from 1 to n is

$$AE_i = \sum_{j=1}^{n} \left[k_j \; \frac{S'_{j(i-1)}}{S_j} \; Z_j PE_i \right]$$

where k_j = coefficient based on soil and plant characteristics in zone j_i

$S'_{j(i-1)}$ = available soil moisture in zone j at morning observation of day i

S_j = total available water in zone j_i

Z_j = adjustment factor for particular soil-drying curve used

PE_i = potential evapotranspiration for day i

In one particular example, Baier and Robertson (1965) divided the soil depth of Mathilda loam soil near Ottawa, Canada, into six arbitrary zones of 5, 7.5, 12.5, 25, 25, and 25 mm water-holding capacities each. The total depth of soil layer under consideration was just over 0.6 m; the top three arbitrary zones were not equal in depth. It was assumed that the available water was approximately 0.15 mm/mm depth of soil in the profile, a total of 100 mm for the entire profile.

The k coefficient expresses the amount of water (in percent of potential evapotranspiration) that can be removed by plant roots from different soil layers during the season. The coefficients must be determined or estimated on the basis of type of cover and density (thus, rooting pattern). If the coefficients add up to 1, actual and potential evapotranspiration will be equal during periods of no water deficiency. Table 5-1 gives some suggested values for different soil layers under a grass cover.

The term $S'_{j(i-1)}/S_j$ suggests a linear relation between actual evapotranspiration and soil moisture content similar to that used by Thornthwaite and Mather (1955b) (Fig. 3-6). Since the soil moisture from the uppermost layer is lost at the potential rate, since the depth of the layers can be varied, and since different forms of drying relations can be utilized in the expression, the versatile budget offers considerable flexibility in handling the problem of drying rate.

TABLE 5-1

Definition of moisture zones and k coefficients
used in versatile budget for permanent grass*

ZONE	DEPTH, MM	AVAILABLE SOIL MOISTURE %	AVAILABLE SOIL MOISTURE MM	k COEFFICIENT
1	0–35	5.0	5.0	0.50
2	35–85	7.5	7.5	0.20
3	85–160	12.5	12.5	0.15
4	160–315	25.0	25.0	0.10
5	315–470	25.0	25.0	0.03
6	470–625	25.0	25.0	0.02

* *With permission, Elsevier Publishing Company, from Baier, 1969, table 1.*

For budgeting purposes, it is necessary to make some assumptions concerning runoff of rainfall exceeding the infiltration rates of the soil. Here, Baier and Robertson followed the results of an earlier study by Linsley, Kohler, and Paulhus (1949) suggesting that the initial soil moisture content chiefly influenced the rate of infiltration. Thus their expression for infiltration involved the term $S'_{j(i-1)}/S_j$ along with the precipitation and several coefficients and constants. It was also assumed that on days with precipitation, the evapotranspiration occurred first at a rate dependent on the soil moisture content at the beginning of the day and that rainfall then occurred later during day i. This would be generally the case with late afternoon convective shower activity.

Baier and Robertson compared the results from both the modulated and versatile budgets with observations by means of Colman blocks during the summer of 1954. Accumulated soil moisture deficits in each soil layer (Fig. 5-3) show that the modulated budget overestimates (if we assume the measured Colman block values are correct) the moisture loss from the top four soil zones (5, 7.5, 12.5, and 25_1) and underestimates the soil moisture loss from the bottom zone 25_3. The versatile budget method appears to overestimate moisture loss in the 12.5 and 25_2 layers and underestimates this loss in the 5 layer. The modulated budget results in slightly too much removal from the upper layers and not enough from the lower layers, while the versatile budget approximates observed conditions more nearly in all layers except 25_2.

Additional studies have been carried out by Baier and his associates since the versatile budget was first proposed, and several minor modifications have been incorporated to provide even more reasonable estimates of soil moisture content (Baier, 1967). The soil moisture available to plants has been divided into six "standard" zones of varying water-holding capacities. Use of standard zones permits one set of k coefficients to be used for a particular crop in any type of soil (see Table 5-1). Then the estimated water uptake during very dry periods has been improved by increasing the rate of water removal from the lower, moist soil layers when the water removal from the upper, drier soil layers decreases. The plant roots seem to become more efficient in removing water from the lower layers when the upper soil layers are dry than when the whole profile is uniformly moist. This adjustment is handled by distributing the k coefficients from the upper dry zones over the more moist, lower zones in proportion to the assumed root distribution. Later versions of the versatile budget have provided even better agreement with measured values than the earlier version.

Many computations for various locations and different years support the conclusion that soil moisture can be determined with considerable precision from climatological data. The climatic approach will permit accurate determination of the movement of water through soils and the amount of storage in any selected layer of the soil, provided the water-holding characteristics of the soil are known.

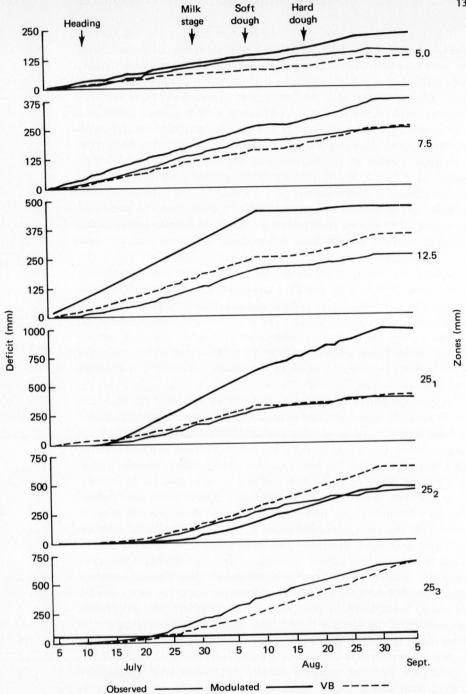

Soil moisture deficits under wheat, per soil layer, accumulated over time, 1954. (*With permission of* Can. J. Plant Sci., *from Baier and Robertson, 1966, Fig. 4.*)

FIG. 5-3

MOISTURE SURPLUS OR RUNOFF

Since the moisture surplus as computed from a bookkeeping procedure represents water that is available for runoff or stream flow, it is possible to obtain estimates of this latter parameter from climatic data alone (see Table 3-6). Where stream gages exist, an accurate check on the climatic bookkeeping procedure is available. If the computational procedure is reliable, it should then be possible to obtain records of stream flow in areas or at times when no stream gage records exist.

Of course, all the moisture available for runoff in a particular month may not be lost in that month; some moisture is often detained on the watershed past the end of the month and will run off in succeeding months. On moderate to large watersheds it has been estimated that only about 50 percent of the available surplus water will run off in a given month, the rest being retained for later runoff.

Measured values of runoff have been obtained from each gaging station on the Delmarva Peninsula (comprising the state of Delaware and the eastern shore areas of Maryland and Virginia) during the period 1949 to 1964 (Mather, 1969b). These data have been compared with the values of computed runoff by using the climatic data from nearby weather stations. The basic data for several watershed areas are summed in Table 5-2, which gives the individual yearly values of measured runoff as well as the computed runoff at the closest climatic station to each of the gaging stations. Correlation coefficients between annual computed and measured runoff for each of the 16 years of data are remarkably high—the values ranging from +0.94 in the case of the relation of computed runoff at Newark, Delaware, to measured runoff in the Christina River at Cooches Bridge, to a value of +0.84 between computed runoff at Princess Anne, Maryland, and measured runoff in Manokin Branch.

Results are less reliable when individual monthly values of computed runoff are compared with the monthly values of measured stream flow. The reasons for this are clear: (1) the influence of intense precipitation causing a variable proportion of the precipitation from an individual storm to run off without entering into the soil can result in considerable disagreement on a short-period basis; (2) the rate of movement of water through the soil and back to the stream may not always be constant but may vary with the season or with other individual climatic or environmental factors; (3) the rate of water release from melting snow is not well understood and can only be approximated in present modelling studies.

In spite of the limitations, results are reasonable. Table 3-6 has already shown the agreement between average monthly values of measured runoff at the Wilmington, Delaware, gaging station on Shellpot Creek based on 19 years of record and the computed runoff based on water budget computations for Wilmington, Delaware, meteorologic data.

TABLE 5-2

Annual values of gaged stream runoff and computed runoff at nearby weather stations, Delmarva Peninsula, 1949–1964 (values in cm depth)*

YEAR	MILLINGTON (CAL. RUNOFF)	UNICORN BRANCH (MEAS. RUNOFF)	NEWARK (CAL. RUNOFF)	CHRISTINA AT COOCHES BRIDGE (MEAS. RUNOFF)	WHITE CLAY CREEK (MEAS. RUNOFF)	BIG ELK CREEK AT ELK MILLS (MEAS. RUNOFF)	PRESTON (CAL. RUNOFF)	FAULKNER BRANCH AT FEDERALSBURG (MEAS. RUNOFF)	PRINCESS ANNE (CAL. RUNOFF)	MANOKIN BRANCH (MEAS. RUNOFF)
1949	33.4	40.8	37.4	38.8	41.5	43.4	25.7		38.5	
1950	27.7	29.6	36.2	34.5	38.4	41.2	25.3		32.0	
1951	35.4	36.3	50.7	47.9	43.2	47.7	32.4	33.8	37.5	
1952	57.6	52.6	72.6	63.5	63.3	66.4	68.6	61.4	57.3	22.6
1953	46.2	44.6	57.1	54.8	55.7	56.4	42.3	43.5	44.7	21.4
1954	26.6	23.7	26.0	22.5	26.9	25.2	29.8	26.3	29.1	13.0
1955	24.4	22.0	41.1	37.6	31.1	32.4	45.1	39.3	49.7	30.5
1956	40.4	30.8	48.0	46.8	39.9	39.7	41.2	42.4	55.2	27.9
1957	40.7	29.3	39.2	41.6	34.4	35.8	39.6	36.9	45.7	22.7
1958	64.7	58.4	66.7	64.9	66.3	66.4	62.6	67.9	66.9	38.0
1959	30.9	27.1	32.6	38.4	34.0	35.3	38.1	50.2	48.5	21.2
1960	53.5	55.3	48.2	53.1	48.4	49.3	49.0	57.2	69.8	29.2
1961	49.9	57.7	40.2	49.4	45.5	48.8	52.6	56.6	66.9	32.4
1962	36.5	32.0	33.7	35.7	32.6	35.6	45.5	39.9	66.1	28.4
1963	35.2	24.4	29.0	26.7	26.1	26.4	40.1	31.4	49.5	23.7
1964	46.1	36.0	33.4	33.6	34.1	32.3	40.7	33.8	52.3	24.9

* From Mather, 1969b

SNOW HYDROLOGY

Both snow accumulation and snowmelt runoff are basic hydrologic problems under a significant degree of meteorologic control. The atmosphere supplies the moisture both for snowfall (provided temperatures are cold enough) and for condensation of water vapor on the snow surface, while meteorological factors control the interchange of energy in the basin snowpack. These energy factors determine, in large measure, the rate and quantity of snowmelt and runoff. The problems of snowmelt and runoff are much more complex than those of accumulation, and it is in this area that the applied climatologist has the opportunity to contribute most significantly.

During the snowmelt period, meltwater from the snow surface will begin to move downward into the lower layers of the snowpack. Since temperatures there are usually below freezing, refreezing of the meltwater will occur with the resulting release of the heat of fusion. This raises the snowpack temperature. Additional heat will also be added to the snowpack both from the air above and from the ground below so that after a period the temperature of the entire pack may approach 0°C. Meltwater continues to be absorbed by the snowpack until the liquid water-holding capacity of the pack is reached and the snow is defined as *ripe*. The density of the snowpack which has been increasing remains fairly constant after the water-holding capacity is reached. Additional meltwater now moves downward through the snow to the ground surface. In this respect the process of water movement through the snowpack has many of the same characteristics as water movement through the soil.

Depending on the infiltration capacity of the soil, and the rate of melting of the snow, the meltwater will either (1) enter the soil and move through it or (2) form a slush layer at the surface that will both contribute to moisture infiltrating the soil surface or to the overland runoff from the area.

Snowmelt can be considered as a heat-transfer process so that an energy budget-type relation can be established to determine the rate of snowmelt at a spot. The U.S. Army Corps of Engineers (1956), in a basic study of the whole snowmelt problem, expressed all the energy terms in the following relation:

$$H_m = H_{rs} + H_{rl} + H_c + H_e + H_g + H_p + H_q$$

where H_m = heat equivalent of snowmelt or amount of heat involved in change of state from ice to water

H_{rs} = absorbed solar radiation

H_{rl} = net longwave radiation exchange between snowpack and environment

H_c = convective or sensible heat transfer from air

H_e = latent heat of vaporization released by conduction

H_g = heat conduction from ground

H_p = heat content of rainwater

H_q = change in energy content within snowpack

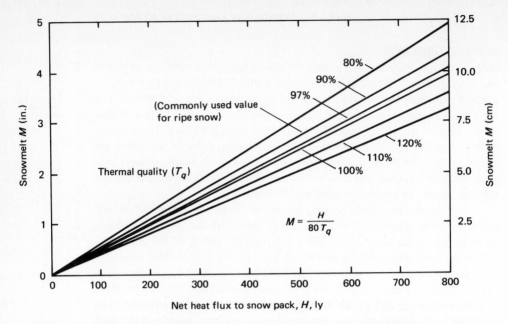

Snowmelt resulting from thermal energy under different values of thermal quality of the snowpack. (*With permission, from U.S. Army Corps of Engineers*, 1956, *plate* 5-1.)

FIG. 5-4

If heat fluxes toward the snowpack are considered positive, then H_{rl} is generally negative for open exposures, H_e and H_q may be positive or negative, H_c is generally positive, and the other terms are always positive. Eighty langleys are required to melt one cubic centimeter of ice at $0°C$, so that the resulting snowmelt in centimeters is $M = H_m/80T_q$, where T_q is the thermal quality of the snowpack. The term "thermal quality" is used to express the fact that snow is different from pure ice at $0°C$. The snow may have a temperature other than $0°C$ and it may have free water in it. Thermal quality is defined as the ratio of the heat needed to produce a given quantity of water from snow to the heat needed to produce the same quantity of water from pure ice at $0°C$. Figure 5-4 illustrates the relation between heat supply and snowmelt for snowpacks with different thermal qualities.

Of the various sources of energy for the melting of snow, convection (H_c), vapor condensation (H_e), and radiation (H_{rs}, H_{rl}) are the most important, followed by the heat from rainwater (H_p). Under most circumstances, conduction (H_g) is a negligible energy contributor to the snowmelt process.

The U.S. Army Corps of Engineers has derived specific physical relations by which each term in the heat budget equation can be evaluated. By applying these relations to the meteorological data collected over the 5-year period 1946–1947 to 1950–1951, they have computed the monthly course of the various heat fluxes for a basin at the Central Sierra Snow Laboratory (Fig. 5-5). The various heat fluxes are expressed in terms of depth

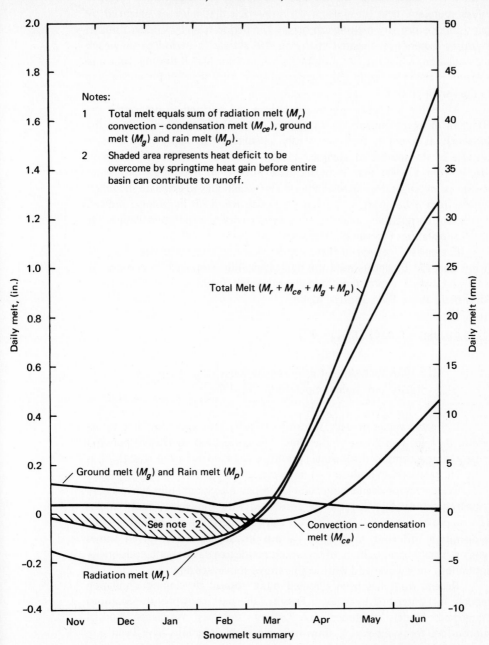

Notes:

1   Total melt equals sum of radiation melt (M_r)
     convection – condensation melt (M_{ce}), ground
     melt (M_g) and rain melt (M_p).

2   Shaded area represents heat deficit to be
     overcome by springtime heat gain before entire
     basin can contribute to runoff.

Total Melt ($M_r + M_{ce} + M_g + M_p$)

Ground melt (M_g) and Rain melt (M_p)

See note   2

Convection – condensation
melt (M_{ce})

Radiation melt (M_r)

Daily melt, (in.)

Daily melt (mm)

Nov   Dec   Jan   Feb   Mar   Apr   May   Jun

Snowmelt summary

Monthly course of various heat fluxes, Central Sierra Snow Laboratory,   FIG. 5-5
averaged for years 1946–1947 to 1950–1951. (*With permission, from U.S.
Army Corps of Engineers,* 1956, *plate* 5-7.)

of melt that would result from the various amounts of energy fluxes. Negative values represent heat losses from the snowpack that must be balanced by gains later before any melting can occur. The period from November through February experiences negative melts on the average, a period of snowpack accumulation, while the flux becomes positive from March through June and ablation is the rule. Rapid melting occurs from April through June according to the computations.

While the individual physical relations can be used to determine the effect of different meteorological, ground, and vegetation conditions on the snowmelt at a point, they are hardly satisfactory to apply over a basin because of the myriad of complex factors involved. Yet it is the snowmelt runoff from a basin that is of ultimate interest in hydrology and water resources investigations. Most of the present techniques to estimate snowmelt runoff from a basin are somewhat empirical, based on hydrologic indices, degree-day correlations, analysis of recession curves, correlation studies, or physical equations (Viessman, Harbaugh, and Knapp, 1972).

Examples of several of these approaches might be instructive. The U.S. Army Corps of Engineers used the hydrologic index approach to develop the following simple expression for daily snowmelt (in inches) over the Boise River basin above Twin Springs, Idaho:

$$Q = 0.00238G + 0.0245 \, (T_{max} - 77)$$

where G = estimated value of daily radiation exchange in open, ly

T_{max} = daily maximum temperature, $°F$, at Boise

Estimated values of daily snowmelt runoff agree with computed values within 0.11 in. two-thirds of the time. The correlation coefficient between measured and computed snowmelt runoff, using data for 1954 and 1955, is 0.95. The relationship is illustrated more clearly in Fig. 5-6.

Degree-day or degree-hour values (temperature summation above a base temperature, usually $0°C$) can be correlated with snowmelt runoff values. The approach is only approximate since many other factors beside temperature influence snowmelt. Various investigators have given values ranging from 0.05 to 0.20 in. of snowmelt per degree day under conditions of continuous snow cover and with temperatures above freezing.

Recent work has been directed more toward developing simulation models for watershed hydrology. Possibly the best known is the Stanford Watershed Model developed by Linsley and his associates. Special subroutines for synthesizing snowmelt runoff hydrographs have been prepared. Their description and use is beyond the scope of the present discussion. However, by making use of the storage and computational facilities of modern high-speed computers, it is possible to achieve remarkably good estimates of snowmelt runoff provided that previous input data involving snowmelt hydrographs and meteorological information have been reliable and representative.

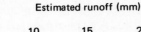

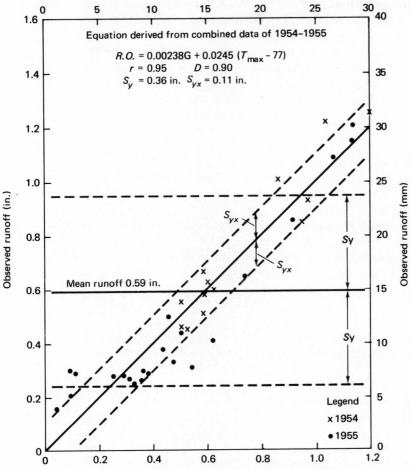

R.O. = Daily generated snowmelt runoff in inches
depth over snow-covered area
G = Daily net allwave radiation absorbed by
snow in open, langleys
T_{max} = Daily maximum temperature, Boise, °F
r = Coefficient of correlation
D = Coefficient of determination
S_y = Standard deviation of observed runoff
S_{yx} = Standard error of estimated runoff

Observed vs. estimated runoff, Boise River basin above Twin Springs, Idaho,   FIG. 5-6
1954, 1955. (*With permission, from U.S. Army Corps of Engineers, 1956,
plate* 6-3.)

PRACTICAL EXAMPLES OF CLIMATOLOGY APPLIED TO HYDROLOGY

The ability to use climatic data to compute evapotranspiration with reasonable accuracy (at least on a monthly basis), to determine soil moisture storage or to evaluate the movement of water through the soil to the water table or to surface streams, when coupled with other measured hydrologic data, permits the quantitative evaluation of many factors that had often been only estimated previously. It is possible to refer to only a few of the many hydrologic studies that have involved an applied climatology approach. However, because of the usefulness of the various approaches in problem solving, several selected examples will be described in more detail here. They, in turn, will suggest other possible uses.

Climatic Water Budget of the Lake Erie Basin

Sanderson (1966) undertook to evaluate each of the factors of the climatic water budget relationship for the Lake Erie basin (precipitation, evaporation, inflow, runoff) for the period 1958 to 1963. Comparing these results with average conditions over the lake basin, she sought a new understanding of the factors responsible for changing lake levels. The water budget technique can provide two possible sets of data: first, quantitative estimates of runoff from the basin around Lake Erie, since 31 percent of this basin area is presently ungaged (some of the gaged records are also listed only as "fair" to "poor" due to the problem of ice obstruction during the winter); and second, quantitative estimates of evaporation from the lake surface itself. In this particular situation, however, Sanderson determined evaporation from the lake (E) as the residual in the following water budget equation for the lake:

$$E = I - O \pm \Delta S + P + R$$

where I = water inflow from upper lakes
  O = outflow to lower lakes
  ΔS = change in storage in lake
  P = overwater precipitation
  R = runoff from surrounding basin

The Thornthwaite water budget was used to compute the basin runoff R, while precipitation was determined by isopleth mapping and planimetering of values. The data on which the isopleths were constructed were values of precipitation at perimeter stations. The other values in the water balance relationship were all measured.

From the various derived values, Sanderson determined the so-called "net basin supply," the water contributed to the Great Lakes system by the individual lake basin. In hydrologic terms, the net basin supply (NBS) may be determined from the relation

$$NBS = R + P - E$$

where R = runoff from basin around lake
  P = overwater precipitation
  E = overwater evaporation

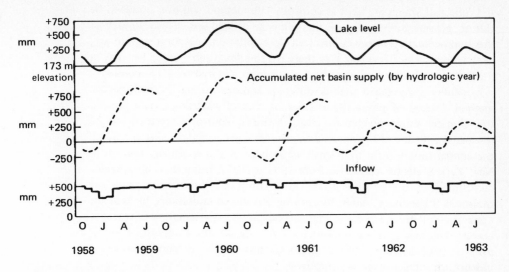

Relation of inflow, net basin supply, and lake levels, Lake Erie basin,   FIG. 5-7
1958–1963. (*From Sanderson, 1966, fig. 33, with permission.*)

Morton and Rosenberg (1959) concluded that the annual change in water levels in the lakes was closely influenced by changes from season to season in the net basin supply.

Figure 5-7 shows a clearcut relation between net basin supply and the water levels in Lake Erie. Inflow from the Detroit River is quite constant, regulated as it is by the large quantities of water stored in the upper lakes system. Lake levels rise for the first 3 years of the study and fall during the last 2 years in accordance with the accumulated net basin supply in each of the years.

The Effects of Reforestation on Water Yield

Land-use change of any sort will result in modification of the factors of the hydrologic balance. Paving over surfaces clearly leads to an increase in runoff of surface water and the more immediate flow of this water to surface streams. Cultivation of previously undisturbed land or the removal of a forest stand also leads to increased volumes of runoff.

Muller (1966), in a study of the effect of reforestation on water yield, has described two different research approaches used to provide information on the effect of reforestation on hydrologic relationships. One, involving the systematic study of forest effects on individual hydroclimatological factors, is called the "physical method." The other, involving study of the overall response of the entire watershed to a change in conditions through comparisons with other similar watersheds which are maintained as controls, is called the "hydrometric method." The hydrometric method will, for example, provide quantitative estimates of water-yield change due to changes in surface cover by comparing several watersheds, some of which are modified and others of which are kept as controls. The method will not, however, be able to provide information on the influence of land-use change on each of the hydroclimatic processes involved (precipitation, evaporation, storage, percolation, surface flow, etc.), so there is a role that each method can play in furthering our understanding of land-use changes and hydrologic responses.

Muller suggested a slightly different approach using the hydrometric method. Instead of physically establishing control watersheds, he employed the climatic water budget to create artificial control watersheds to be compared with the changing conditions on real watersheds. Muller computed the annual runoff from four small watersheds in the Allegheny uplands of New York State for each year from 1935 to 1957, using the Thornthwaite climatic water budget model. He then compared these values with the measured stream flow values. In carrying out the computations, he kept the same value of storage in the soil throughout, thus not attempting to take into account any change in vegetation cover. As a result, the computed values of runoff reflect the runoff that would occur due to climatic conditions alone, independent of any changing land cover.

Measured runoff tended to be greater than computed runoff at the beginning of the study when forests covered 20 to 50 percent of the watersheds; measured runoff was less than computed in the latter years when forests covered 80 to 90 percent of three of the watersheds (Fig. 5-8). The differences between measured and computed water runoff were tested for statistical significance, and the results from three of the four watersheds proved significant. Calculating the linear regression lines and the regression coefficients, Muller was able to obtain the average yearly decrease in runoff due to reforestation or other changes in land cover over the period of investigation. While Muller's study goes further to suggest certain hydrologic consequences if reforestation continues, these conclusions need not concern us here.

The Water Budget of Delaware Bay

Quantitative determination of flows in large water bodies, such as bays or estuaries, without gaging stations has always created difficult problems for hydrologists. When the mouth of the bay is wide and currents move both ways, often at the same time at different levels, it has been nearly impossible to obtain more than subjective estimates based on records from the gaged streams flowing into the estuary.

A climatic water-budget bookkeeping approach allows us to obtain a quantitative estimate of net flow at any place within the estuary system, although it still cannot provide information on the actual volume of flow in each direction. An example of the application of such a technique is available from a recent study on Delaware Bay, located between Delaware and New Jersey in the eastern United States, by Mather, Swaye, and Hartmann (1972).

The Delaware estuary includes the whole area from the last stream gaging station (at head of tide) on the Delaware River at Trenton, New Jersey, to a line from Cape May, New Jersey, to Cape Henlopen, Delaware,

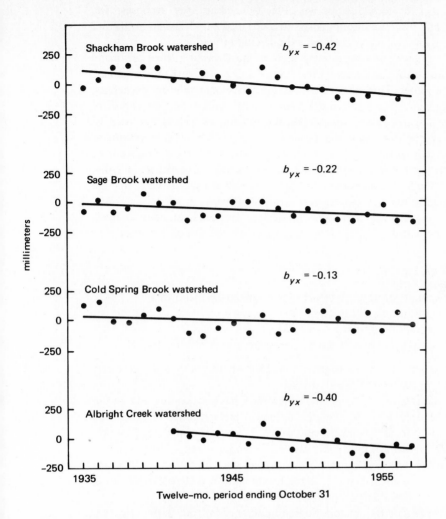

Differences between calculated and measured water yield at experimental watersheds on the Allegheny Plateau, 1935-1957. (*From Muller, 1966, fig. 6, with permission.*)   FIG. 5-8

where the estuary terminates in the Atlantic Ocean. This distance is 216 km (134 mi). The estuary varies in width from 200 m at Trenton to some 900 m at Philadelphia. Its widest extent is 43 km while its mouth is 19 km wide at Cape May–Cape Henlopen. The total drainage area is nearly 34,000 km², of which nearly 18,000 km² lies above the head of tide at Trenton.

Knowledge of the net flow in the estuary is of importance for many reasons. There is, of course, a relation between estuary flow and salinity conditions in the estuary which is of vital importance to the shellfish industry. The movement of the salt line up and down the estuary is of importance to those urban centers which must rely on estuary water for municipal uses or to industries who use it for their own purposes. The movement and dispersal of pollutants must also respond ultimately to the net balance between freshwater outflow and saltwater inflow.

Unfortunately, in any estuary such as the Delaware, there are many sources and sinks of water, thus complicating the task of obtaining quantitative values. There are, of course, the environmental water exchanges, the additions due to precipitation on the estuary surface, and the runoff of water from the land areas surrounding the estuary, as well as the losses of water by evaporation from the estuary surface. Other water additions or losses can occur through the bottom of the estuary and there can be fresh- or saltwater movement in either direction at the mouth of the estuary. Finally, there are a number of man-made additions and withdrawals of water that are more difficult to identify and evaluate quantitatively.

If it is assumed that we can neglect the unmeasured man-made additions and withdrawals, we can set up a balance accounting for other gains and losses.

Net flow = $RO + P - E \pm U$

where RO = runoff of water to estuary from surrounding land area

P = precipitation onto estuary surface

E = evaporation from estuary surface

U = underflow to or away from estuary

This last term is often neglected because we do not have the necessary information to evaluate it quantitatively.

For the Delaware estuary, monthly runoff has been computed by using climatic data from some 30 climatic stations in the watershed of the estuary below Trenton. Runoff from the area above Trenton was obtained from the stream gage at Trenton itself. The values of computed runoff from the land area below Trenton were plotted on maps, and lines of equal runoff drawn. Planimetering the areas, it was possible to obtain the total volume of runoff from this watershed area by months.

Next, the data on precipitation and evaporation from stations near the estuary were plotted on maps, and lines of equal precipitation and evapotranspiration were drawn over the estuary itself. While it is recognized that values determined at land stations might not hold over the water surface, no other reasonable technique was available. The limited extent of the water surface and the shallow nature of the estuary should act to keep its response fairly close to that of the surrounding land areas. Planimetering the areas between isolines and multiplying by the mean value between adjacent isolines provided values of the volume of precipitation and evaporation over the estuary surface.

Table 5-3 provides sample monthly values of the various inflows and outflows (neglecting underflow and any man-made diversions) for 1968. The seasonal pattern of each factor is in accord with the expectations—land runoff is a maximum in winter and early spring, precipitation a weak maximum in the summer period, and evaporation a marked summer maximum. The balance of these factors or the net flow at the mouth of the estuary shows a significant change through the year from values over 1,100 m^3/sec per month in March to less than 140 m^3/sec per month in late summer and fall.

Table 5-4 gives just the monthly net flow figures for a 20-year period starting in 1949. The month of July 1955 had so little precipitation and runoff from the land that there was actually a *net* inflow of saltwater to the estuary, 6 m^3/sec. The very dry period in the mid-1960s is quite marked with annual values dropping to 240 m^3/sec. The highest annual flow, over 870 m^3/sec, was found in 1952.

The calculated values of net flow are quite consistent with other available meteorologic and hydrologic information. For example, significant correlation coefficients have been found between monthly flow values and chloride content at various places in the estuary. Studies are currently under way to refine the flow values by including information on man-made diversions of water.

The Annual Cycle of Stored Soil Moisture

During the 1950s oceanographers working with the seasonal changes in the heights of sea levels concluded that a vast quantity of water, some 5.0×10^{18} g, was removed from the oceans between fall and spring each year and returned, of course, to the oceans between spring and fall. This resulted in an annual variation in mean sea level of about 1.4 cm when spread evenly over all the oceans of the world.

TABLE 5-3

Monthly water budget of the Delaware estuary, 1968
(all values in m^3/sec)*

1968	RUNOFF AT TRENTON	RUNOFF FROM BASIN DOWN-STREAM FROM TRENTON	PRECIPITA-TION ONTO ESTUARY	POTENTIAL EVAPOTRANS. FROM ESTUARY	NET FLOW AT MOUTH
Jan.	195	219	48	2	460
Feb.	280	182	21	2	481
Mar.	473	565	102	18	1122
Apr.	336	330	29	40	655
May	442	343	85	64	806
June	576	224	78	105	773
July	203	116	25	120	224
Aug.	126	54	40	114	106
Sept.	135	29	23	78	109
Oct.	126	19	55	47	153
Nov.	277	39	80	21	375
Dec.	306	79	51	2	434
Average					475

* *From Mather, Swaye, and Hartmann, 1972.*

van Hylckama (1956) investigated this problem from the climatic viewpoint by considering where this amount of moisture must go when it leaves the oceans. Clearly, little moisture is stored in the atmosphere at any time and the change in atmospheric storage from winter to summer (considering both hemispheres together) is quite negligible. Storage in rivers, lakes, and other surface water reservoirs does not change significantly through the year. Therefore, the only place where this amount of water can be held is in the soil itself.

In many areas of the world, evaporation of moisture is less than precipitation during the winter months and greater than precipitation during the summer periods (see, for example, the water budget diagrams in Figs. 4-2 and 4-3). The water budget at a place allows us to obtain quantitative values of the seasonal course of surplus and deficit and to determine the actual amount of moisture stored in or on the soil at any time. van Hylckama utilized the computations of the water budget at over 14,000 stations around the world to compute the amount of storage in and on the land by months of the year. From these values, he determined the monthly moisture storage in and on the land by $1°$ squares of latitude and longitude; these values were later summed by $10°$ squares and by latitudinal belts (Table 5-5).

In the Northern Hemisphere, maximum land storage is found in the late winter season (February to April) in midlatitudes, following the period of adequate precipitation and greatly reduced evapotranspiration. Maximum land storage in low latitudes (0 to $35°$) occurs in late summer and fall (August to October), following the rainy season. Evapotranspiration will not vary greatly through the year in low latitudes so that storage depends on the availability of precipitation, while in midlatitudes, precipitation varies less through the year and storage depends more on the seasonal pattern of evapotranspiration. Land areas in low latitudes in the Northern Hemisphere

MONTH	1949	1950	1951	1952	1953	1954	1955	1956	1957
Jan.	17	5	9	14	12	4	4	3	5
Feb.	13	7	13	10	12	7	6	9	8
Mar.	8	12	12	14	14	9	11	12	8
Apr.	8	10	12	18	14	7	7	13	11
May	7	6	4	12	11	6	3	7	4
June	2	4	3	6	4	1	2	4	2
July	1	2	3	4	2	0	0	3	0
Aug.	1	2	2	3	2	1	10	1	0
Sept.	1	2	3	3	1	1	3	2	1
Oct.	1	1	2	1	1	1	10	3	1
Nov.	1	6	10	7	2	5	8	5	2
Dec.	4	10	11	12	7	6	3	9	8
Av.	5.3	5.6	7.0	8.7	6.8	4.0	5.6	5.9	4.2

* From Mather, Swaye, and Hartmann, 1972.

are less extensive than in midlatitudes so that the pattern of storage found in midlatitudes dominates.

The seasonal patterns of storage will be just reversed in the Southern Hemisphere. Thus, the winter season of maximum land storage in midlatitudes is from August to October, and the late summer and fall period of storage in low latitudes occurs from February to April. But in the Southern Hemisphere, the maximum amount of land is found in low latitudes so that the pattern of storage found in those latitudes will be reflected most clearly in the hemisphere totals.

If the land areas of the globe were equally distributed between Northern and Southern Hemispheres, one would balance the other and there would be no net change in storage over all the land areas from winter to summer. However, in actuality, there is much more land in midlatitudes in the Northern Hemisphere and low latitudes in the Southern Hemisphere, and so the patterns of storage conditions in those latitudes (which happen to go hand in hand) control the seasonal course of storage for the whole globe. Hemispheric totals given by months at the bottom of Table 5-5 indicate the influence of these belts of larger land areas.

Approximately 25×10^{18} g of moisture are stored in and on the upper layers of the soil (the only part of the storage that changes appreciably from season to season) in every month of the year. This monthly figure, according to van Hylckama's study, varies from a maximum of 29.44×10^{18} g in March to a minimum of 21.77×10^{18} g in September. The difference between the late winter storage and the late summer storage is 7.6×10^{18} g, which is in quite good agreement with the earlier figure of about 5.0×10^{18} g of water

TABLE 5-4

Monthly total net flow of water at mouth of Delaware estuary ($100 \ m^3$/sec) from climatic water balance*

1958	1959	1960	1961	1962	1963	1964	1965	1966	1967	1968
7	5	8	3	6	3	8	2	2	6	5
5	6	11	12	5	3	7	6	6	5	5
17	9	7	15	12	15	12	7	8	11	11
19	9	12	14	12	8	12	6	6	9	7
10	3	6	8	4	4	5	2	5	8	8
4	1	4	4	2	2	2	1	2	3	8
3	2	3	3	1	1	1	0	0	3	2
4	1	2	2	1	1	0	1	0	7	1
2	1	9	1	1	1	1	0	1	3	1
4	3	4	2	1	0	1	1	2	2	2
6	5	3	2	5	3	1	1	2	4	4
4	8	3	3	3	3	2	2	5	9	4
7.1	4.4	6.0	5.8	4.4	3.7	4.3	2.4	3.2	5.8	4.8

TABLE 5-5

Moisture detained in and on the land areas of the world by 10° belts and by months ($\times 10^{16}$ g)*

LATITUDE	JAN.	FEB.	MAR.	APR.	MAY	JUNE	JULY	AUG.	SEPT.	OCT.	NOV.	DEC.
60–70°N	325.3	345.7	364.3	377.2	355.6	257.4	198.8	177.6	182.3	243.0	274.5	299.6
50–60	485.5	558.2	548.7	558.0	475.9	393.0	318.9	283.7	285.3	330.7	396.8	444.2
40–50	397.3	437.8	472.7	458.5	397.6	333.0	267.7	225.5	222.7	247.5	296.3	353.0
30–40	269.2	296.5	313.3	307.0	275.0	234.6	215.9	196.5	180.1	175.7	193.2	230.5
20–30	158.8	157.6	155.5	159.4	174.5	201.3	280.8	249.9	238.2	202.1	177.0	165.0
10–20	93.8	73.3	55.5	46.8	55.8	92.4	143.7	204.2	212.7	185.0	147.0	117.7
0–10	215.5	191.8	189.6	212.3	245.7	273.4	297.6	316.9	318.4	307.0	281.5	246.7
0–10°S	333.3	368.2	400.3	411.6	378.1	335.9	277.6	238.8	217.4	222.0	254.5	288.9
10–20	223.9	257.2	267.2	227.8	181.6	150.1	119.7	98.0	81.0	75.7	102.0	151.7
20–30	84.8	87.9	88.7	85.0	82.8	84.7	81.2	76.2	74.0	72.5	73.3	77.4
30–40	48.1	43.9	43.8	48.4	58.4	70.5	83.6	82.7	82.2	78.0	72.7	66.2
40–50	21.7	18.2	18.3	20.1	22.0	23.9	25.4	25.5	38.1	30.2	25.8	22.5
50–60	5.2	4.6	4.8	5.0	5.1	5.3	5.5	5.5	5.4	5.3	6.7	5.9
60–70	28.3	23.3	21.3	24.8	27.8	31.3	33.4	34.8	39.3	42.4	45.3	30.3
North total	1945.4	2060.9	2099.6	2119.2	1980.1	1785.1	1723.4	1654.3	1639.7	1691.0	1766.3	1856.7
South total	745.3	803.3	844.4	822.7	755.8	701.7	626.4	561.5	537.4	526.1	580.3	642.9
Grand total	2690.7	2864.2	2944.0	2941.9	2735.9	2486.8	2349.8	2215.8	2177.1	2217.1	2346.6	2499.6

* From van Hylckama, 1956, table 16, with permission.

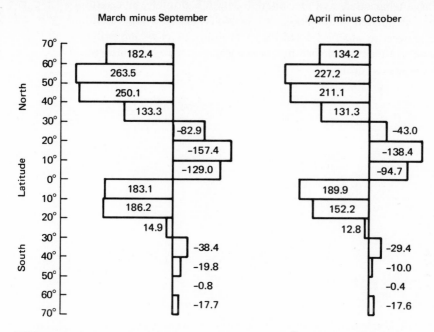

Differences in moisture detention on the land by 10° belts of latitude, March minus September and April minus October. All values are in 10^{16} g. (*From van Hylckama, 1956, fig. 10, with permission.*)   FIG. 5-9

removed from the ocean during the late winter season and replaced in the ocean by early fall.

Figure 5-9 illustrates the significant latitudinal pattern of difference in storage between late winter and late summer. The diagram clearly shows the large quantity of storage in middle to high latitudes in the Northern Hemisphere in late winter, reinforced by the excess of storage in tropical latitudes in the Southern Hemisphere at the time when the precipitation belts are south of the equator.

CONCLUSIONS

There is a close relationship between meteorologic factors such as precipitation and evaporation and different hydrologic and soil responses. This chapter has attempted to explore some of these relationships by means of actual case studies involving soil moisture storage, stream runoff, and lake levels. Only a few examples could be included here, but they should provide some indication of how climatic data can be applied to studies of various hydrologic problems. The hydrologist will, in some cases, have more and better data available to him and so he will not have to make the assumptions necessary in these examples. The examples do show, however, that even with readily available temperature and precipitation data it is possible, by careful manipulation of limited information, to obtain reasonable and useful quantitative answers to practical problems. Availability of more and better data will permit refinement of such hydrometeorologic studies and increase the value of the answers achieved.

CLIMATE AND AGRICULTURE: PHENOLOGY, DROUGHT, AND SCHEDULING PROBLEMS

6

The great need to increase agricultural production in recent years has led to a succession of significant books and monographs in the field of agricultural meteorology and climatology (see Wang, 1963a, 1972; Waggoner (ed.), 1965; Chang, 1968; for example). These works have been concerned with the influence of different physical parameters such as radiation, moisture, wind, and humidity on plants and animals, with energy and moisture balance studies, and with the role of short-period weather events on the development or yield of plants and animals. The emphasis has not generally been on the use of weather forecasts in agriculture.

The present and the following chapter attempt only a general survey of agricultural climatology. The role that climatic factors play in such basic problems as plant development (phenology), crop scheduling, irrigation, frost protection, diseases, and pests will be discussed. The relation of all the weather factors to the ultimate goal of agriculture—yield or production—will be considered from several viewpoints. Finally, there is a brief discussion of the problems of economic decision making in agriculture, a new field with vast potential and relevancy in this period when increased production from an ever-dwindling land area is vital. For more detail, the reader is referred to one of the aforementioned works.

In fulfilling its obligation "to establish and record the climatic conditions of the United States," the Weather Bureau recognized the need to standardize instrumentation and observation sites as well as type of observation and form of statistical presentation. Unfortunately, the standardized climatological stations are usually neither situated nor equipped to measure weather factors at the places and times that are critical for plants. For example, the temperature regime in a standard instrument shelter at standard height above the ground gives no measure of the range of temperature conditions at different heights within plant stands even at the observation site, let alone at a place some distance from the site.

The climate of a region as determined by means of the standardized observations is more or less of an abstraction (Thornthwaite and Mather, 1954). Actually, any region is made up of many local climates called "microclimates." They are the ones that must concern the farmer, the agronomist, and the biologist. These local climates vary greatly among themselves. For example, the climates of adjacent north- and south-facing slopes may resemble in many respects the standard climates of places hundreds of miles away to the north and south. Furthermore, the climate in a standard weather shelter is very different from that within a few inches of the ground in the open. Nighttime temperatures are lower and daytime temperatures are higher near the ground than a few feet above it.

There are many examples of failure to solve important agricultural and biological problems because of the inadequacy of our knowledge concerning local climates. Each problem requires study from the climatic as well as from the biologic side. Special exposure of instruments, sometimes the construction of special instruments, always the special handling of observational data are required. The task is not one for the biologist or the climatologist alone, but for both in collaboration.

CLIMATE AND CROP SCHEDULES

For more than 200 years, scientists have been attempting to determine the relationship between plant development and climate. The eighteenth-century French scientist, Rene Antoine de Réaumur, inventor of an alcohol thermometer and a temperature scale (still in use in scattered areas) having $0°$ as freezing and $80°$ as the temperature of boiling water, may well deserve the title of father of agricultural climatology. He was the first to determine quantitatively the amount of heat necessary to bring a plant from germination to a given stage of development. He did so by summing the mean daily temperature of the air for each day between one stage of development and the next. Réaumur (1735) felt that the sum of these daily temperatures from one particular stage of development to the next should be approximately constant for the same plant from year to year. This degree-day technique is still used today to express the relative heating or cooling needs of a climate.

Réaumur's original work on temperature summation created considerable interest and stimulated many similar studies. While Réaumur speculated that the temperature sums from one stage of development to the next should be constant for a particular species, later studies showed that actually these sums were quite variable from year to year and from one climatic region to another. The validity of the scheme was still accepted, however. Later modifications which were suggested preserved Réaumur's principle of temperature summation and involved changing only the manner of obtaining the sums.

The work of Réaumur and his successors carries a special lesson for us today. Recent demands for more useful information on scheduling the development and maturity of plants have led to a reinvestigation of factors influencing crop development; the temperature summation or heat-unit system of determining maturity, originally suggested by Réaumur, has received renewed attention (Thornthwaite, 1952c).

A heat unit is defined as one degree of temperature above the base growing temperature for the particular crop. Peas have a base growing temperature of about $40°F$ and sweet corn about $50°F$. The heat units of a day are obtained by subtracting the base temperature from the mean temperature of the day. The number of heat units needed to mature a crop is obtained by summing the daily heat units from the date of planting to the ideal harvest date.

The method has been widely tested and many deficiencies have been revealed. Despite these, genuine improvement in harvesting certain crops has resulted. Even though the temperature summations are different for different places, it is possible to achieve useful values for each location. The fact that the temperature summation differs from one year to another is a serious handicap. It prevents the reliable forecasting of harvest dates. But it is still possible to work out planting schedules which fix the harvest order of individual fields and prevent harvest pileups even though the harvest dates are not exactly known. The deficiencies now being found in the "heat-unit theory" are precisely the ones that had been pointed out in the decades following Réaumur's formulation of the principle of temperature summation in 1735.

PHENOLOGY

Phenology may be defined as the relation of climate to periodic biologic activity in either plants or animals. Although phenological observations have been made for hundreds of years, most of these have been merely of dates of planting, fruiting, budding, flowering, and leaf fall of different types of plants or particular events in the life of animals. Correlation studies with various factors of climate have been limited by the fact that many days elapse between each successive stage. Daily observations of plant development, for example, reveal more clearly any correlation between climatic factors and plant development that might actually exist. Many plants, however, do not grow in a manner permitting daily development to be readily determined.

The technique of making observations of daily plant development is based primarily on observation of the leaf formation. Not only is the presence or absence of a leaf (or "node" as it is called) of significance in determining development, but also the shape, character, and condition of the most recently formed leaf are important in providing information on the amount of development which has occurred since the last full leaf formed. In these observations, the physical size of the leaf or the plant itself or the increase in length or overall size since the last observation is not used; such measures of growth are more related to fertility, seed selection, and other factors independent of climate.

Studies have shown that plant development (progress toward maturity) is essentially uninfluenced by external environmental factors such as fertility, soil type, cultivation practice, or soil moisture content. As long as moisture and nutrients are sufficient to maintain the plant in a growing condition, its rate of development seems to be determined by climatic conditions and, more especially, by the available energy. Thus it is development rather than growth that must be used as an indicator of the effect of climate on the plant.

To simplify and standardize the procedure for making observations, detailed instructions have been prepared (Higgins, 1952). Careful drawings of the stages of development of the plants reduce the subjective elements in the readings and make it possible for even untrained observers to recognize the various daily developmental stages.

Phenological Observation in Peas

As the pea plant grows, successive nodes or pairs of leaves develop along the stem. Each node represents a definite stage of development of the plant. The development from one node to the next involves a series of readily discernible changes in the plant, so that it is possible at any time to say what fraction of the development has occurred from the last node to the one next to come. Numbering the nodes serially permits the use of a single figure to indicate the development stage of the plant at any time.

In order to record the development of the plant beyond the last full node, 10 substages of development have been identified. The stages, indicated by tenths of a node, are shown in Fig. 6-1. Higgins (1952, pp. 5-6) has described the 10 stages as follows:

0.0   This drawing shows a completed node at the tip of a plant at which point occurs a mature leaf composed of four leaflets and a tendril. Between the lower part of the leaflets is an immature stem which supports a small, tightly closed leaf bud.

0.1   Bud begins to develop. It increases in size and the tendril unfolds from between the first pair of leaflets.

0.2   Second pair of leaflets which are held closely together begins to show between the first pair.

0.3   Second pair of leaflets and tendril elongate.

0.4   Second pair of leaflets separate. Elongation of these and of the tendril takes place.

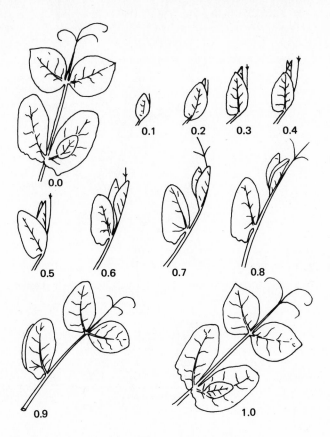

Stages of garden pea leaf development. (*From Higgins*, 1952, *fig. 2, with permission.*)   FIG. 6-1

0.5   Second pair of separated leaflets and tendril elongate.

0.6   Second pair of leaflets begins to separate from the first pair.

0.7   Second pair of leaflets becomes completely separated from the first pair.

0.8   Second pair of leaflets begins to unfold and becomes separated further from the first pair. The first pair of leaflets remains tightly closed.

0.9   Second pair of leaflets unfolds completely while the first pair begins to unfold.

1.0   Both pairs of leaflets have fully expanded and between the first pair of leaflets is a tightly closed leaf bud. This is a completed node.

Figure 6-2 is a sketch of a pea plant at a developmental stage between the eighth and ninth nodes. From the description of the internodal

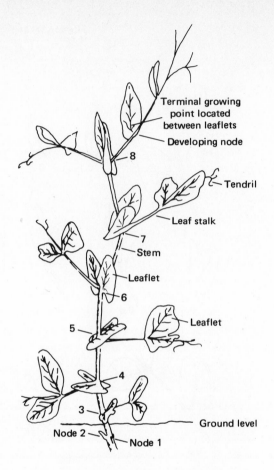

Vegetative development of the garden pea. (*From Higgins, 1952, fig. 1, with permission.*)   FIG. 6-2

development provided by Higgins, one might conclude that about 0.8 of a new node has developed. The plant would be described as at stage 8.8. Recognizing the stage of development of a plant to tenths of a node is relatively simple; different observers do not vary appreciably in their individual answers. Normally 10 or more plants are tagged and read daily in order to eliminate individual variations. An average rate of development from day to day is thus determined. This daily development rate may be related to daily weather conditions.

In a study of the daily leaf development of *Tephrosia vogelii* Hook f., Higgins and Decker (1971) found that maximum daily temperatures were more highly correlated with leaf development than were minimum, mean, or range of temperature. Average daily leaf development was some 20 percent greater at Solomons, Maryland, on the lower Chesapeake than at Oakland, Maryland, in the mountains of the western part of the state in mid-July. The mid-July rate at Solomons was more than twice as great as was the rate found in mid-April at Salisbury on the Delmarva Peninsula. Potential leaf yields for various areas could be forecast as the result of a relationship between plant development stage and harvested yield obtained earlier at Glenn Dale, Maryland.

Scheduling of Crop Harvests

Knowledge of the daily development of a plant and its relation to various climatic factors has made possible the application of scheduling techniques in farming never before possible. As a result, the farmer has more control over his farm operation. Moreover, the scheduling technique offers a means by which the farmer can obtain a climatic forecast of his future operations that may prove more useful than would be possible if he relied on long-range weather forecasts.

While advising Seabrook Farms, a large grower and packager of frozen vegetables in southern New Jersey, on irrigation, Thornthwaite began a detailed study of the various problems involved in pea growing and harvesting (Thornthwaite, 1953b). Seabrook Farms was divided into a number of divisions or smaller farms in partial competition with each other. While each division was told how many acres of peas to plant, there was little advice on how to space the planting or harvesting of the peas. Thus each farm planted as many acres of peas at one time as possible. The result was a big glut of peas arriving simultaneously at the factory. To ease the harvest as much as possible, peas were picked before they were mature, during their mature period, and after maturity, night and day. Peas stood in trucks in the factory yard for many hours waiting to be processed. A large number of peas had to be thrown away because they were too young, too old, or had remained unprocessed for too long a time.

Thornthwaite recognized that climatology could help this problem. Starting first with the real bottleneck, which was the number of acres of peas that could be handled by the processing plant in any one day, and given the information on the length of time the factory needed to process peas, Thornthwaite supplied the climatic information of how much energy from the sun was required for the pea to develop from germination to maturity. Using climatological normals, he developed a climatic calendar which expressed the length of each day of the year in terms of the energy available for plant development on that day.

He found, for example, that about 7,500 units of energy were available during the average year at Seabrook. By the end of March, energy units were accumulating at a rate of 10 per day. By the middle of May about 30 units of energy were accumulating each day, and the peak accumulation of 53 units per day occurred near the end of July. The rate of accumulation of energy may be expressed by the S-shaped curve of Fig. 6-3 where the units of energy accumulated are called "growth units," after Thornthwaite. In actuality, these are units of development (progress toward maturity) rather than growth (increase in physical size).

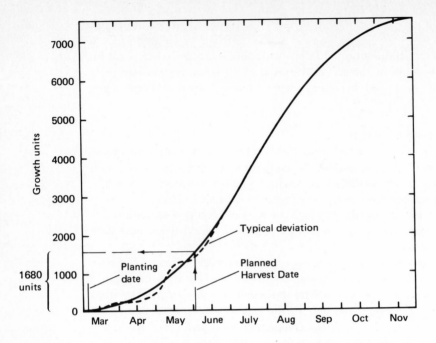

Growth curve at Seabrook, N.J. (*With permission of the Operations Research Society of America, from Thornthwaite, 1953b, fig. 1.*)   FIG. 6-3

Thornthwaite now converted the civil calendar with 24 hr in each day into a climatic calendar based on the number of "growth" units accumulated on the average each day. Days in April are about one-fifth the length of days in July. With the information on when to harvest and on the amount of energy required to bring a particular plant from germination to maturity, Thornthwaite was able to count back on his climatic calendar to determine the date of planting. For harvest on successive days in June, it was necessary to plant peas anywhere from 3 to 6 days apart in March and April. Substitution of different pea varieties with different energy requirements for maturation allowed successive planting in March or April and successive harvesting in June. Occurrence of particularly warm or cold growing seasons result in harvest either earlier or later than originally planned, but since all fields are affected in the same way, the sequence of successive harvesting of just the number of acres that can be handled by the processing plant is not affected.

The climatic calendar technique of scheduling the planting and harvesting of crops, now used extensively on a wide variety of crops, was reduced by Thornthwaite to a simple circular slide-rule device (Fig. 6-4). The outer linear scale gives the "growth" or development index. Actual day lengths for various crop-growing areas in latitudinal belts across the United States are shown on the moveable inner scales. A table accompanying the rule gives the number of "growth" units to maturity for a large number of crops. The inner calendar scale is set so that the date of planting in the particular zone is

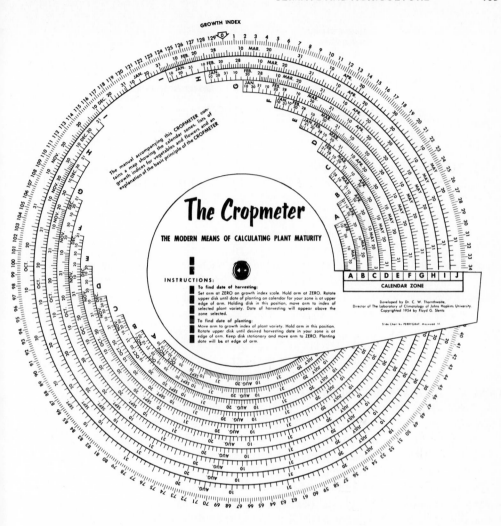

The cropmeter slide rule for calculating plant maturity, developed by C. W. Thornthwaite. (*Reprinted with permission of C. W. Thornthwaite Associates, and Floyd G. Slentz, copyright 1954, Floyd G. Slentz.*)

FIG. 6-4

opposite the zero on the growth index scale. The date of harvest in that zone is then read opposite the number on the growth index scale representing the number of accumulated energy units for maturity of the particular crop.

In addition to permitting crops to be picked at their peak of maturity, crop scheduling has increased the efficiency in the use of labor and equipment by smoothing the flow of produce from the fields to the factory. This has resulted in reduced overtime labor and has permitted the farming operation to be carried on within the limits imposed by the capacity of the

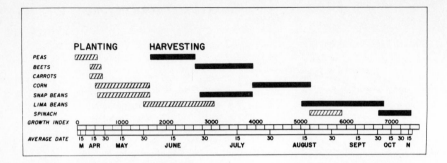

Hypothetical planting and harvest schedule for several crops at Seabrook Farms, Seabrook, N.J. (*With permission of the American Meteorological Society, from Seabrook, 1953.*)

FIG. 6-5

processing plant and the available farm equipment. Figure 6-5 schematically shows the relation between planting and harvesting times for several crops at Seabrook Farms. The crops are manipulated in such a way that only a manageable amount of planting and harvesting is scheduled for any given day. Harvest schedules are so arranged that, in general, the first harvest of a crop begins after the last harvest of another crop, e.g., peas and beets, beets and corn, snapbeans and corn. Farm and factory equipment can then be fully utilized for each crop, and conversions of equipment can take place after the end of one harvest and before the start of the next.

RELATION OF WATER USE AND PLANT DEVELOPMENT

A relationship between climate and plant development has been achieved through detailed observations of peas and other crops planted successively through the growing season. A relationship between climatic factors and the water needs of plants had also been found by growing various crops on evapotranspirometer tanks (Chap. 2). Thornthwaite suggested that plant development could be thought of in terms of units; for example, a pea plant at a given stage of its development toward maturity will have accumulated a certain number of identifiable units of development. The pea plant can, thus, be used as a climatic yardstick with the nodes marking various subdivisions. The fact that transpiration and development are related and are both influenced by temperature in the same way provides a new interpretation of development units. A development unit can be defined as either (1) the interval between successive nodes on a pea vine or (2) the amount of water needed for transpiration (Thornthwaite and Mather, 1954). Since this latter quantity is directly available from potential evapotranspiration, it is a more useful scale. It is easy to convert from one system to the other.

Thus defined, a development unit is the amount of plant development occurring while a unit amount of water is transpired. One hundred development units correspond to 1.0 cm depth of transpired water (Thornthwaite, 1952*b*). This unfamiliar concept provides a single solution to the two important agricultural problems of scheduling supplementary irrigation and crop harvesting. Figure 6-6 shows the march of plant nodal development, measured water use (potential evapotranspiration), development units, and daily incoming solar radiation through the year at Seabrook, New Jersey. All four curves follow the same general pattern but there are individual differences. Net radiation would have probably provided a closer relation to the other factors than solar radiation if the data had been available. Measured potential evapotranspiration and pea node development both lag behind insolation in spring although the factors are all in fair agreement in fall. The reason for the spring lag is probably found in the need for some of the solar energy to go into heating the soil and evaporating water from very moist soils in spring. Not all the energy is directly available for plant development.

DROUGHT AND IRRIGATION SCHEDULING

Drought occurs whenever the supply of moisture from precipitation or stored in the soil is insufficient to fulfill the optimum water need of plants.[1] At first, the failure of the climate and soil to provide sufficient water to the vegetation cover for evaporation and transpiration (evapotranspiration) is relatively unimportant, and this early drought is usually overlooked. Before long, however, continued failure of the water supply to equal the water need results in some wilting of plant leaves and a reduction in growth and yield. Often even these periods of drought go almost unnoticed by farmers since the plants have not been killed and yields are only reduced from some potential value which is seldom ever achieved at any time. It is only when the water supply fails to satisfy the water needs for a fairly long and continuous period that the existence of drought and the potential danger of a crop failure are recognized. To many, this is now the time to bring out the irrigation equipment in the hopes of preventing a total crop failure.

Thornthwaite (1947; *Newsweek*, 1949) described four types of drought which result from a failure of the water supply to satisfy needs. He called these permanent, seasonal, contingent, and invisible drought. Permanent drought is found in the desert areas where in no season is precipitation equal to the water needs. Plants are adapted to the dry conditions and can survive only by elaborate means of controlling transpiration losses and by rapid growth and reproduction following the limited rains that do occur. Agriculture is impossible without irrigation through the whole crop season. Regions of seasonal drought have well-defined rainy and dry seasons. These droughts can be expected each year and result from large-scale seasonal circulation changes. Agriculture is possible during the rainy season or with the use of irrigation in the dry season. Contingent droughts result from the fact that rainfall is irregular and variable. These droughts are characteristic of

---

[1] This is only one of many definitions of drought. The literature on the subject is extensive, and the number of quite different definitions of drought is large. The definition should be determined, in large measure, by the use intended.

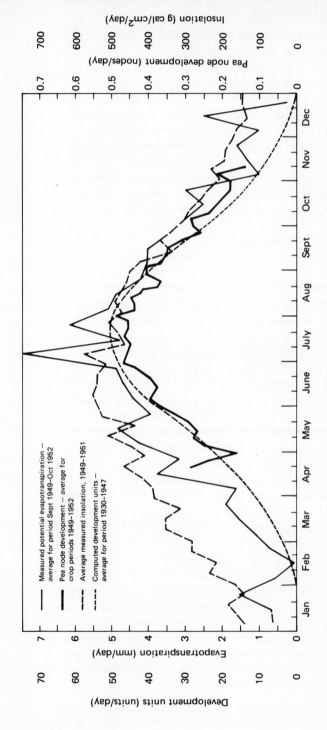

Annual march of solar insolation, plant development, the development index, and potential evapotranspiration at Seabrook, N.J. *(With permission of the American Meteorological Society, from Thornthwaite and Mather, 1954, fig. 3.)*

FIG. 6-6

subhumid and humid areas as a result of the occurrence of significant periods without rain. They can occur at any season but are usually more severe during periods of greatest water need. They are serious because of their unpredictability and because without supplemental irrigation, periods of crop failure can result.

These three types of drought are fairly evident by the wilting of crops or the lack of much vegetative growth or appreciable yields. The fourth type of drought, invisible drought, is less easily recognized. Invisible drought can occur at any time, even during periods with precipitation, when the daily supply of moisture fails to equal the daily water needs of the plants. As a result, there is a slow drying of the soil and plants fail to grow at their optimum rate. To the casual observer, there is no appearance of drought for plants continue to grow, little or no wilting is observed, and yields appear to be "normal." The frequent rains lull the farmer into a false sense of security that growing conditions are satisfactory. He does not realize that the failure of plants to obtain quite all the water they need reduces yields below those potentially possible. Invisible drought is almost entirely confined to the humid regions of the country where frequent rains will seem to argue against the need for an irrigation agriculture. It is this invisible drought which is most insidious in reducing potential farm yield and which can be most successfully combated with supplemental irrigation. At the same time, the economic needs or feasibility of supplemental irrigation to combat invisible drought is the most difficult to establish.

The Palmer Crop Moisture Index[1]

Palmer (1965) described a different approach to the measurement and classification of meteorological drought. Defining drought as a significant reduction of available moisture below that required for the near-normal operation of the established economy of a region, Palmer deviated from the older concepts of drought as periods of time with rainfall less than certain specified amounts. The Palmer definition would suggest that a semiarid region might not necessarily have a drought since the existing economy of the area would have already adjusted to the low levels of precipitation normally received; a drought would exist only if the supply of moisture was such as to affect adversely those levels of activity. Thus it argued for a "deviation from normal experience" as the criterion for drought, permitting different degrees of dryness for drought to exist in different areas. Severity of drought is considered to be a function of the length of the period of abnormal moisture deficiency as well as the magnitude of this deficiency.

To determine the actual moisture deficit at a place, Palmer divided the soil into two layers and assumed that 1 in. depth of water can be stored in the upper layer. This water is used at the potential rate until depleted; then water

---

[1] The author is indebted to W. C. Palmer for reviewing the following material and suggesting a number of valuable additions.

will be removed from the lower layer of the soil. In this respect, the approach is similar to the modulated soil moisture budget of Holmes and Robertson (see page 135). Water is lost from the lower soil layer at a decreasing rate dependent on the amount of moisture actually in storage. Recharge by precipitation occurs in the upper soil layer only until this layer reaches field capacity. After that, recharge of the lower layer can begin. No runoff occurs until both layers reach field capacity. Soil moisture loss from each layer can be expressed as

$$L_s = S'_s \quad \text{or} \quad PE - P \quad \text{whichever is less}$$

$$L_u = (PE - P - L_s)\frac{S'_u}{AWC} \quad \text{for } L_u \leqslant S'_u$$

where L_s, L_u = moisture loss from surface and underlying layers, respectively
S'_s, S'_u = available stored soil moisture, surface and underlying layer, respectively, at start of month
PE = potential evapotranspiration for month
P = precipitation for month
AWC = combined available water capacity in both soil layers

In carrying out the hydrologic accounting, it is desirable to start at a time when the two soil layers are at field capacity (or actual measured values of soil moisture storage are available); the available water-holding capacity of the soil must be known. Values of potential evapotranspiration are then compared with the precipitation by means of a bookkeeping procedure using the expressions for L_s and L_u given previously. As demand for water exceeds supply, these relations provide values of the amount of water stored in each of the two layers in the soil. Accompanying instructions indicate the disposition of any excess precipitation that might occur.

Palmer defines a number of other terms that are necessary in later computations. These include

PR (potential recharge): amount of moisture needed to bring the soil to field capacity

PL (potential loss): amount of moisture that could be lost from the soil if no precipitation occurred in the month

PRO (potential runoff): total amount of soil moisture storage available

α (coefficient of evapotranspiration): ratio of average actual water loss to average potential evapotranspiration for a month (when both ET and PE are 0, α is considered to be 1.0)

β (coefficient of recharge): ratio of the average recharge to the average potential recharge

γ (coefficient of runoff): ratio of the average runoff to the average potential runoff

Δ (coefficient of depletion): ratio of the average total moisture loss from the soil to the average potential total loss

CAFEC values: value of the particular parameter that would have been climatically appropriate for the conditions of the time and place being examined. (These values are indicated by means of a caret, ^, over the particular parameter.)

$$\hat{ET} = \alpha PE \qquad \hat{R} = \beta PR \qquad \hat{RO} = \gamma PRO \qquad \hat{L} = \Delta PL$$

These relations are then combined in the following expression to obtain a measure of the amount of precipitation necessary in a month to maintain "normal" evapotranspiration, runoff, and soil moisture storage considering antecedent moisture conditions

$$\hat{P} = \hat{ET} + \hat{R} + \hat{RO} - \hat{L}$$

Two additional terms defined by Palmer are:

K (climatic characteristic): empirically derived weighting factor which depends on average moisture supply $(P + L)$, average moisture demand $(PE + R + RO)$, and the long-term mean of the absolute values of the precipitation anomaly $(P - \hat{P})$ (Palmer, 1965, eqs. 26, 27). K tends to be large in arid regions and small in humid regions.

Z (moisture anomaly index): product of the climatic characteristic K and the precipitation deficiency d, where $d = P - \hat{P}$.

Palmer considered that drought severity can be adequately expressed by four distinct classes which he called mild, moderate, severe, and extreme drought. He assigned a value of -1.0 as the boundary marking the start of mild drought, -2.0 separating mild and moderate drought, -3.0 separating moderate and severe drought, and -4.0 separating severe and extreme drought.

Palmer now had the problem of reconciling the effect of intense short-period deficits as opposed to less intense but long-lasting deficits. By considering maximum rates of accumulation of the Z values (the moisture anomaly index) at different stations, he concluded that Z-value accumulation at the rate of -12.0 for a 1-month or -85.0 over a 60-month period would constitute extreme drought conditions. He then developed the following expression for severity of drought:

$$X_i = \sum_{t=1}^{i} \frac{Z}{0.309\, t_1 + 2.691}$$

This expression provides only a partial solution because a single very wet or dry month in a long series of opposite months is unrealistically reflected in the value of severity achieved. Palmer handled this problem by including each month in the index on an increment basis. As a result, the

effect of duration is only indirectly included but the approach eliminates
other, more serious problems. By substituting into the previous expression for
drought severity, Palmer achieved the final expression for severity as

$$X_i = X_{i-1} + \frac{Z_i}{3} - 0.103X_{i-1}$$

where the subscript i refers to the current month.

The values of X will range from greater than +4.0 (very much wetter
than normal) to greater than -4.0 (extreme drought) with values from +0.49
to -0.49 constituting a near-normal class.

There are many unfamiliar terms and definitions in the Palmer drought
index. The steps of the computation process are logically ordered, however,
so that a value of drought severity is not difficult to determine for any month
or other time period. Actually, the computations can be divided into four
subroutines as follows:

1   Computation of the probability that a drought or wet spell has ended

2   Computation of the probability that a wet spell has begun

3   Computation of the probability that a drought has begun

4   Computation of the actual severity of any wet spell *or* any drought that
has already been established

Not all these subroutines need to be evaluated for each month. During
long periods of drought only subroutine 4 is used since only seldom in such a
period is a month found that might require computations of the probability
that the drought has ended.

Palmer has investigated the possibility of using weekly values of
precipitation and evapotranspiration in the computation of the drought
index. He concludes that while the weekly analysis provides more detail and
allows better determination of the timing of events, the overall results are
very similar to those obtained from the monthly computations and the latter
are considerably easier to obtain.

Later, Palmer (1968) developed an agricultural drought index based on
successive weekly values of the computed abnormal evapotranspiration
deficits $(ET - \hat{ET})$. Generalized maps of this measure of the effects of
abnormally dry weather on growing crops (Fig. 6-7) are a regular feature of
the *Weekly Weather and Crop Bulletin* issued by the National Oceanographic
and Atmospheric Agency.

CLIMATIC APPROACH TO IRRIGATION SCHEDULING

If a farmer proposes to supply supplementary water to his crops, he must
have some practical means of determining how much water to use and when
it is needed for his particular farm. Common practice among farmers is to
watch the plants for signs of moisture deficiency as a basis for supplying
water. This is not very satisfactory for by the time the plants begin to show

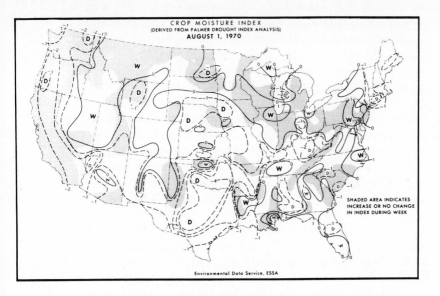

Distribution of Palmer crop moisture index over the conterminous United States for week ending Aug. 1, 1970. For growing rather than mature vegetation. (*From* Weekly Weather and Crop Bull., *NOAA*.)

FIG. 6-7

some signs of water need they are already suffering and yield has been reduced correspondingly. Instead of watching the crop for indications of drought, some investigators suggest watching the soil instead. One investigator has stated that the only known way to be sure that soil moisture is present in readily available form is by frequent examination of the subsoil by the use of a soil auger. In recent years, several devices have been developed to be installed permanently in the soil to provide a continuous indication of the amount of moisture remaining. Among these devices are elements made of gypsum fiber glass, and nylon in which the electrical resistance varies with moisture. While useful, these devices often do not provide the information that is needed. They provide a sample at only one spot, which may or may not be representative of a wider area, and they assume that the response to moisture of the sensor is similar in all respects to that of the soil so that the measured values will represent conditions in the soil. This assumption is seldom satisfied.

The climatological approach to the problem of determining soil moisture content, while indirect, is simple and practical. It avoids the errors involved in calibration and exposure of measuring instruments and can provide average values over good-sized areas, depending on the climatic data used. One climatic approach was originally suggested by Thornthwaite (1945*a, b*) and expanded in more detail in later papers by both Thornthwaite and Mather (1955*a, b*) and Mather (1954*b*).

In the climatological approach, the moisture in the soil is regarded as being a balance between what enters it as a result of precipitation and what leaves through evaporation and transpiration. An irrigation schedule can be set up as a bookkeeping procedure. The moisture in the soil may be regarded as a bank account. Precipitation adds to the account, evapotranspiration withdraws from it. We merely need to keep track of the evapotranspiration and restore by irrigation whatever is not promptly returned by precipitation.

When the moisture content of the soil is at field capacity or above, any water that is added to it by precipitation is lost by downward percolation. This gravitational water is only detained briefly, the period depending on the permeability of the soil and the amount of the gravitational water. When the soil moisture is below field capacity, precipitation first brings the soil moisture storage up to that level. The amount of water that can be stored in the root zone of the soil depends on its depth and on the soil type and structure.

Evaporation from a moist soil immediately begins to lower the moisture content of the soil. As the soil dries, the rate of evapotranspiration diminishes. At first, evapotranspiration goes on at nearly the maximum rate from all soils, but by the time 25 mm of water has been removed, the rates from different soils begin to differentiate. When one-half of the water is gone, the rate of evapotranspiration falls to one-half of the potential rate and plants begin to suffer from drought. With a constant rate of potential evapotranspiration of 5 mm/day, the half-rate would be reached after 7 days in coarse sand but not until after 37 days in fine-textured soil.

An irrigation schedule is a natural outgrowth of this method of computing soil moisture. One can set up limits below which the soil moisture will not be allowed to fall for the particular crop and depth of root zone in question. Then by keeping daily account of how much water has been lost from the soil, it is possible to know exactly when the predetermined level of soil moisture depletion is reached and just how much to irrigate to bring the moisture level back to a safe value. Shallow-rooted crops will have to be irrigated more frequently but with smaller amounts of water than will deeper-rooted pastures or orchards. If irrigation is scheduled by keeping continuous account of the soil moisture, no great moisture deficiency can develop in the soil to limit growth, and there will be no over-irrigation to damage both soil and crop and to result in a wasteful misuse of water.

Figure 6-8 indicates three hypothetical irrigation programs for fields near Seabrook, New Jersey, in 1954. During the 2-month period covered by the example, only one 60-mm irrigation was required on the deeper-rooted orchards, although a considerably greater degree of deficit was permitted to develop in the soils of the orchards. Two 40-mm irrigations were called for on the pasture soils during the May–June period while five 20-mm irrigations were needed to supply sufficient water for the shallow-rooted crops. Clearly, different totals of deficit developed under the three different irrigation programs. Each program was adapted to the depth of rooting of the vegetation and, thus, to the amount of deficit possible for each particular vegetation type before reduction in yield became appreciable.

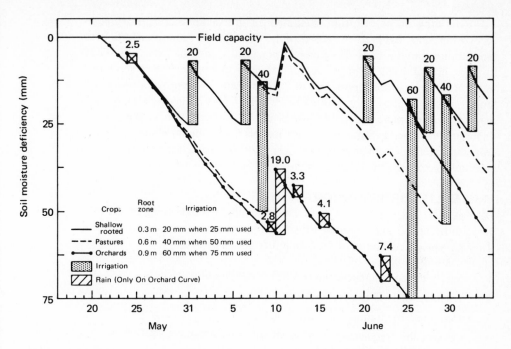

Irrigation schedule for three different types of crops, Seabrook, N.J., 1954.   FIG. 6-8
Assume soil holds 75 mm of water per 0.3 m depth of soil at field capacity.
Rain on May 21 brought soil up to field capacity so that computations of
deficit begin at 0 on that day. (*From Thornthwaite and Mather, 1955b, fig.
3.1.*)

Evaluation of hypothetical irrigation schedules, using past weather data,
provides considerable information on the possible need for irrigation and may
supply factual data on the size and degree of use of a planned irrigation
system. For example, such a hypothetical irrigation schedule was computed
for the period 1948 to 1954 for fields in the vicinity of Memphis, Tennessee.
The schedule called for the application of 40 mm of irrigation every time the
moisture deficit in the soil reached 50 mm. Table 6-1 gives the results of the
number of irrigations and the quantities of water that would have had to
be applied each year. The table shows that some irrigation was needed in each
of the 7 years, ranging from four 40-mm applications in the wet years of
1949 and 1951 to 12 applications in 1953. The table also provides
information on the quantity of water needed per acre, from which one can
determine the total needed water supply for an irrigated farm or the number
of acres that can be irrigated with a given, but limited, water supply. From
the actual daily data it is also possible to determine the spacing of the
irrigations through the year and to plan size of pipes and pumps, number of
laterals and sprinklers, and other factors of the actual irrigation system in

order to permit irrigation of all fields under the anticipated climatic water demands.

The program of irrigation illustrated for the Memphis area, developed for cotton and pastures, is based only on climatic conditions. Some modification on the basis of economic practicality may be necessary. For instance, economically, it might be more feasible to irrigate 60 mm depth of water every time a moisture deficit of 75 mm occurs. This program will give longer periods of appreciable soil moisture deficit (and reduced yield), but it may result in a more profitable overall use of the irrigation system. It is important to remember that any irrigation plan must be developed for the crop and soil in question and must be a compromise between the climatic and economic factors which are involved in irrigation scheduling.

PROBABILITY OF IRRIGATION NEEDS

Baier et al. (1968, 1972) have developed a simplified bookkeeping procedure for estimating the likelihood of irrigation requirements from climatic data. They have prepared a series of four computer programs that

1   Verify the completeness and condition of all input data cards

2   Calculate daily latent evaporation

3   Determine the frequency of supplemental irrigation requirements

4   Convert the frequencies into depths of supplemental irrigation water requirements

After the daily value of potential evapotranspiration has been calculated from the available meteorological data, it is used to calculate the daily soil water budget. The amount of water available to change the budget from the previous day (ΔW_n) is

$$\Delta W_n = P_n - (CU)(PE_n)$$

where P_n = precipitation on day in question

CU = consumptive use factor

PE_n = daily value of potential evapotranspiration

TABLE 6-1

Irrigation summary for area around Memphis, Tenn., 1948–1954*

YEAR	NO. OF 40-MM IRRIGATION APPLICATIONS	WATER APPLIED	
		MM	1,000 LITERS/ACRE
1948	7	270	1,080
1949	4	150	617
1950	5	190	770
1951	4	150	617
1952	9	350	1,388
1953	12	450	1,850
1954	10	380	1,542

* From Thornthwaite and Mather, 1955a, table 1.

The consumptive use factor is actually the ratio of the consumptive use of water by a crop to the potential evapotranspiration. If the crop is actively growing, completely covers the soil, and is supplied with all the water it needs, the consumptive use factor will be 1.00. If the crop is spaced in rows (with gaps between rows) in such a way that energy can reach the lower portions of the plant, then the consumptive use factor may exceed 1.00. With short plants not covering all the surface, or with plants not actively transpiring, the consumptive use factor will be less than 1.00.

The amount of readily available soil moisture, S_n, on the nth day is the sum of the available soil moisture from the previous day S_{n-1} and the ΔW_n, as long as this total does not exceed the available water-holding capacity of the soil. If the water-holding capacity is exceeded, a water surplus develops and percolation and/or runoff of excess water occurs. If $S_{n-1} + \Delta W_n$ exceeds 0 but is less than the water-holding capacity, then there is water available in the soil and no irrigation requirement for that day. If that total is less than 0, then the irrigation requirement IR for that day will be given by the relation

$$(IR)_n = -(S_{n-1} + \Delta W_n)$$

The total irrigation requirement for any period of days can be obtained by summation of this expression for each day of the period.

Because soils and vegetation covers vary markedly even within short distances, the computer programs to evaluate irrigation requirements have been prepared for various water-holding capacities, a range of consumptive use factors, and many different probability levels. Table 6-2 lists a series of

TABLE 6-2

Supplemental irrigation requirements for different storage capacities and consumptive use factors at Ottawa during first week of August*

STORAGE CAPACITY	CU FACTOR	PROBABILITY, %							
		25	50	75	80	85	90	95	99
		in.	in.	in.	in.	in.	in.	in.	in.
1.00	0.25	0.00	0.00	0.00	0.00	0.00	0.00	0.00	0.00
	0.50	0.00	0.00	0.11	0.19	0.28	0.40	0.57	0.89
	0.75	0.00	0.24	0.54	0.62	0.71	0.82	0.99	1.30
	1.00	0.20	0.54	0.89	0.97	1.08	1.20	1.38	1.74
3.00	0.25	0.00	0.00	0.00	0.00	0.00	0.00	0.00	0.00
	0.50	0.00	0.00	0.00	0.00	0.00	0.00	0.04	0.20
	0.75	0.00	0.00	0.28	0.41	0.56	0.74	1.01	1.52
	1.00	0.00	0.38	0.81	0.92	1.04	1.20	1.43	1.87
5.00	0.25	0.00	0.00	0.00	0.00	0.00	0.00	0.00	0.00
	0.50	0.00	0.00	0.00	0.00	0.00	0.00	0.00	0.00
	0.75	0.00	0.00	0.00	0.00	0.00	0.27	0.69	1.49
	1.00	0.00	0.14	0.68	0.82	0.98	1.17	1.46	2.01

* With permission of Can. Agr. Eng., from Baier and Robertson, 1967, table 2.

irrigation requirements for the Ottawa, Ontario, Canada, area for the first week of August under different soils (varying storage capacity) and vegetation cover (*CU* factor) conditions. The effect of soils and cover can easily be seen. Eighty percent of the time on soil having a 3-in. storage capacity and with a complete cover of actively transpiring vegetation, the irrigation requirements will be 0.92 in. or less. One year in five they will exceed this amount. On the average, they will equal 0.38 in., while one year in ten they will exceed 1.20 in. Irrigation requirements will exceed 0.97 in. one year in five with a complete vegetation cover on a soil with only 1-in storage capacity, while on a soil with 5 in. of storage capacity, requirements will only exceed 0.82 in. one year in five. The greater the storage capacity, the less will be the irrigation requirements. And, of course, the more actively transpiring the vegetation cover, the greater will be the irrigation requirements for the same probability value.

The computer programs provide basic information for evaluating water resources in relation to agriculture not only at a point but also over geographical areas. Information based on specific storage and consumptive use factors can easily be plotted on maps and the geographical pattern of irrigation requirements determined. Irrigation-need data would be of great use in planning irrigation enterprises, in determining equipment needs, in formulating costs to establish an irrigated farm, or in studying the feasibility of supplemental irrigation in a particular area.

The rapid increase in the use of irrigation in agricultural production in the United States in the past few decades has resulted in a greatly increased demand for water. The areas of greatest need for irrigation, of course, correspond almost exactly with the areas of most limited water resources. Thus the agricultural demand for water in many arid and semiarid areas has severely reduced available supplies and limited any other possible development of these areas.

Evaluation of some form of climatic water budget on a daily basis provides detailed figures not only on the irrigation need but also on the water deficit and water surplus and the day-to-day storage of water in the soil. Significant water deficits, resulting in reduced crop yields, are found nearly every year even in a so-called humid part of the country. The daily climatic water budget can provide quantitative answers to the questions of when and how much to irrigate, resulting in the achievement of more efficient irrigation practices. Such computations will not only counteract the damaging tendency to overirrigate or to irrigate solely by custom or long-standing routine (Hudson, 1962), but they will also allow us to begin to analyze the present distribution of irrigation agriculture to see whether it is rational and if it results in the most efficient use of our limited water resources.

Drought is a real problem in nearly all parts of our country. It limits agricultural yields everywhere. Drought can be combated by irrigation. Since future demands for food and fiber require that we maintain a high and efficient agricultural production, it will ultimately be necessary to further increase our irrigation agriculture. Because of limited supplies in the arid and semiarid areas, because of the high cost of water and of land management for irrigation, because of the priority of competing demands for such water, and the potential for increasing yields in the more humid East, it is evident that serious thought must be given to the further growth of irrigation agriculture in the humid and subhumid areas rather than to increasing our already great demands for water by increased agricultural development in the arid and semiarid areas of the West.

CLIMATE AND AGRICULTURE: FACTORS CONCERNED WITH HEALTH AND YIELD

7

Essentially all phases of plant or animal life from germination or birth through growth or development to maturity, to final yield or productivity, and to reproduction, are influenced to some degree by the environmental factors of weather and climate. But this influence may vary during the life of the plant or animal. Weather factors may play a significant role during one stage of development yet hardly have any effect at another time; moisture may be important at one time, while temperature or humidity may be significant at another time; high temperature may be needed at one time, while low temperatures are important at another time; climatic factors may have more influence on secondary factors such as diseases or pests than they do on the plant or animal directly. Agricultural climatology is, thus, a complex subject requiring detailed observations and skillful interpretation and analysis.

FROST DAMAGE PROTECTION

Protection of crops from the dangers of frost is a significant and practical exercise in applied climatology. Agriculturalists define two general types of frost: white frost occurs when light, feathery ice crystals form on the vegetation as a result of sublimation; black frost results in no surface ice formation although low temperatures cause injury to plant tissues. Frost damage does not occur at a specific critical temperature; injury depends in part on the sensitivity of the plant itself and its particular height above ground. Some years ago, the U.S. Weather Bureau used to report the dates of first and last killing frost, although this practice has now been replaced by one which merely lists first and last dates of certain low minimum temperatures. While by definition first or last killing frost referred to the frost that brought killing damage to 50 percent or more of the native vegetation in an area, it was a difficult definition to apply because of the need to estimate a percentage kill over an appreciable area as well as to identify the so-called "native vegetation."

Frost normally occurs when the temperatures near the earth's surface drop sufficiently below $0°C$ for freezing to occur. Early fall or late spring frosts occur as a result of radiation or advection conditions. Radiation frost requires cold, calm, clear, and fairly dry nights, while advection frost requires the downslope movement of cold air into hollows, valleys, or other low-lying areas. With radiation frost, air temperatures are already near the frost point, and it is the additional cooling provided by increased nocturnal radiation with negligible mixing of the surface air layers with warmer air above that causes the frost to develop. With advection frost it is accumulation of a pool of cold air that has drained from higher-lying areas that results in local temperatures dropping below the freezing point.

Wang (1972) has provided an interesting discussion of possible ways to combat both types of frost. Three general methods of prevention are suggested: (1) through the application of heat to the plants and environment from available heat sources; (2) through reduction of heat loss from the vegetation and environment; or (3) through a combination of the first two methods.

Several techniques exist by which heat may be supplied to the vegetation and environment. The most common is to increase the turbulent mixing of the lower atmosphere by fans or propellers. In most cases when frost develops near the ground, the cold ground layer of the atmosphere is overlain by warmer air layers. In other words, an inversion of temperature exists through the lowest layer of the atmosphere. If the warmer air aloft can be mixed with the colder surface air by means of turbulent exchange, the surface temperature might be kept sufficiently above the frost point to prevent serious frost damage.

Wind machines are common in citrus orchards in Florida, Georgia, and California. In more northern areas, airplanes and helicopters have been used to protect wheat from damage from frost. One helicopter can protect a much larger area [50 to 70 acres (20 to 28 ha)] than can a single wind machine [15 to 20 acres (6 to 8 ha)], but the cost factor favors the wind machines, which have been used for 50 years.

Stationary fans produce jets of air in fixed directions, resulting in the cold surface air being pushed to lower elevations to be replaced by the warmer air from above. Multiple or rotating fans increase the efficiency of

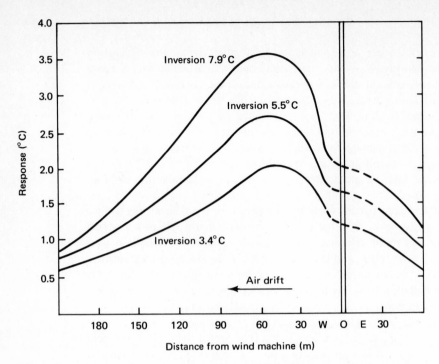

Magnitude and areal extent of warming response under three inversion conditions at a 12-m level from use of a 90-hp single-engine wind machine. (*Reprinted with permission from Schultz, 1962, copyright 1962, the Pergamon Press, Ltd.*)

FIG. 7-1

mixing of the air over a larger area. For example, in a citrus orchard propellers on a dual-engine wind generator (60 to 80 hp) located on towers at 30 to 35 ft (9 to 10 m) above the ground and rotating slowly so that any given direction is covered every few minutes create strong, low-level gusts of air to about 150 ft (45 m) in all directions from the towers. While the actual gusts cover less than 2 acres (0.8 ha) of the orchard, Wang (1972) reports that such a single system can protect some 15 to 20 acres (6 to 8 ha) of orchard depending on the height and intensity of the inversion and the degree of ground cooling.

Schultz (1962) studied the use of a single-motor wind machine in Riverside, California, and showed that its effectiveness depended on the intensity of the inversion at the height of the wind machine itself. He prepared a diagram showing how temperature change is related to distance from the wind machine under different inversion conditions (Fig. 7-1). In his experience, maximum temperature rises occurred some 60 m from the machine itself; a temperature increase of at least 0.5°C was found at a distance of 200 m.

Schultz suggested that with a minimum shelter temperature of 28°F (-2.2°C) or above, no frost protection is necessary for ripe or near-ripe fruit. At 25 to 27°F (-2.8 to -4°C), the wind machine would provide sufficient protection. At 24°F (-4.4°C), protection is possible with a wind machine if a medium or strong inversion exists, while at 23°F (-5°C), a combination of wind machines and small heaters will probably be necessary to provide adequate protection.

A second method of adding heat to the crop utilizes flooding or spray irrigation techniques. Flooding has long been employed to protect cranberry plants. While effective for certain other plants, it is, of course, restricted to species that can withstand immersion in water.

Water contains large quantities of latent heat which are released as the water cools or as the liquid water is converted to ice. (Cooling one cubic meter of water one degree Celsius releases some 0.28×10^5 cal.) Release of this latent heat may be sufficient to prevent the formation of frost. Wang (1972) has listed the following prerequisites for effective protection by flooding: (1) short-growing vegetation unaffected by flooding; (2) adequate water supply as well as low rates of losses by percolation and runoff; and (3) sufficient time to provide opportunity for flooding. Needless to say, these conditions are present in only a limited number of cases.

Sprinkler irrigation has wider applicability; it does not require flooding the field, the equipment can be easily moved, and smaller quantities of water are required. Protection against frost is provided by (1) the release of latent heat as the water cools and freezes; (2) the interception of terrestrial radiation by the increased water vapor in the air; and (3) the increased soil heat capacity of the wet soil, permitting more storage of heat during the day. Some investigators (de Zeeuw, 1958) feel that spraying is the most effective method of frost protection, surpassing both wind machines and heaters. All agree that the rate of water application should be kept to a minimum; Businger (1955) has suggested that 0.12 in. (3 mm)/hr is sufficient for air temperatures of 23°F (-5°C).

A third method of obtaining additional heat is by increasing daytime heat storage in the soil or reducing nighttime heat loss from the soil. Many investigators have successfully developed techniques to increase the absorptivity of soil by reducing the surface albedo through the addition of such things as coal dust. Reflectors (or, in many cases, just white-painted walls) have been used to increase the amount of solar heating of the soil surface. The thermal capacity of the soil can be increased by irrigation of the soil. Finally, a mulch (to be discussed later) or shelterbelt can be employed to reduce the outgoing terrestrial radiation and, thus, keep the soil warmer during the evening hours.

There are, of course, other ways to reduce nocturnal heat loss. One of the oldest is by means of orchard heaters designed to add small quantities of sensible heat to the lowest air layers through the combustion of fuels. Some turbulent mixing of the air is also created by the convective currents established by the heaters. If a protective blanket of smoke is formed over the area, the outgoing long-wave radiation will also be reduced somewhat. Wahlberg (1950) has found, however, that combined use of 10 to 12 heaters per acre along with a wind machine provides better protection at lower cost than using either device alone.

The creation of artificial fogs to reduce outgoing radiation has been tried in several areas with only limited success. Much research in this technique of frost protection remains to be done.

One new technique still under active study is the use of foam as a protection against frost (Desjardins and Simonovitch, 1968). Foams have several properties that make them very attractive for frost protection: (1) they last only a limited number of hours before they break down and essentially disappear, thus eliminating the need to remove the protective material by hand; (2) the foam is made of a solution of hydrolyzed protein or amino acids which provides an additional source of nitrogen for the vegetation when the foam breaks down; and (3) tests have shown that the foam is an excellent insulator, with nocturnal temperatures in the foam some 10 to 12°F (6 to 7°C) above those at the same height over unprotected surfaces.

Possible drawbacks include (1) colder temperatures just above the foam due to the lack of upward heat transfer through the foam and the radiative and evaporative cooling at the foam surface (this might limit foam use to low-growing plants that can be completely covered); (2) the need to spread the foam under and around plants for partial support (the relatively high density of the foam might otherwise crush unsupported plants); (3) the foam is less effective under windy conditions associated with advection frost due to a lack of stability in wind; and (4) the relatively high cost of the foam.

The use of foam as a protective agent against frost has limited applicability at present for low-growing, high-value crops over limited areas. However, with experience its value and use as a protective device may increase. It can never replace heaters or wind machines in many applications, however.

THE INFLUENCE OF MULCHES ON PLANT MICROCLIMATE

The word "mulch," possibly derived from the German *molsch* meaning soft or rotten, is now applied to any substance spread over a soil surface to protect, in some way, the growing plants or the deeper soil layers from unwanted influences. Originally used in connection with rotting straw and other similar materials, mulch now refers to such substances as paper, plastic film, other soil layers, rocks, wood chips, or peat moss.

The use of mulch greatly influences the microclimate of the local area where the mulch is applied by (1) changing the disposition of the radiant energy and (2) modifying the water budget of the surface. As more economical materials have been developed in recent years, study of the climate-influencing properties of mulches and of their role in applied agroclimatology has increased.

Waggoner, Miller, and De Roo (1960) have employed an energy balance approach to study the effect of different mulches on the local microclimate. Neglecting the advective or horizontal exchanges as well as the small amount of energy used for photosynthesis and the temperature changes within the plant, the energy balance can be written

$$R_i + R_0 + W + A + G = 0$$

where R_i = incoming radiation

R_0 = outgoing radiation

W = energy lost in evapotranspiration

A = vertical energy exchange by conduction and convection with the air

G = energy exchange by conduction in the soil

In their study, Waggoner et al. measured the radiation terms by means of hemispherical and net radiometers, W was determined from the water loss from small soil-filled aluminum cans, G was measured by means of heat flow transducers at 1 cm depth in the soil profile, and A was not measured but rather obtained as the residual in the expression. The reasonableness of the estimate was checked against the vertical profile of air temperatures as measured above the various mulched surfaces.

Waggoner and his associates have provided data on the disposition of energy over a number of different surfaces including black and translucent plastic film (38 and 152 μm thick, respectively), aluminum film bonded to polyethylene film (each film 152 μm thick), paper, and straw surfaces. In each experiment, a bare soil surface was maintained as a reference. The basic findings, relative to the results from the bare soil plot, may be summed briefly. The opaque black film resulted in little warming of the soil in the day but did increase its temperature by several degrees at night. The translucent film, with many water droplets forming on the underside, resulted in a warming of the soil by many degrees during the day and by several degrees at night. With the reflective aluminum film the soil was several degrees cooler in the day and several degrees warmer at night.

In all cases, the film reduced the evaporation of water and thus conserved heat in the soil. This additional water, available for plant growth, would be a useful plus resulting from the application of mulches. The paper and hay mulches, because of the increased air movement through the mulch itself, were not as effective in conserving water as the film mulches.

The major influence of the mulches depended in large measure on the trapped air between the mulch and the soil surface. Thus, if the black film had been in perfect contact with the ground, the increased temperature of the highly absorbent film would have been conducted to the upper soil layers and increased their temperature considerably (as has been seen many times with the use of carbon black or soot films having close connection with the soil surface). Instead, the insulating air layer between the film and the surface served to keep the soil surface cooler under the black film. The air layer did not serve as an insulating blanket in the case of the translucent film since the solar radiation could penetrate to the soil surface but the translucent film prevented outgoing long-wave radiation (the greenhouse effect) and thus increased soil temperature appreciably.

The influences of the different mulches on both temperature and evaporation have been summed up in Table 7-1 in a qualitative fashion. The plus sign stands for an increase (double plus, a marked increase), the minus for a decrease, and the 0 for no appreciable change in the tabulated quantity.

CLIMATE AND PLANT AND ANIMAL DISEASES [1]

Many environmental factors influence the development and spread of pests and diseases. Significant among these are temperature, humidity, wind, rainfall, dew, cloudiness, radiation, and frost penetration into the ground. At the present time, studies of the relation of these factors to plant diseases are quite unsophisticated, often because of the lack of good environmental data. For example, mean shelter temperatures and monthly rainfall amounts are often used to provide some indication of the energy and moisture relations at crop heights in fields some distance away.

Radiation, as well as soil moisture and soil temperature data, would seem to be more closely related to the spread of many pests and diseases than would shelter temperatures and rainfall, but these data are seldom measured routinely. Since it may be necessary to relate climatic factors with information on plant health over a period of years before relationships are discovered, the backlog of soil or radiation data needed is almost always lacking.

The aim in crop-disease or crop-parasite studies is to identify the nature of the relationship and the role of external factors such as weather or climate so that forecasts of impending outbreaks can be made sufficiently far in advance to permit the effective use of control measures. Wang (1972) has divided pest and disease forecasts into three categories, depending on the

[1] The material in this section dealing with the work of Grainger (pp. 187–191), Ollerenshaw (pp. 191–192), and Hogg (pp. 193) has been adapted with permission from Taylor, "Weather and Agriculture" (1967), Pergamon Press Limited.

TABLE 7-1

Summary of temperature and moisture effects of different mulches (+ for increase, − for decrease, 0 for no change)*

	BLACK FILM	TRANSLUCENT FILM	ALUMINUM FOIL	KRAFT PAPER	HAY
Ground heat storage	−	+	− −	−	− −
Midday soil surf. temp.	0	+ +	−	−	− −
Night soil surf. temp.	+	+	+	0	+
Mean soil temp. (−3 cm)	0	+	0	−	−
Diurnal range temp. (−3 cm)	−	+	− − −	−	− −
Soil moisture conservation	+ +	+ +	+ +	+	+

* With permission of the Connecticut Agricultural Experiment Station, New Haven, from Waggoner, Miller, and DeRoo, 1960, p. 30.

length of time involved in the forecast. His long-term warnings refer to periods several months in advance; medium-term warnings refer to a period more than 2 days but not over a month in advance; short-term warnings cover periods less than 2 days in advance.

The relation between climatic factors and plant diseases can be direct and straightforward, as in the blowing of spores from one infected field to an adjacent one, or it can be complex and interrelated, as in the case when intermediary hosts are necessary for the development of the parasitic organisms (see the later discussion of liver fluke disease). Climatic factors may influence all or part of the relationship, either favorably, preventing the actual development of the virus or spores, or unfavorably, resulting in rapid multiplication or reproduction of the virus or parasite so that disease can spread quickly. We are only beginning to understand the complex relationships of climatic factors to plant diseases. As our knowledge advances, some of our present conclusions will need to be revised. A case in point is our increased understanding of the factors affecting potato blight to be discussed below. Detailed discussions of several interesting relationships illustrating the nature of the recent climate-disease studies are included in the following paragraphs. These should provide suggestions for directing future research studies into many other disease relationships that still need to be investigated.

1 Climate and Potato Blight[1]

One of the earliest of all investigations of climate-plant disease relationships involved the development, spread, and intensity of the fungus *Phytophthora infestans* which causes late potato blight. The early work on this relationship (van Everdingen, 1926) involved the specifications of favorable weather conditions for the development and spread of the fungus. These conditions involved

1   No less than 4 hours of dew at night

2   Minimum nocturnal temperature not below 10°C

3   Following day mean cloudiness 8/10 or greater

4   Rainfall greater than 0.1 mm in 24 hr following the night with dew

Lacking observations of dew and cloudiness, information on dew point temperatures and sunshine hours was substituted.

In the Netherlands, an outbreak of potato blight followed the occurrence of these weather conditions within a 2-week period. However, later application of these limiting climatic values to other areas such as

[1] I am indebted to Dr. John Grainger who reviewed the original of this section and who made several suggestions for including new material on blight. He graciously prepared Fig. 7-2 to illustrate the relation between weather factors and c_p/R_s ratio during the potato-growing season.

southwest England resulted in too many forecasts of possible blight. The rules were not sufficiently sensitive for the British environment. Conditions 3 and 4 did not necessarily ensure that the potato leaves would remain moist for a long enough period of time for the blight organism to develop.

Beaumont, who worked on this problem for many years, first considered adding one more criterion to the Dutch rules, namely, that the relative humidity in midafternoon of the day following the night with dew should not be less than 75 percent. This, he felt, would ensure fairly moist conditions on the plant leaves throughout most of that day. Later work led Beaumont (1940, 1947) to the formulation of a simple temperature-humidity rule that now has widespread applicability. The rule states that the first outbreak of potato blight follows in 7 to 21 days after a so-called "critical period" defined as air temperature not less than 10°C and relative humidity not below 75 percent for a period of at least 48 hr.

Such a definition leaves itself open to modification with the identification of so-called "near critical periods" in which possibly 1 hr or so of the 48 might have conditions of lower temperature or humidity. Application of the Beaumont rule to other areas and years has shown that it works reasonably well but it still does not identify all periods of blight outbreak. More significantly, there are many times when it identifies a possible outbreak and none occurs. In spite of this, recording instruments (temperature and humidity or wet and dry bulb recorders) have been developed expressly for the purpose of identifying "Beaumont periods," as those 48-hr periods of particular temperature and humidity have come to be called.

Bourke (1953) modified the Beaumont rules for the Irish situation by specifying the need for a period of 12 hr with temperatures above 10°C and relative humidity 90 percent or above. The 12-hr period is only justified if the vegetation remains moist for at least another 4 hr; otherwise the specification calls for 90 percent relative humidity or higher for 16 hr. Bourke (1957) also specified "blight weather" in terms of a synoptic forecast map.

Grainger (1967) listed all these various rules and then sought to identify the common factors among them. The Dutch rules specified dew at night and cloudy, rainy conditions the next day in order to keep moisture on the leaves. As modified by Beaumont, afternoon humidity must remain high to ensure no drying of the leaves after the night with dew. Later modifications called for a fairly long period with over 90 percent relative humidity. This also emphasized the need for moisture on the plant leaves. It is known that dry blight spores such as those likely to initiate a new attack will begin to grow only in water and thus need a period with free water on the potato leaves. Grainger (1967) suggested that the necessary condition was really one or more periods of water deposition within a 48-hr interval that together total about 12-hr duration. One wet period should be separated from the next wet period by a period of dryness no greater in length than the preceding wet period. The spore needs water and cannot survive if the period of dryness is severe enough to permit desiccation. All the earlier rules concerning dew, high humidity, rain, and cloudiness were related to the real need for free moisture on the leaves. Because of the ease with which the Beaumont period can be recorded instrumentally, it is still used in many places in lieu of actual observations of free water on the plant surface. There is even a self-calculating recording instrument that can be calibrated for either Beaumont or Bourke periods (Grainger, 1955).

Grainger's studies went further than earlier ones for he identified one other salient fact to explain why the Beaumont rules often overestimated the opportunities for blight to develop. In considering the physiological "disease potential" of the potato, he found that potato blight easily kills sprouts on tubers early in the season. In late maturity the blight can also occur in epidemic form on the potato shoot. However, in the middle stages of development, the potato is not particularly receptive to blight fungus. Thus the potato has a changing "proneness" to blight through its life cycle. Grainger suggested that the fungus, as it attacks the potato plant, must obtain energy and food (carbohydrate, nitrogen, and mineral matter) from the plant itself. When the plant only has little extra food, and particularly, only meager "spare" carbohydrate to supply energy for the fungus, the ability of the fungus to grow and become of epidemic proportions is strongly limited. Grainger (1956, 1962, 1968) evaluated the ratio C_p/R_s, where C_p is the weight of total carbohydrate in the whole plant and R_s is the "residual" (carbohydrate-free) dry weight of the shoot. He found that when the ratio C_p/R_s was less than 0.5 no disease occurred, while between 0.5 and 1.0 possibilities of blight were only slight. Epidemic conditions were possible only if the C_p/R_s ratio was greater than 1.0 and rising during mature growth (Fig. 7-2). The plant was indeed supersensitive to blight in the early stages of growth when a small shoot weight (R_s) was linked to a large tuber weight (C_p). C_p/R_s was then over 10.0 and often high above that value (Fig. 7-2). Fortunately periods of suitable weather (e.g., Beaumont periods) occur but rarely during this highly receptive early phase; the temperature is usually too low.

Grainger established one additional rule calling for the first appearance of blight to follow the first period of suitable weather (e.g., Beaumont period) to occur only after the C_p/R_s ratio of the potato has risen above 1.0. Use of this rule in concert with the earlier Beaumont rules brings order and relevancy to blight forecasting. Beaumont periods occurring when the C_p/R_s ratio is less than 1.0 are nonproductive of blight because the host plant does not have enough carbohydrate for the growth and reproduction of the blight spores. While weather conditions may satisfy all available blight rules, there is a physiological barrier to the development of blight because of the low C_p/R_s ratio. Eliminating such nonvalid Beaumont periods makes the next weather period recorded with a C_p/R_s ratio greater than 1.0 an effective indicator of the first possibility of blight (Fig. 7-2). C_p/R_s measurements involve tissue drying and chemical analysis, which may not always be quick enough for use in blight forecasting. Grainger and Rutherford (1963) found that significant correlations existed between the fresh weights of the above-ground parts of a potato plant and R_s and between the below-ground parts and C_p. A very

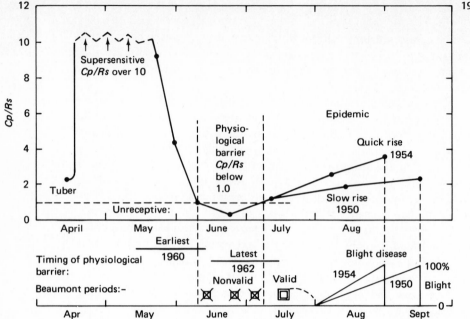

Typical graph of C_p/R_s changes in main-crop-potato "Kerr's Pink" in west   FIG. 7-2
Scotland, annotated to show (1) that periods of suitable weather (Beaumont
periods) occurring during the unreceptive physiological barrier phase must be
regarded as nonvalid; (2) that first Beaumont period to occur after C_p/R_s has
risen above 1.0 is valid and forecasts the first appearance of blight by a
suitable interval which permits effective control measures to be applied; (3)
that quick rise of C_p/R_s in the epidemic phase is related to rapid blight
development and vice versa; (4) that since the physiological barrier phase of
agricultural main-crop potatoes can occur in practice as early as shown for
1960 or as late as given for 1962, its timing and duration can only be
established by measurements of the C_p/R_s ratio. (*Diagram prepared by J.
Grainger and used with kind permission.*)

rapid test that is accurate enough to define the duration of the physiological
barrier phase (C_p/R_s below 1.0) is now used in Scotland and has proved
adequate in practice over several years.

Blight is weather sensitive but it is also sensitive to the condition of the
host plant. If the host is not in a condition to receive and feed the fungus, no
amount of favorable weather will initiate an epidemic. After the blight
appears, high amounts of moisture may not be needed, for the spores
continue to germinate and grow under most normal weather conditions. The
rate of spread of the epidemic, once blight is present, appears to be more
related to the C_p/R_s ratio than to favorable weather.

Weather may, however, still have a final limiting effect. Freshly
liberated blight spores can germinate to a sufficient degree in air at high
humidity, and they do not then need free water (Glendinning, MacDonald,

and Grainger, 1963). One warm day of continuous bright sun nevertheless dries the spores so rapidly that they again need free water to germinate. If such a day occurs at an early stage of a blight epidemic, it is sufficient to arrest further progress, and spraying should be postponed to the next Beaumont period.

C_p/R_s has been found to indicate disease potential to both fungal and bacterial pathogens in over 40 host-parasite relationships (Grainger, 1968). Details and timing of potato growth vary from year to year because of weather; they are different in warm areas and cold, and they vary according to whether big seed tubers are used, or small "eyes" or even rooted stem cuttings. C_p/R_s measurements, however, can always be used to reveal changes in host receptivity, which can then be considered along with weather in any new disease-forecasting venture.

2 Climate and Liver Fluke Disease

The adult parasite of the liver fluke lives in the bile ducts of many animals, the most important being sheep, cattle, and rabbits. Eggs produced by these adults are deposited on pastures in the feces of the host animal; the eggs hatch and produce a free-swimming larva which must contact a host snail (*Lymnaea truncatula*) within 24 hr or it will die. Development and reproduction occurs within the snail, and young flukes leave the snail to encyst on vegetation. The fluke can remain here for many weeks if necessary before the foliage is eaten by the final host and the flukes enter the liver to grow and reach maturity.

Ollerenshaw (1967) studied the importance of climate in the life cycle of the liver fluke and identified various stages in the whole cycle that are particularly climate sensitive. For example, the intermediate host snail (*L. truncatula*) is necessary before the life cycle of the parasite can be completed. But this snail is found in soil that is not too acid, saturated, with even a little free water on its surface. These conditions might be found on poorly drained land, around springs or water holes. The occurrence of such areas is, of course, closely related to general climatic conditions as well as to the actual weather conditions in any particular year. The snail grows well under proper soil and moisture conditions but it withdraws and becomes inactive under dry conditions; such adverse conditions rarely terminate its existence, however.

Outbreaks of liver fluke disease are not necessarily related to the population density of snails. There are always some individuals present, and while the number of snails is responsive to the occurrence of wet or dry seasons, epidemics have occurred even with relatively small populations of snails.

Ollerenshaw (1967) has shown that other aspects of the life cycle are more climate sensitive than the influence of climate on the snail itself. Moisture and temperature seem to be the limiting conditions. For example, there will be no development of the parasite either in the egg or in the snail unless the temperature exceeds 10°C. Moisture is even more limiting since fluke eggs cannot survive desiccation and the eggs cannot hatch or the larva enter the host under dry conditions. Development in the snail proceeds normally only if the snail is active (i.e., under moist, nonacid conditions). Moisture is also necessary as the parasite leaves the snail and finds its way to vegetation. However, once encysted on the vegetation, the parasite can survive for some time under dry conditions.

Ollerenshaw (1967) has pointed out that in Great Britain favorable moisture and temperature conditions must last for at least 3 months for the parasite to go from egg to its encysted state on the vegetation. Since temperature is limiting in late autumn, winter, and early spring, the main development of the parasite occurs from May to October.

The need for three favorable months before infection of the foliage can occur results in two possible periods of infection of sheep and cattle during the year. First, eggs hatching in early summer and developing during the summer in the snail infect vegetation in late summer or early fall (August to October), causing outbreaks of liver fluke disease in stock in late fall and winter. Second, eggs hatching in late summer and fall do not complete their cycle in the snail until spring because of low temperature. The parasites pass onto the vegetation in summer and cause stock infection in late summer and early fall.

Comparison of the geographical pattern of the disease in Great Britain with the distribution of temperature and moisture emphasizes the climatic limitations specified by Ollerenshaw. Low temperatures in Scotland restrict the development of the parasite and cause a minor but fairly constant fluke problem every year. In eastern England, the climate is dry, transpiration exceeding precipitation in 3 or 4 months from May to October. This dry weather prevents the life cycle of the parasite from being completed so that the disease is rare. In western England, the climate is more conducive to parasite development. In those years when moisture conditions are favorable, epidemics of liver fluke disease may develop. When precipitation falls behind evapotranspiration, the parasite has difficulty developing and outbreaks of the disease are more limited and less serious.

3 Climate and Airborne Plant Diseases

Spores of various plant diseases, as well as disease-carrying insects, are often transported great distances by the general circulation of the atmosphere. Research to identify the source area of the disease, in an effort to forecast future outbreaks, requires reference to upper-level circulation maps and the construction of airflow trajectories. A trajectory is defined as the actual path described by an isolated portion of the air. It is obtained by tracking the movement of the air backward over any previous period of time to determine where the particular parcel of air now over the area was located yesterday, and each previous day of interest. Lacking sufficient upper-level winds, the needed information is often approximated by the geostrophic wind. This often causes certain errors in both distance and direction, but it may be the only technique possible.

Hogg (1967) used the trajectory technique to identify the source of black stem rust of wheat in southwest England. For each of the late spring and early summer months (May to July), he determined the surface, 700 mb, and 500 mb trajectories at 12-hr intervals. Based on his hypothesis that only the air from France, Spain, or North Africa could bring the rust spore, Hogg only graphed the frequencies of trajectories from those areas for the years 1947, 1948, 1950, and 1954. The years 1947 and 1950 were severe rust damage years, while 1948 and 1954 were very light damage years.

Analysis of the graphs shows a marked variation in the trajectories with years. There is definite agreement between years with severe epidemics of black rust and trajectories from the Iberian Peninsula. In these years there were several periods when the trajectories at all levels seemed to correspond. These might be particularly significant times for rust development. In the years with little or no black rust many fewer trajectories from Iberia were noticed. This by itself does not provide proof of the hypothesis, but it does suggest the need to pursue this line of investigation further.

Craigie (1945) also noted the relation between wheat stem rust in the plains provinces of Canada and wind movement. Rust first appears in the Red Valley of Manitoba, spreading westward to Saskatchewan and Alberta where it appears about a month later than in Manitoba.

Significant correlations between diseases and weather or climate factors are extremely useful in predicting the occurrence or spread of diseases and in warning the farmer in advance of the need to spray, dust, or take other defensive actions. Such action, when outbreaks are not imminent, is wasteful; if spraying or dusting occurs too early, later vegetative growth will still be susceptible. To be most effective, the defensive action should occur at just the time that the disease outbreak is beginning; accurate timing is essential.

CLIMATE AND ANIMALS

Cold-blooded animals, whose temperatures usually adjust fairly well to the environment, can move from one spot to another, seeking favorable climatic conditions, or they can hibernate or remain protected during periods of stress. Warm-blooded animals can produce heat through metabolism and activity or lose it through radiation or sweating. While most do not have the complete heat-regulating mechanism that man does, they do have some control over their heat balance and bodily responses to thermal stimuli.

Since animals can move, one particular microclimate is not necessarily significant for them. The applied climatologist must be prepared to study the meso- and even the macro-climate since these will influence the response of the animal. Because they have sensory and nervous systems, other aspects of the atmospheric environment including acoustical and optical properties, electricity, and pressure may be of importance. Thus the range of environmental conditions that impinge on the animal and that may influence its reaction are far more complex than in the case of plants.

Studies of the effect of climate on animals have generally involved investigations under controlled conditions where temperature, humidity, wind, or other factors can be regulated. One of the reasons for this is that the animals themselves have to be somewhat confined, if routine and systematic observations of respiration, body temperature, blood composition, and endocrine activities are to be studied. More important is the need to control some of the many variables if there is to be hope of understanding the influence of the others.

Animal-climate studies have usually involved the tolerance of the species to limiting climatic conditions, the reproductive activity, milk or meat production, or the effect of parasites and other disease organisms. In many cases, it is the response to high temperature that is more limiting than cold temperature so that temperature-tolerance studies are often confined to extreme maximum temperatures. For example, cattle are generally tolerant to cold temperatures, but above about 27°C their rate of weight gain or milk production begins to suffer. Thus cattle studies are almost entirely confined to the effects of tropical or subtropical conditions.

Pigs, lacking efficient sweat glands, must control their temperatures by shade or wet conditions. Studies on pigs have, therefore, been generally directed toward controlling high temperature and radiation, or low humidity. In sheep and goats, studies have been mainly directed toward the relation of wool production to climate (especially the effect of shearing on sheep) and toward reproductive activity and fertility.

Climate-Comfort Relations

The body temperature of most European dairy cattle increases as ambient air temperatures rise above 21°C (Wang, 1972). As temperatures exceed 27°C, appetite decreases with a consequent falling off of milk yield. Light intensity may also influence the loss of appetite, and so consideration has been given to pasturing during periods of darkness under tropical conditions.

Zebu cattle have a greater heat tolerance than do European breeds, possibly because their body form permits greater heat dissipation. Respiration rate increases with increasing temperature up to some limit, resulting in increased heat dissipation. After this limit is passed, respiratory rate declines. Sweating and characteristics of the coat are further areas of consideration in studies of increased or decreased heat dissipation.

As in the case of humans, one climatic factor alone seldom places severe limitations on animals. Thus a single-factor approach is not warranted. Rather it is a combination of several factors such as temperature, humidity, and wind that must be investigated in relation to animal reactions if the influence of climate is to be understood.

Kibler and Brody (1954) and Brody (1956), among others, have studied the response of animals to combinations of heat and humidity. Using three breeds of cattle (European, Indian, and Santa Gertrudis), all confined to climate-controlled chambers with constant temperatures, wind, and humidity, the "comfort zone" and the response to uncomfortable conditions were determined for each of the breeds. European cattle had a "comfort zone" between 30 and 60°F (-1 to 15°C), while Indian cattle were comfortable between 50 and 80°F (10 to 27°C). This higher heat tolerance probably resulted from lower heat production, greater surface area per unit weight, and shorter hair.

When the temperature was raised to 105°F (40°C), the Indian cattle had a rectal temperature of 105°F (40°C), Jersey cattle had 106°F (41°C), and Holstein cattle had 108°F (42°C), which approached the lethal limit. European cattle were much more cold tolerant and increased their heat production by only 10 percent or so when the air temperature dropped to 9°F (-13°C), well below their comfort zone. Zebu cattle increased heat production some 60 percent when the air temperature dropped from their comfort zone of 50 to 9°F (10 to -13°C). All cattle seem to need protection against heat, although some are better equipped at heat regulation than others; all breeds seem fairly cold tolerant and need only adequate food under such conditions along with protection against severe weather such as blizzards, hail, and driving rain.

Humidity seems to have only little effect on cattle in the field although some reaction has been found in climate-controlled chambers. Kibler and Brody (1954) did prepare a comfort diagram (Fig. 7-3) for cattle, quite similar to one available for humans, describing the effect of different combinations of temperature, humidity, and wind speed on cattle well-being. The "zone of normal rectal temperature" was determined using a wind speed of 0.5 mph (0.2 m/s), although it was shown that normal rectal temperatures could also be maintained at higher air temperatures and humidities if the wind speed exceeded 6 mph (2.7 m/sec) to increase heat dissipation.

Temperature, light intensity, and possibly day length, all seem to influence the reproduction and health of rams and ewes. Dutt and Simpson (1957), for example, blamed high summer temperatures for the low rate of conception of Southdown rams. They felt that this rate could be improved by keeping the rams at colder (7° to 9° C) and more moist (70 to 80 percent relative humidity) conditions. Light does increase the production of eggs in chickens and intensify sexual activity in birds, although few studies on light effects on cattle, sheep, or hogs have appeared.

Priestley (1967) has considered the heat balance of animals and shown how it is related to their size and degree of insulation. He points out that most animals do not sweat or pant and, therefore, do not lose their radiation heat load through evaporation of water. Rather, outgoing radiation and conduction-convection must play an increasing role in the maintenance of their thermal balance. The well-studied sheep emphasizes the insulating value of its wool coat. Very little heat is conducted inward since wool is an excellent insulator. The sheep will lose the greatest amount of heat by longwave radiation. Temperatures at the end of the wool fleece will rise to very high levels (60°C to over 80°C), increasing outward radiation. Possibly as much as 60 percent of the heat loss will occur by outgoing reradiation and 30 or so percent by conduction-convection. These percents will vary appreciably

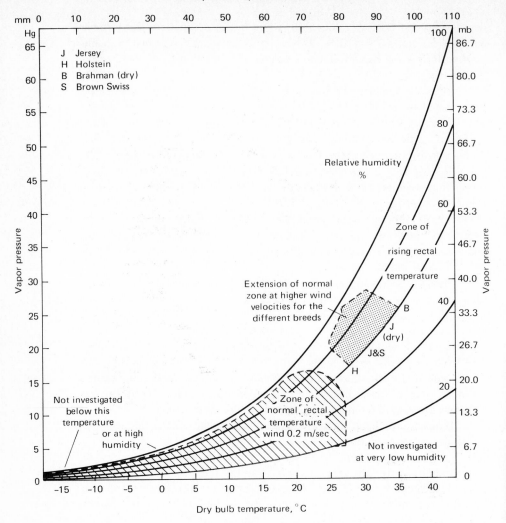

Comfort zone for dairy cattle as affected by environmental temperature, humidity, and air movement. (*With permission of the Missouri Agricultural Experiment Station, from Kibler and Brody, 1954, fig. 1.*)

FIG. 7-3

under windy conditions. Only about 5 percent of the total heat load on the sheep is conducted inward through the fleece.

Size is also important in the process of convection heat loss. The rate of heat loss by conduction depends, in part, on the temperature difference between the surface and the air blowing over it. But as the surface becomes larger, the air moving over it becomes modified by its presence. As a result smaller differences in temperature between the surface and the air exist. The rate of convection loss decreases, therefore, as the surface area of the body increases. Priestley points out that a lamb would lose more heat by conduction-convection than would its larger mother. As smaller body sizes are encountered, convection plays an increasingly important role in heat loss until a size is reached at which extreme surface temperatures can no longer occur, so that the need for an insulating coat or layer is eliminated. Many insects have no need for any insulation from high temperatures nor mechanisms for evaporative cooling. They are convectors rather than radiators of excess heat, maintaining their heat balance by large convective heat losses.

The relation of the biological activities of birds to temperature, light, wind, or other climatic conditions is sometimes quite difficult to document. A number of investigators have studied the phenomenon of bird migration and attempted to relate it to various climatic stimuli. Climate or weather may not be the only control, and, even if it is, one might question which climatic conditions are important. Certainly temperature would seem to be significant but what temperature should be considered—the temperature in the area of departure, the temperature at flight elevation, the temperature at the destination, or the temperature effect on the food supply of the bird?

A number of investigators have suggested that changing temperature rather than some absolute value of temperature might be the factor for initiation of migration. Robbins (1949) felt that the migration of the Canada goose paralleled the northward advance of the $35°F$ $(1.7°C)$ isotherm. However, others have reported a much more rapid migration seemingly unrelated to the slow movement of a particular isotherm. Logic might suggest that food supply would be a critical factor for all migrating animals. They should not migrate unless their own food supply is inadequate or unless climatic conditions are favorable for the production of adequate food along their migration path or in the area to which they are going. Such conditions may be correlated with climatic conditions in their original home.

Those animals which do not have a wide tolerance to climatic conditions must maintain thermal balance by frequent movement, by venturing forth just at certain times of day or night, or by creating just the right environment for themselves in their living quarters. Priestley (1967), reporting on an original and thorough study conducted by H. J. Frith and begun in 1947, showed how the Australian mallee fowl builds and maintains an incubator for its eggs in which the temperature varies less than $2°C$ over a 6-month period. The mallee fowl builds a nest on the ground, in the spring of the year, consisting of soil mixed with wet leaves and twigs to provide heat by decomposition. The nest is a mound some 3 m in diameter and $\frac{1}{2}$ m high. During the incubation period of almost 8 months, the nest is tended by the male bird whose beak is quite temperature sensitive.

By opening the egg chamber almost daily early in the season, excess heat is lost. As the summer heat increases, the egg chamber is opened more slowly and left protected from the radiation by some 15 to 30 cm of soil.

Digging in the nest area is done only about once a week in the very hottest part of the year. Toward fall, the digging by the fowl to open the egg chamber occurs later in the day, and the sun's rays are used to bring needed heat to the eggs.

Investigators built two test mounds, a "control" mound built entirely of soil, and an "artificial" mound built entirely similarly to the real mound but not tended thereafter. Temperature records from the artificial and control mounds and from a nest maintained by a mallee fowl show the ability of the fowl to control his environment within a $2°C$ range for a 6-month period by opening and closing his nest, while temperatures increased $15°C$ in the control mound during the incubation period and decreased by $11°C$ during the first 3 months of the incubation period in the artificial mound. The ability of the mallee fowl to maintain a stable nest temperature is an example of practical applied climatology seldom equaled in the world of humans.

CLIMATE AND AGRICULTURAL YIELD

Investigators have long tried to find precise relations between agricultural output, as expressed in terms of crop yield or the production of such items as milk or honey, and simple, readily available weather data such as temperature and rainfall. In many cases these studies are still the only ones possible since more refined measures of the climatic environment are not available. In recent years, however, with improved instrumentation, it has been possible to begin to relate agricultural output to such factors as soil temperature or soil moisture content, to radiation, or to derived climatic indices that reflect how well the vegetation is utilizing the energy and water potential for growth and development. Use of these more active climatic factors has resulted in the achievement of significant correlations between climatic factors and yields. The correlations have often been meaningful enough to be of predictive value. Forecast of crop yields based on relations to measured weather variables are now possible several months in advance for many agricultural commodities. The possibility of severe intervening weather, the sudden development of blights or other diseases, or the inability to harvest at the right time due to labor, mechanical, or environmental factors may influence the accuracy of the forecasts, yet in many areas of agriculture, considerable guesswork in terms of national or regional yields has now been eliminated from farming.

Questions involving the relation of yield to climate or weather are among the most complex in applied climatology. For example, weather and climate over the whole crop-growing season may be less important in ultimate yield than the weather of a particular period during the development of the crop. Monthly or seasonal data may obscure the real relationship. Yield of different crops may be expressed in terms of amount of fruits, vegetative growth, or tubers; conditions favoring vegetative growth may not at all favor the production of fruits or of tubers. Thus different weather factors may be significant within just a single agricultural area. Weather varies markedly from place to place, and conditions as measured in a Stevenson screen may not be representative of the weather in a field even a few miles distant. This is particularly true for precipitation, but it can hold as well for radiation and even temperature, especially if topographical differences are involved.

Three selected types of climate-yield studies will be discussed in the following pages. They cover significant areas of the subject presently under investigation and provide a feel for the current direction of applied climatology studies in the field of agricultural output. In general, the studies may be considered under the following headings.

Climate and agricultural production. Yield vs.:

1   Primary weather factors (such as temperature or rainfall individually)

2   Combined primary weather factors

3   Secondary (derived) climatic factors (such as soil temperature or soil moisture content)

1 Yield Relations with Individual Primary Weather Elements

Hurst (1967) has shown a good correlation between honey production in Great Britain and monthly temperatures (especially in July and August). The relationship is simple and direct: the higher the temperature, the greater the activity of the bees and the greater the production of honey. However, monthly temperatures need not be closely related to the distribution of sunny, hot days that are particularly favorable for honey production. Thus, in 1970, Hurst reported on the relation between honey production and number of days each month with maximum temperature over certain limits (20, 25, and $30°C$). Actually there were not enough days with temperatures over $30°C$ so he used only the first two limits. Averaging 8 years at Kew with very high honey production (greater than 28 kg per hive each) and comparing these results with the 8 years of very low honey production (less than 8 kg per hive each), he found a direct relation with maximum temperatures (Table 7-2).

Nearly the same relation was found at six other stations located in both northern and southern areas of Britain. In every instance, it was the July and August temperatures which were most significantly related to honey production.

TABLE 7-2

Relation between maximum daily temperatures
and honey production, Great Britain*

HONEY PRODUCTION (AVERAGE 8 YRS)	DAYS WITH TEMP. OVER 20°C		DAYS WITH TEMP. OVER 25°C	
	JULY	AUGUST	JULY	AUGUST
34.9 kg	23.7	23.3	9.6	6.5
5.7 kg	11.6	13.9	1.5	1.9

* *From Hurst, 1970,* Meteorol. Mag., *no. 1172, vol. 99, p. 76, with permission of the Controller of Her Majesty's Stationery Office.*

By omitting 1940, which had a very warm June period along with high honey production, Hurst showed that there was not a single high honey year with less than 21 July days with maximum temperatures over 20°C and no low honey production years with more than 18 such hot days. August results were similar but not as marked.

Smith (1968) studied the relation between milk yields and climatic factors in Britain and obtained a simple but direct forecasting aid based only on June rainfall. He found that the annual milk yield for the period April through March for herds in England and Wales can be predicted 9 months ahead of time (by the end of June), based on the April–June milk production data and the total June rainfall over the country. The mean percentage error is 0.31. When only the April milk production and rainfall data are used (11-month forecast), the mean percentage error rises to 0.53. Clearly, a much more valuable statistic is obtained by including two additional months of data; even so the length of the lead time (9 months) is still long enough to be useful. Fourteen years of data were available, and the milk production figures were based on a little under 1 million dairy cows. The data, thus, were from a large enough sample to make the result significant.

Smith noted that over the 14 years of record, milk production had increased with improvements in dairy husbandry and a gradual change to higher-yielding cattle breeds. This trend had to be removed first before any relation to rainfall could by sought.

Smith suggested the main meteorological influences on milk yield might be

1   The earliness of the spring (reflected in April production)

2   The summer rainfall (unpredicted but suggested by the June rainfall)

3   The rainfall during the early haying season since this is related to the quality of the main winter food

The June rainfall figures are quite important in providing at least partial answers to 2 and 3 above. If June rainfall is low (below 50 mm), the quality of the hay is more apt to be good and milk production large. If June rainfall is high (above 75 mm), hay quality is poor and winter milk production is low.

The milk production forecast formula developed by Smith is

$$Y = 0.25J + 0.20A + 0.13n - 0.0065R + 10.45$$

where J = April–June milk yield
A = April milk yield
n = number of years after 1953 (to correct for yearly trend in improved milk production due to other than climatic factors)
R = rainfall factor based on June rainfall, mm

R is the difference in the rainfall from a base value of either 50 or 75 mm; thus it is a negative value for a deficiency of rainfall below 50 mm, zero if the rainfall is between 50 and 75 mm, and a positive value for rainfall in excess of 75 mm.

Sugarcane yields have been related to climate and weather factors in several quite complex ways. Actual production of dry matter is related to weather factors during the growing season, while the weather conditions during the few weeks just before harvest time are closely related to sugar content of the juice. In this latter instance, minimum temperatures are inversely related to sugar content. The lower the minimum temperature during the 2-week period before harvest, the higher will be the sugar content of the juice.

Researchers at the Hawaiian Sugar Planters' Association (1963, 1969) have found an interesting correlation between sugar yield and the diurnal temperature range—the larger the range of temperature, the greater the yield. In Hawaii there seems to have been a trend toward a larger diurnal temperature range in the 30 or 40 years from 1900 to nearly 1940 and a trend toward a decreasing daily range since the early 1940s. If the diurnal temperature range is being influenced by some man-made factors, such as pollution or changes in surface conditions, then this relation may be of usefulness in predicting sugar yields in the period ahead.

The Sugar Planters' Association research has also established a relation (correlation coefficient +0.80) between solar radiation and tons of sugar produced per acre. A number of investigators have noted that with densely growing crops, resulting in little energy wastage, there will be a close relationship between total radiation and dry matter production. Sugarcane is such a crop so that in years with adequate soil moisture the total yield can be well predicted by radiation alone. Knowledge of such an influence should aid management in making decisions to irrigate and fertilize in years with high radiation amounts in order to meet the needs of a predictably heavy crop yield.

In studying the yield of tomatoes in Delaware, Wang (1963b) found a good relation between certain rainfall factors and yield. He determined the relative maximum rainfall, R_R, during the flowering period and defined it as the highest accumulation of weekly rainfall in the particular growing season. By plotting scatter diagrams of the relative maximum rainfall (for periods ranging from 1 up to 6 weeks) for a series of years against yield in those years, Wang found that when R_R was too low during flowering, yield decreased. Yield also declined when the relative maximum rainfall was too high. The result was a parabolic relation between relative rainfall maximum and yield (Fig. 7-4). Using 4-week rainfall figures, he found an optimal yield of about 12 tons/acre (26,900 kg/ha) with a rainfall of about 4.6 or 4.7 in. (120 mm) (Fig. 7-5). When the 4-week relative maximum rainfall dropped below 2.4 (60 mm) or exceeded 7.1 in. (180 mm), yield dropped below the mean yield line of 9 tons/acre (20,000 kg/ha). Plotting the relative maximum rainfall amounts for various length periods (1 up to 7 weeks), Wang obtained

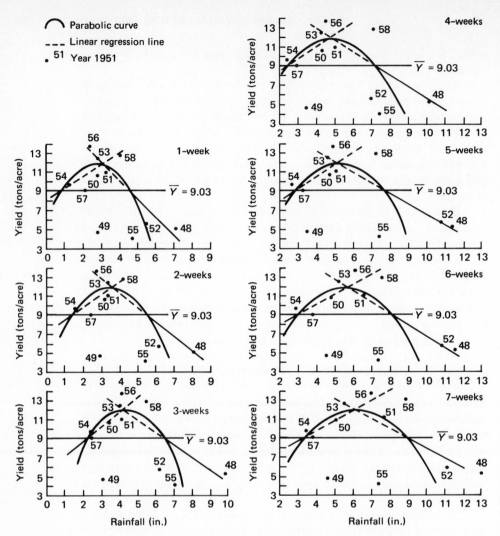

Weekly relative maximum rainfall vs. yield of tomatoes during blossoming period, central Delaware, 1948–1958. (*With permission of the American Society for Horticultural Science, from Wang, 1963b, vol. 82, p. 437; also Wang, 1972, "Agricultural Meteorology," Milieu Information Service, San Jose, Calif., p. 311.*)

FIG. 7-4

cumulative curves for the upper and lower limits of optimum yield [greater than 11 tons/acre (24,600 kg/ha)] (Fig. 7-6).

Wang (1963b) also applied a relative minimum rainfall parameter (R_r, the lowest daily or weekly rainfall amount relative to the crop-growing season) to the study of sweet corn, peas, tomatoes, cucumbers, and other crops. He reported that a time period of 3 to 5 weeks of precipitation seems to correlate most significantly with crop yield. The correlations obtained were both positive and negative depending on the crop and its developmental

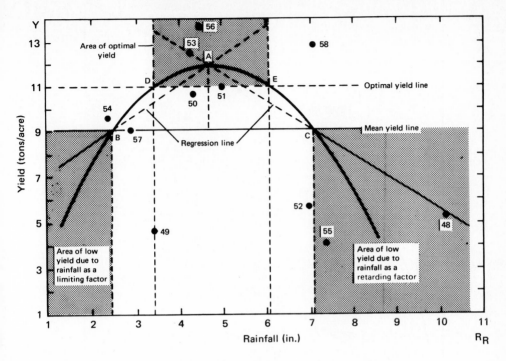

Four-weeks relative maximum rainfall vs. yield of tomatoes during blossom-   FIG. 7-5
ing period, 1948–1958. (*With permission of the American Society for Horticultural Science, from Wang, 1963b, p. 438; also Wang, 1972, "Agricultural Meteorology," Milieu Information Service, San Jose, Calif., p. 311.*)

stage. For example, for Tendermost and Tenderblonde sweet corn in Wisconsin, he found an expression of the form

$$Y_d = -2604(3R_r) + 2161$$

where Y_d = departure of yield, lb/acre

$3R_r$ = relative minimum rainfall over a 3-week interval in growing period from planting to tasseling

However, for Victory Golden sweet corn in the same area, the relation was

$$Y_d = 3056(3R_r) - 3627$$

using a growing season from silking to harvest. In the first case the correlation coefficient was −0.79, while in the second case the coefficient was +0.68. Thus, during different stages of development, low amounts of rainfall can be either favorable or unfavorable to later yields.

Wang suggested that the optimum temperature range for tomatoes lies between 66 and 83°F (19 and 28°C). If mean daily temperatures outside that range occur (mean daily temperatures less than 66°F or more than 83°F),

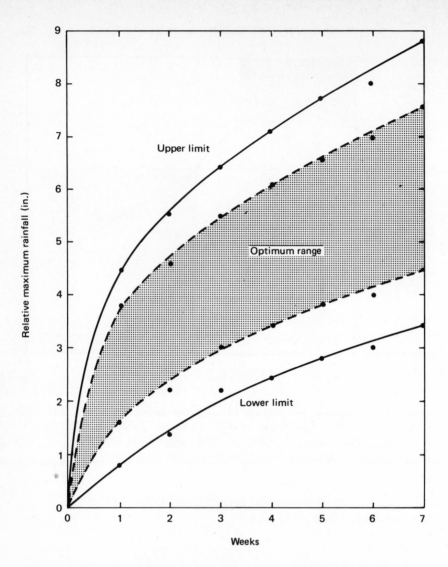

Relative maximum rainfall in successive weeks for optimum and above-mean   FIG. 7-6
tomato yield in central Delaware, 1948–1958. (*With permission of the
American Society for Horticultural Science, from Wang, 1963b, p. 439; also
Wang, 1972, "Agricultural Meteorology," Milieu Information Service, San
Jose, Calif., p. 312.*)

reduced yields are to be expected. Using the percent frequency of
temperatures over 83°F during the reproductive period, he obtained a
significant negative correlation with yield ($r = -0.90$) along with the following
regression equation:

$$Y = -0.29T_f + 13.29$$

where Y = yield of tomatoes, tons/acre

T_f = temperature frequency over 83°F, given in percent

Gangopadhyaya and Sarker (1965) related wheat yield to rainfall distribution. Any rainfall over and above the average for a period of about a month preceding sowing and during germination is found to be beneficial to yield, while extra rainfall during tilling is harmful. Additional rainfall more than a month prior to sowing did not improve yield and, in fact, might be detrimental. Approximately 75 percent of the total variation in wheat yield can be explained by rainfall distribution.

One clear conclusion is that if the additional rainfall does not occur during the favorable period prior to sowing or germination, the farmer should supply it by irrigation to improve final yields.

2 Yield Relations with Combinations of Primary Weather Factors

Slightly more sophisticated than the single-parameter studies are those that involve two or more of the basic meteorological parameters. Several of these are cited for illustrative purposes in the following paragraphs.

Francis (1967) has studied the relation between California quail populations and various weather or climate conditions, using data for a 14-year period in San Luis Obispo County. He found that the annual changes in the productivity of quail could be accounted for almost entirely by only two climatic factors—rainfall and temperature—if one knew also the percent of adults among the females in the hunting sample of the previous year. This latter factor is, of course, of importance since adult females will produce young more efficiently than will subadult females.

Francis found a positive correlation between quail production and seasonal rainfall for the period September 1 to April 29. However, in developing the actual relation, McMillan (1964) found that a more significant moisture factor was the soil moisture content at the end of April; the reason for the relation to rainfall was because of the high correlation between rainfall and soil moisture content (up to some maximum value). McMillan's expression was

$$Q = 0.021A + 0.929M - 0.120P - 0.975$$

where Q = quail production as determined by ratio of subadults to adults in hunting season

A = percent of adult females taken in previous fall's hunt

M = soil moisture content, in., at end of April (calculated from Thornthwaite system)

P = precipitation, in., from September 1–April 29

While a simple correlation of rainfall with quail production is positive, the partial correlation coefficient in a multiple regression is negative because of the strong positive correlation of rainfall with soil moisture content. When precipitation exceeds the amount needed for optimum soil moisture, production decreased.

Francis found that only once out of 18 sets of data tested did the observed ratio of subadults to adult birds differ from the computed ratio by an amount greater than the sampling error of the observed data and the standard error of estimate of the computed ratio. In that year, the birds taken in the hunting season in the different age groups were not sex differentiated so that an estimate had to be made. This could have contributed to the poor results for that particular year. Francis felt that the predictive value of the regression equation would offer a useful research and management tool.

Sen and his associates (1966) carried out a detailed investigation of the various factors affecting tea yield in the Assam Valley of India. First, to identify the weather variables with the most profound influence on tea yield, Sen et al. obtained mutual correlation coefficients between the 15 climatic variables R_1, R_2, R_3, T_1, T_2, T_3, D_1, D_2, D_3, H_1, H_2, H_3, and S_1, S_2, S_3. (R = rainfall, cm; T = mean temperature, °C; D = difference or range in temperature between maximum and minimum; H = average daily relative humidity, %; S = average number of sunshine hours per day; subscripts 1, 2, 3 = periods January to March, April to June, and July to September of the current year, respectively. In a later aspect of the study the investigators added the subscript 4 to cover the September to December period.) Identifying the variables that were mutually uncorrelated and thus possibly independent, Sen et al. fitted, by the method of least squares, a multiple regression of the yield of the early, main, and late tea crops on these independent climatic variables. The independent variables were later removed one by one by backward elimination until only the most critical variables influencing tea yield remained. These were found to be (1) rainfall and mean temperature in the January to March period (influencing both the early and main crop), (2) rainfall during the July to September period (influencing the main crop), and (3) rainfall and temperature in October to December (influencing the late crop).

In the crop-growing area, the January to March period is generally dry and cold, April to June is the period of early rains, the main rains come from July to September, while the October to December period is again usually dry. An increase in rainfall up to about 18 cm in the January to March period was beneficial; rainfall above that amount did not increase later yield. An increase in rainfall with high temperatures was more significant than an increase with cold temperatures. In the October to December period, the rate of increase in tea yield with increase in temperature tended to slow as rainfall increased.

The study by Sen and his associates involved routinely available monthly climatic statistics investigated by a regression-type statistical analysis. While significant results were obtained, it is possible that weather data selected for phenological events rather than months, and covering additional parameters, might have improved the significance of the results.

Thompson (1970) has investigated both climatic and technologic factors involved in the production of soybeans in the central United States (Ohio through Iowa) by using a multiple curvilinear regression analysis. A linear time trend was used to determine the influence of technological change so that it could be removed, permitting the effect of weather to be identified quantitatively. Highest yields of soybeans are associated with air temperatures that are above normal in June but below normal in July and August. At the same time for best yields, precipitation must be near normal in the September to June period and above normal in the months of July and August.

Clearly, average July and August temperatures result in a water demand that cannot be satisfied by normal precipitation. Cooler temperatures or above-normal precipitation result in reduced moisture stress and so increase yields. While Thompson uses the basic parameters of temperature and rainfall, it is the derived parameters of soil moisture content or moisture stress that are of prime significance.

3   Relation of Yield to Secondary or Derived Climatic Elements

Hurst and Smith (1967),[1] seeking a simple method to express the relative dryness of a summer, suggested a parameter which they called "grass-growing days." It is defined as the number of days from April through September during which the soil moisture deficit does not exceed 2 in. (50 mm). Smith (1956) had earlier shown that when deficits are less than 2 in. (50 mm) hay transpiration is closely correlated with hay yield.

Using daily values of measured rainfall and evapotranspiration as calculated from weather data and corrected for sunshine, Hurst and Smith determined a daily water budget from which they obtained the soil moisture deficit for each day of the growing season. The southwestern and north-western portions of England as well as Wales all experienced more than 170 grass-growing days on the average while the area around the mouth of the Thames in southeastern England has less than 130 such days.

The figures of milk production by sections of the country are quite closely correlated ($r = -0.91$) with the number of grass-growing days lost due to high soil moisture deficits. Soil moisture is more intimately related to grass growth than to either precipitation or temperature, or both together. Use of the derived parameter provides a more significant relationship with hay yield as well as with related agricultural factors such as milk production.

Earlier, deWit (1958) had shown that transpiration and dry matter production were closely related. If crop production is primarily expressed in terms of vegetative growth, maximum yields are usually obtained when water is available at the so-called potential evapotranspiration rate. This quantity of water may not be actually required by the plant each day in the growing season, for some vegetation needs less water during certain stages of development. Also with low plant density, water is not used at the potential rate.

---

[1]   Material adapted with permission from Taylor, "Weather and Agriculture" (1967), Pergamon Press Limited.

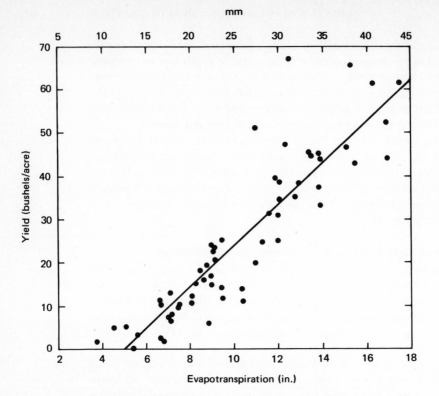

Relationship between wheat yield and evapotranspiration in tanks, FIG. 7-7
1922-1952. (*With permission of the Agricultural Institute of Canada, from Staple and Lehane*, 1954, *p.* 557.)

Chang (1968) has pointed out that lettuce, spinach, radishes, cauliflower, celery, alfalfa, cotton, tobacco, sweet potatoes, bananas, and peaches all produce higher yields when the soil moisture content is maintained above 50 percent of field capacity. Yields of tomatoes, snapbeans, barley, sugarbeets, Irish potatoes, peanuts, pears, and apples appear not to be significantly affected by increases in soil moisture content above 50 percent, according to Chang.

Staple and Lehane (1954) investigated the relation between water use and yield of spring wheat in Swift Current, Canada, both as grown in tanks and in the open field. They found that approximately 5 in. (125 mm) depth of water was necessary for the establishment of the plants; the relation between yield and evapotranspiration was linear in the tanks and curvilinear in the field-grown wheat (see Figs. 7-7 and 7-8). More than 30 years of data were available for study.

Chang et al. (1963) have emphasized the need not only to study the water and yield relationships during the entire growing season of the crop but also to compare relationships found in different areas and different years. He suggested that the ratio of actual to potential evapotranspiration would be a

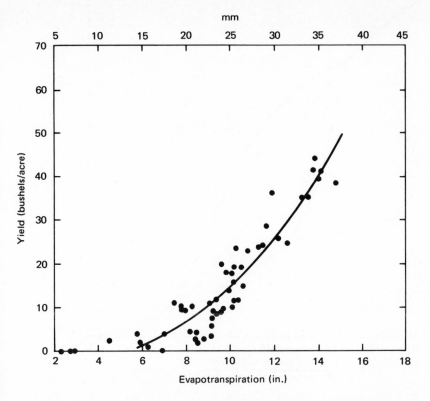

Relationship between wheat yield and evapotranspiration on field plots.   FIG. 7-8
(*With permission of the Agricultural Institute of Canada, from Staple and Lehane,* 1954, *p.* 558.)

more useful indicator than actual evapotranspiration alone. He has used this ratio to obtain a generalized relationship to yield (Fig. 7-9). When actual water use approaches potential, at least in vegetative crops, actual yield approaches potential. When actual evapotranspiration is only 0.1 or 0.2 of potential, actual yield achieves about the same fraction of potential yield. In the middle portion of the curve, large variations in results can occur, depending on the distribution of rainfall, the timing of droughts, or the efficiency of irrigation. This middle period is the critical period in the life of the plant. Irrigation at the right time and in the right amount can result in a doubling of yield according to the different suggested relationships of Fig. 7-9. It is not the absolute totals of precipitation or temperature that are important, but rather the availability of water to meet the plants requirements at the time they are most significant. This is best determined through a comparison of actual use and potential need (AE/PE) or an evaluation of soil moisture storage or deficit.

Mather (1968a), in a study of the future direction of irrigation agriculture, related corn and soybean yield in southern New Jersey to the

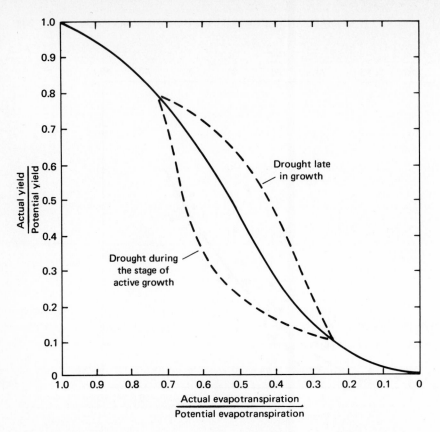

Generalized relationship between yield and adequacy of water application. FIG. 7-9 (*With permission of the Agron. J., from Chang, Campbell and Robinson*, 1963, *p.* 452.)

accumulated soil moisture deficit ($PE - AE$ as determined from the Thornthwaite climatic water balance). Using data on seasonal deficits rather than for specific periods in the life cycle of the plant over a 15-year period, he obtained significant negative correlation coefficients with both sweet corn ($r = -0.85$) and soybean ($r = -0.75$) yields (Fig. 7-10). Although his study was preliminary in nature and did not attempt to correct for the increase in annual yields due to improved technology, it still showed the important role that soil moisture deficit plays in the ultimate yield of a crop. Refinements in technique employed by later investigators (see following paragraphs) would have resulted in further improvements in these correlations and regression equations.

More sophisticated studies have appeared recently relating yield to factors of soil moisture deficit and moisture stress. For example, Mack and Ferguson (1968) and Baier and Robertson (1967b) both made use of modifications of a soil-moisture bookkeeping procedure to obtain actual and potential evapotranspiration and from them to derive values of soil moisture

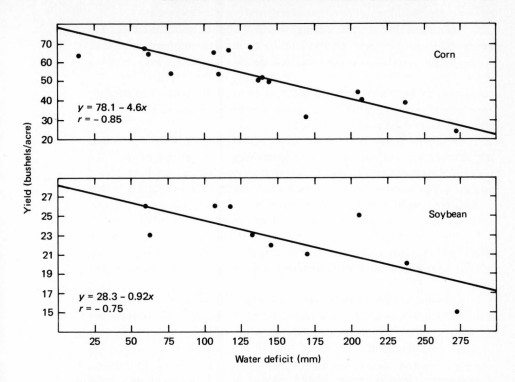

Relation between corn and soybean yield and water deficit, Seabrook, N.J.,   FIG. 7-10
1947–1961. (*With permission of the Association of Pacific Coast Geographers, from Mather, 1968a.*)

storage with time and depth within the zone of rooting of the crop. All this information is obtained from the basic climatic statistics of temperature and precipitation.

Baier and Robertson (1968) then applied these soil moisture estimates rather than the basic statistics of temperature and rainfall to the problem of predicting wheat yield. Using wheat yields from 39 plantings across Canada over a period of 5 years, they found that yield was more closely related to soil moisture than to precipitation or maximum and minimum air temperature. Rainfall, by itself, was not at all useful for yield estimation, while minimum temperature was somewhat better than maximum temperature on the basis of a multiple correlation analysis.

They then divided the crop-growing season into five periods (planting to jointing, jointing to heading, heading to milk stage, milk stage to soft dough, and soft dough to ripe), and the soil into six zones (from the surface downward) containing the following percent of the total water-holding capacity in the plant-root zone: zone 1, 5.0 percent; zone 2, 7.5 percent; zone 3, 12.5 percent; zone 4, 25.0 percent; zone 5, 25.0 percent; zone 6, 25.0

percent. This permitted studying the influence on yield of soil moisture in different depths of the soil at different stages in plant development.

Soil moisture content during the planting to jointing stage was unimportant in final yield. In the jointing to heading and heading to milk stages, simple correlation studies revealed significant relations between yield and soil moisture in each of the six depth zones. As the crop developed, the correlation between yield and soil moisture in the lower portion of the root zone profile increased while the relation between yield and soil moisture in the upper part of the profile decreased.

Of all the zones and developmental periods studied it was found that the value of soil moisture in zone 3 (containing 12.5 percent of the total water-holding capacity) during the jointing to heading development period was most significantly correlated with final wheat yield ($r = 0.68$). Higher correlations would have been obtained if the soil moisture content during other selected development periods and depth zones had also been included.

The same two investigators (Baier and Robertson, 1967b) further related soil moisture content to several components of wheat yield, including number of heads per unit area, number of kernels per head, and 1,000-kernel weight.

Considering each of the three yield components as dependent variables, Baier and Robertson developed a regression model in which the independent variables were the soil moisture values per zone and the crop-development periods. They found that several of the soil moisture variables, such as the moisture content in all rooting depths during the jointing to heading developmental period, were highly related to all three of the yield components and thus to final wheat yield. Determining yield components did not provide any better estimates of total grain yield than they achieved in the previously described study of soil moisture vs. total yield. The study did provide more insight into the influence of soil moisture at different depths and developmental stages on factors contributing to ultimate yield, however.

Mack and Ferguson (1968) made use of a modulated soil moisture budget (originally developed by Holmes and Robertson, 1959, and later modified to become the versatile soil moisture budget of Baier and Robertson, 1966; see Chap. 5) to provide quantitative estimates of soil moisture content at any time during the wheat-growing season in Canada. They determined actual evapotranspiration (AE), and moisture stress ($PE - AE$) from the soil moisture budget and related various combinations of these derived factors to wheat yield. They found that wheat yields from experimental plots were more closely correlated with what they called "moisture stress" ($PE - AE$: $r = -0.83$) than to the other tested parameters: seasonal precipitation ($r = 0.62$); potential evapotranspiration ($r = -0.59$); the ratio AE/PE ($r = 0.78$); degree-days ($r = 0.59$).[1]

---

[1]   t should be pointed out that moisture stress as so defined ($PE - AE$) is equivalent to moisture deficit as defined earlier by Thornthwaite and used by Mather (1968a) in his study of corn and soybean yield just reported.

The authors considered the effect of moisture stress at different developmental stages on grain yield. From regression analysis, Mack and Ferguson reported that for each centimeter of moisture stress from emergence to harvest, grain yields were reduced an average of 156 kg/ha. For each centimeter of moisture stress from fifth-leaf to soft-dough stage and from soft-dough to maturity, final yield was reduced by 311 and 69 kg/ha, respectively. Thus moisture stress is more important to final yield at one time in the life of the crop than at another time.

Nix and Fitzpatrick (1969) also studied the relation between moisture stress and grain yield, using the ratio of actual to potential evapotranspiration (AE/PE) at particular periods in the development of the crop. They agreed with both Baier and Robertson and with Mack and Ferguson that moisture stress at a certain "critical" period (ear emergence through anthesis to early grain filling) was more highly correlated with yield than stress at other times. The AE/PE ratio accounted for 60 to 83 percent of the variation in yield in the wheat and grain sorghum varieties tested in Australia.

From these studies, some investigators have argued that no real genetic differences in drought resistance exist among wheat varieties but rather that different varieties reach the "critical" developmental period at different times. Thus, moisture stress conditions will vary, resulting in better yields for some and poorer for others. Nix and Fitzpatrick suggest that grain yield is largely influenced by the date of anthesis and the soil moisture condition at that time. Two given varieties may actually have the same moisture demands so that neither is more "drought resistant" than the other, but yield in the one that flowers at a time of less moisture stress will be greater.

Nix and Fitzpatrick suggest the need to consider such things as irrigation at selected times in the development of a crop in order to reduce critical moisture stress periods, manipulation of planting dates in relation to weather possibilities, reduction in water removal during vegetative stages through control of plant density and stand geometry, manipulation of the fallow-crop-rotation practices, as well as selection of species with lower energy requirements prior to flowering. All these techniques can result in a more favorable soil moisture situation at the time of anthesis and, thus, lead to increased yields.

Soil temperature may be more expressive of energy conditions for crop development and yield than air temperature. Many plants, especially those with underground tubers, have soil temperature limits. Allmaras, Burrows, and Larson (1964) studied the influence of soil temperature on dry matter production of corn and found a direct and almost linear relation between temperature at the 10 cm depth and dry matter production between 15.6 and 27.4°C. Above the latter temperature, dry matter production decreased again (Fig. 7-11).

Temperature in the upper layers of the soil is vital to agriculturalists, and much effort has been expended in trying to control soil temperature in order to maintain optimum growing conditions for plants. Surface soil

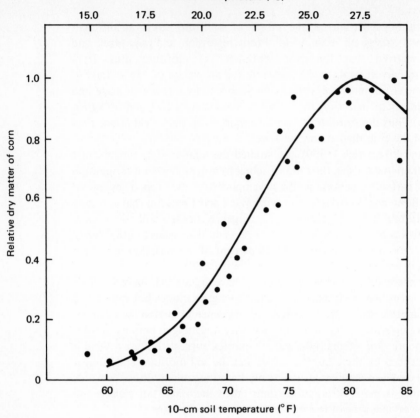

Relative yield of dry matter in young corn plants as affected by average 10-cm soil temperature. (*With permission of the Soil Science Society of America, from Allmaras, Burrows, and Larson, 1964, p. 274.*)   FIG. 7-11

temperature can be raised by ridge planting if followed by shallow cultivation to aerate the soil. Other techniques which generally involve influencing the amount of absorbed short-wave radiation or the quantity of reradiated terrestrial radiation include the use of mulches; changing the absorptivity of the ground by cultivation, compaction, irrigation, or drainage; controlling the rate of evaporation; or varying air temperature by means of wind machines or shelterbelts. The effects of some of these treatments have been discussed earlier.

DECISION MAKING IN AGRICULTURAL CLIMATOLOGY

The foregoing studies of climatic effects on plants and animals are ultimately directed toward the key problem in agriculture: how can our knowledge be put to profitable use to aid the farmer in making economically wise decisions? Just because we know the relation between soil moisture and hay production, or hay production and milk yield, this information serves us little unless we utilize it in some decision-making process that will aid in the development of a more rational farm program. Improved technology in recent years, while increasing yields, has also greatly increased the farmer's fixed costs and influenced his profit margin. Farm machinery and irrigation equipment are expensive items and decisions on the extent of their use need sound economic justification. In recent years there have been a growing number of studies on economic aspects of weather forecasts, or of climatic control, and on the whole field of economics and risk taking in industry and agriculture. Weather and climate data are seldom given in a form directly usable in economic decision making, and few farmers know what questions to ask of the applied climatologist in order to be able to obtain fundamental information on the effect of alternate courses of action. We are only beginning to explore the whole field of weather and farm management decisions.

Duckham (1967),[1] in an informative discussion of decision making in agriculture, lists some of the basic information needed to quantify relationships among weather events, their biological and/or their operational effects, and the financial results of these biological or operational effects. Answers are needed to a number of pertinent questions, such as,

1   Is the enterprise or operation significantly weather dependent? If so, why, and to what weather factors at what stage or time, and to what degree? If a direct answer is not possible, is it still possible to quantify, empirically, the relationships between weather and both biological and operational effects? If these relations can be quantified, financial effects can usually be estimated without too much difficulty.

2   How reliable are the timing and intensity of the particular weather events and are they predicted on the basis of probability analysis or by actual weather forecasting?

3   What options are open either to avoid or to exploit weather events whose probabilities are generally known?

4   If questions 1 to 3 can be answered adequately, is it possible to express the answers in financial terms as part of a decision-making matrix?

Duckham illustrates the whole process by means of a partly hypothetical example of dairy farming on pastures in southeastern England. To provide a starting place for his decision-making matrix, Duckham sets up the following three possible operational programs, clearly all weather dependent: (1) very high stocking rates, very high nitrogen application, with irrigation, (2) high stocking rates, high nitrogen, without irrigation, and (3) medium stocking rates, low nitrogen, without irrigation.

---

[1] Material adapted with permission from Taylor, "Weather and Agriculture" (1967), Pergamon Press Limited.

On the basis of climatological data, Duckham determines the frequency of wet and dry years in 10 years (3 wet, 3 dry, and 4 fairly normal years). Milk yield in years with normal moisture is fairly well known, but milk yield in wet and dry years is more questionable. The yield effects of nitrogen additions, of timing of dry periods, or of irrigation applications are more difficult to quantify and so assumptions must be made. Based on these assumptions, as well as on estimates of the price of milk and costs for irrigation and fertilizer applications, Duckham shows that under wet conditions greatest return is from high stocking and no irrigation (program 2), while under normal and dry years, the greatest return results from very high stocking and irrigation (program 1). The mean annual profit expected over a 10-year period including three wet and three dry years is £11/acre under program 1, £4/acre under program 2, and £8/acre under program 3. The lower return from program 2 results from losses during the dry years when high stocking rates without irrigation are used.

Analyzing the stocking programs in relation to the various types of farmers, Duckham feels that each program may have its place. Program 1 requires management skill, considerable capital is risked in number of cows and irrigation equipment, and intensive farming is anticipated. Risks are great but so are returns for the skillful farmer. Program 2 has great profit instability since it has maximum return under wet years but losses under dry years. If purchase of food is economically possible during dry years, program 2 would be more feasible but large swings in profit would still exist. Program 3 has relative stability since profit per acre is the same in all years. Opportunities to increase profits through excellent management do not exist but risks are also minimized. Thus program 3 is excellent for a farmer not wanting to take risks or without the good credit needed for risk taking.

Because of the many unknowns this approach can offer little more than suggestions for decision making at present. The exercise is important, however, for it begins to focus attention on the questions to be asked, the quantitative information needed, the various alternative programs to be evaluated, and the measurements required. It is a large and challenging field for applied climatologists.

Kolb and Rapp (1962) attempted to evaluate the economic usefulness of forecasts in the raisin industry. Actually, their study was confined to the raisin-drying portion of the harvest, since this is most sensitive to weather conditions. The quality of the raisins is influenced by the length of the grape-growing season. Grapes harvested too soon have low sugar content and produce poor raisins, although such grapes will dry before any real risk of rain which might damage the drying raisins. Grapes left on the vines until their sugar content is higher have a considerably greater chance to be damaged or lost completely by rains during the drying season.

The drying period for grapes runs from August 20 through October 5. Three to four weeks without rains (or less than 2.5 mm) are needed. Grapes not set out to dry before September 15 are usually sold for other purposes because of the greatly increased chances of rain through September and October.

Assuming that the raisins can be protected by rolling the paper trays on which they are drying at a cost of $6/ton, while rain will cause a complete loss of the raisins valued at $200/ton,[1] the "operational risk" or cost/loss ratio becomes 6/200 or 0.03. Jorgensen (1949) showed that the average probability of rain for each day from September 1 to October 20 is 0.0222 (actually it varies markedly from 0.005 on September 14 to over 0.035 on October 1). Since the average chances for rain are less than the cost/loss ratio, the decision to protect or not protect the grapes might logically be made not to protect.

Based on the changing probability of rain and the need for a 4-week drying period, it is possible to determine the expected losses from rain as a function of the date drying begins. This shows an expected loss of just under $100/ton if drying starts on September 5, rising to almost $140/ton if drying begins on September 10, and $170/ton for a drying start on September 15. This is the economic reason for the suggestion that any grapes not in the drying process by then should be sold for other purposes.

Until this point, the study involved no weather forecast. Clearly, if perfect forecasts were possible so that protection would only be undertaken when there was a forecast of rain, costs for protection would drop (to about $3.70/ton/season) and grapes could be kept on the vines until high quality was assured. Lacking perfect forecasts, however, use of available forecasts results in some improvement in the results over those possible without forecasts. The expected cost (C) if the farmer protects every time there is a forecast of rain can be written as

$$C = 6N_r + 200\, N_{nR:R}$$

where N_r = number of rain forecasts made

$N_{nR:R}$ = number of times it rains when no forecast of rain is made

Obtaining quantitative information from this expression involves some assumptions about the skill of the forecaster. Kolb and Rapp prepared a table (Table 7-3) indicating the expected cost for a ton of raisins based on the number of rain forecasts made and the number of rains actually occurring when no rain was forecast. The probability of one rain in 28 days (4-week drying period) is 28 × 0.0222 or 0.62. Assuming a loss of $200/ton with this rain, the expected cost of never protecting the drying raisins would be 0.62 × $200 = $124/ton. The table provided by Kolb and Rapp is time independent as it does not take into account the rising probability of rain as the drying season progresses.

The right-hand column provides the expected cost to the farmer if he protects the trays of raisins every time rain is forecast. The other two columns show the number of forecasts of rain and incorrect no-rain forecasts during a 28-day drying period. The values in the left-hand column indicate a

[1] If many raisins are lost due to heavy rains, the price for the remaining raisins will rise and offset monetary losses to some extent. This is not considered in the example.

decreasing frequency of forecasting rain which, of course, is matched with an increasing frequency of missed rain forecasts in the middle column. Substituting these values in the previous cost equation results in the cost data in the last column. For the Fresno County area, a forecast of rain three or four times in the 28-day drying period should result in no times of missed rain forecast and provide the highest return on the raisin crop.

These two brief examples of economic decision making in agricultural climatology only suggest a small portion of the field open to the applied climatologist. It is a most relevant subject which has just been touched. It must receive increased attention in the future as farmers seek to maximize profits not only by using better farming techniques and better seed, but also by reaching better management decisions.

TABLE 7-3

Expected costs for 28-day drying period with forecast and nonforecast rain*

NO. OF TIMES RAIN FORECAST	PROB. OF RAINS OCCURRING WHEN NO RAIN FORECAST	EXPECTED COST/TON IF PROTECTED WHEN RAIN FORECAST
6.65	0	$ 39.90
3.59	0	21.54
2.44	0.07	28.65
1.39	0.17	42.34
0.86	0.25	55.16
0.48	0.33	68.88
0.00	0.62	124.00

* With permission of American Meteorological Society, after Kolb and Rapp, 1962, table 2, p. 11.

CLIMATE, CLOTHING, AND HUMAN COMFORT

8

Man is a complex mechanism; he produces heat in proportion to his activity and operating efficiency, and he must have heat-regulating facilities to ensure the adequate dissipation of that heat since his body can tolerate only a small change in temperature. Any upset in the heat balance of the body through exposure to very cold or very hot environments, or any breakdown in the heat-regulating mechanism of the body itself through disease or malfunction, may result in significant changes in the body's temperature, with consequent severe stress. The human heat balance, influenced as it is by the climatic environment, becomes an area of great concern to applied climatologists.

MAINTENANCE OF CONSTANT BODY TEMPERATURE
An inanimate dry object, not in the sun, will come to the same temperature as its surroundings if given time. If the object is exposed to the atmosphere and, at the same time, is wet, its temperature will tend to be lower than the air around it because of the heat removed in the evaporation process. Within the object, the temperature distribution will not be uniform but rather there will be a gradient toward the cooler, evaporating surface. While temperature is sufficient to characterize many environmental relations in the case of a dry inanimate object, temperature, humidity, and air movement are all necessary in the case of a wetted object.

If our object is a man or animal actively producing heat, environmental relationships are further complicated. The object produces its own heat and thus tends to raise its temperature above the surroundings. A temperature gradient is established from the interior to the surface which, of course, reinforces the gradient resulting from any surface evaporation. However, because of the internal heat production, the surface temperature may either be above or below that of the surroundings. A stabilized body temperature (homeothermy) is possible only when the receipts of heat by internal production and from the external environment equal the losses to the external environment by radiation, evaporation, and conduction-convection.

The three climatic factors of temperature (radiation), wind, and humidity are all involved in the relation of man with his environment. It is the combined effect of these climatic elements that determines man's state of balance or imbalance with the surroundings. But these elements all interact with one another so that changes in one are always accompanied by changes in the others. Each of these changes is accompanied by changes in the body's heat loss or heat production. As a result, the maintenance of homeothermy becomes a very complex process.

HEAT GAINS AND LOSSES

Evaluation of the heat balance of the body requires an accounting of all the body's heat gains and losses. If these do not remain essentially in balance, a danger signal is flashed and the human must change his rate of heat production or seek a modified climatic environment before serious stress develops.

Gains of heat to the body involve three possible areas of concern: conduction-convection, radiation, and metabolic heat production. Losses of heat from the body also involve three areas of concern: conduction-convection, evaporation-convection, and radiation. The general heat-exchange relationship between man and his environment may be expressed as

$$S = (M - W) + (\pm R \pm C - E)$$

where S = rate of heat storage in body

M = rate of total metabolic energy production

W = rate of external work performed, expressed as its heat equivalent [Forbes (1949) has indicated that about 10 to 15 percent of total energy production is utilized in occupations involving moderate to heavy manual labor]

R = net rate of heat exchange by radiation between man and his environment

C = net rate of heat exchange by conduction-convection between man and his environment

E = rate of heat loss by evaporation-convection to the environment

The total metabolic energy production (M) represents the net rate of conversion of potential energy into external work and heat by the body. It can be divided into four categories.

1   "Basal" metabolic rate: the rate at which energy is liberated under standard resting conditions (average value is 40 kg cal/m² hr).

2   Extra metabolism of food alimentation: the rate at which energy is liberated as a result of taking food. Usually 10 to 15 percent of the caloric value of ingested food will appear as heat in the following 3 to 4 hr.

3   Extra metabolism of exercise: the rate at which energy is liberated as a result of physical activity. Activities such as walking continually at a speed of 3 to $3\frac{1}{2}$ mph (5 km/hr) result in 100 to 140 kg cal/m² hr of extra metabolic heat production. Violent exercise may result in over 300 kg cal/m² hr extra heat production for short periods. Long-term increases in heat production (for 24 hr or more) due to exercise do not normally exceed 130 percent of the basal rate for fairly sedentary workers and 300 percent of the basal rate for heavy manual laborers.

4   Extra metabolism of shivering: the rate at which energy is produced as a result of shivering and muscle tensing. Adolph and Molnar (1946) have shown that shivering may produce as much as 170 kg cal/m² hr for a 2-hr period and up to 100 kg cal/m² hr for a 4-hr or longer period.

Average energy production for native peoples in cold and hot climatic environments without much in the way of clothing is estimated in Table 8-1.

Ten to fifteen percent of this energy production goes to perform external work (W), while the rest is produced as heat. Thus the total heat production by native peoples in a cold climate as a result of their daily activities might be about 230 kg cal/m² hr. Since man may be approximated by a cylinder 1.8 m in height and 0.3 m in diameter having a vertical surface area of about 1.75 m² (the horizontal surface area can be disregarded), this results in a total heat production by man of approximately 400 kg cal/hr in cold climates. It is this heat supply that must be augmented by radiation and conduction-convection to maintain the body's thermal equilibrium in cold climatic environments.

TABLE 8-1

Estimated average metabolic energy production for native peoples

METABOLISM	AV. ENERGY PRODUCTION KG CAL/M² HR	
	COLD CLIMATES	HOT CLIMATES
Basal	40	40
Extra:		
of food alimentation	Negligible (< 10)	Negligible (< 10)
of exercise	120	60
of shivering	100	0
Total metabolic energy production	260	100

Native peoples in hot climates have a total metabolic energy production of approximately 100 kg cal/m^2 hr, assuming that they reduce the extra metabolism of exercise by 50 percent to 60 kg cal/m^2 hr and eliminate the metabolism of shivering. This results in a heat supply of the order of 175 kg cal/hr for a 1.75 m^2 man that must, in addition to radiation and conduction-convection heat gains, be dissipated by various avenues of heat loss in hot climates.

Possible heat gains and losses by evaporation, conduction-convection, and radiation to balance the metabolic heat production of the body are outlined in Table 8-2.

THE ROLE OF CLOTHING

The foregoing discussion has generally considered man as a nude or seminude individual. In our modern society, it is seldom that man approximates this condition (except at the beach) so we must, perforce, be concerned with the influence that clothing plays in the relation between man and his surrounding environment. The problem becomes complex because of (1) the variety of fabrics and styles possible; (2) the effect of body movements on effectiveness of different clothing assemblies; (3) the effect of clothing assemblies on the different avenues of heat exchange between man and his environment; and (4) the lack of equilibrium conditions so that we are faced with interpreting the interplay of always-varying conditions in the environment, in human heat production, and in the interspersed clothing assemblies (through changing assemblies or through bodily actions).

Clothing can be considered to play two major roles: (1) protecting the wearer from adverse environmental conditions whether they be climatic (radiation, temperature, wind, or moisture primarily), mechanical (terrain effects or climatic-related conditions such as snow, ice, blowing dust), or biological (insects, reptiles, or other animals, irritation or thorny plants, or against the effect of other humans in the environment through the use of helmets or arms); and (2) satisfying some individual or societal psychologic function relating to decoration, ego building, or modesty. It is interesting to note that while the protective function of clothing was probably the first reason for its use, the psychologic function is now more significant in our continued use of clothing. With efficient heating and cooling systems in home, car, and office, with cars and other means of rapid transportation from place to place, with foodstuffs conveniently available to us at the corner supermarket, modern man is seldom forced to venture out into an alien environment for too long. Clothing for protection, except under special and limited conditions, plays a less significant role than does clothing for decoration, modesty, and self-satisfaction. This is, in a way, unfortunate because as we pay less attention to the protective aspect of clothing, or its role in moderating climatic stress, we place more reliance on mechanical means (heating and air-conditioning systems primarily) for maintaining the proper heat balance between man and his environment. Increased concern for the protective role of clothing would undoubtedly result in a more economic adjustment to stressful environmental conditions. (For example, the addition

TABLE 8-2

Heat exchange processes of the body

HEAT GAINS	HEAT LOSSES
1 Condensation of moisture onto the body directly from the atmosphere seldom occurs since the body would have to be at a lower temperature than the moist air. Can be disregarded	1 Evaporation of moisture from perspiration or applied moisture depends on percentage of body covered by moisture (which may vary from less than 10 percent to 100 percent) and the vapor pressure of the air. Evaporation cooling will be retarded by clothing (evaporation from wet clothing will only indirectly cool the body) and speeded by convection which will remove moist air from vicinity of body
2 Heat conduction-convection toward body: (*a*) By contact with objects at above skin temperatures: effective where external surfaces—ground, buildings, etc.—are warmed above skin temperature and will thus conduct heat to cooler body (*b*) From air whose temperature is above skin temperature: effective in very hot environments where heat is added to body from hot environmental air. Rate of heat addition influenced by convection (air movement)	2 Heat conduction-convection away from body: (*a*) By contact with objects having below-skin temperatures. Generally more effective than item 2(*a*) under Gains (*b*) To air whose temperature is below skin temperature. In still air, heat transfer either to or away from body is slow but the process speeds appreciably with any air motion
3 Heat gain by radiation to body: (*a*) From sun either as short-wave radiation direct or scattered from clouds, sky, atmospheric particles, or terrain. Amount of surface presented by man depends on angle of sun and posture of man. Amount absorbed by man depends on color of skin and clothing: white colors reflect about 90 percent of visible and 60 percent of short infrared; black colors reflect 15 percent or less of visible and 40 percent of short infrared (*b*) Thermal radiation (long-wave) from hotter surroundings. Color is of no importance in long infrared radiation exchanges although certain polished metal surfaces will result in high reflectivities. Other surfaces are almost completely absorptive—no reflectivity	3 Heat loss by radiation from body: (*a*) To colder surroundings. Net long-wave infrared radiation loss to environment due to higher skin temperatures. (*b*) Net long-wave infrared radiation loss to "sky." Often neglected as an avenue of heat loss, this radiation loss depends on the temperature difference between the skin and the so-called effective atmospheric temperature. Without atmosphere, temperature of outer space ($-273°C$) would result in very rapid radiation heat loss. With atmosphere, temperature of "sky" becomes some intermediate value. In warm, humid regions this temperature is near ground temperature, but in hot, dry environments it can be somewhat lower and so constitute the only channel of radiation loss

of a sweater to an individual during a short period of cold conditions is economically more justified than turning up the thermostat and warming a whole house to conditions that would be comfortable without a sweater.)

The present discussion is only concerned with the role of clothing as protection against the climatic environment since this is where the applied climatologist can contribute to the clothing problem. Other aspects of clothing are important but are clearly beyond the limits of this study.

The effect of clothing is to interfere in some way with the different avenues of heat exchange between man and his environment by radiation, convection, conduction, and evaporation. Various formulas and relations have been derived from theoretical considerations and actual measurements on human subjects (or replicas of humans). These provide estimates of the order of magnitude of heat gains and losses by the different avenues of exchange. A brief review of these relations follows.

Radiation (R)[1]

The radiation term in the heat-balance relation for man involves both the receipt of short-wave solar radiation and the exchange of thermal or long-wave radiation between the human and the surrounding environment. All bodies radiate to a certain degree. The net rate of heat exchange by radiation between man and his surroundings is merely the difference between the radiation emitted by the body and that received from the sun, sky, and surrounding terrain. It is proportional to the differences in the fourth power of the respective temperatures.

For an individual within a closed environment, such as in a room, the radiation-exchange relation becomes somewhat more complex because of the need to include the effect of any clothing assembly, body geometry, the various gains and losses of heat by radiation to different objects with different temperatures, and the possible effects of movement by the individual or by the air itself (which might change surface temperatures). In this situation, a quantity known as the mean radiant temperature (MRT) of the environment has been suggested. The MRT is nothing more than the mean temperature of all the surfaces surrounding the individual, weighted on the basis of their emissivities. Since the environment is hardly ever homogeneous but rather is made up of many different objects with different temperatures and emissivities, the MRT is experimentally determined from knowledge of air movement and the temperature of a Globe thermometer, an ordinary mercury-in-glass thermometer inserted within a blackened copper sphere whose diameter approximates 10 to 15 cm.

Belding (1967) has provided the following formula for computing the MRT from knowledge of the Globe thermometer temperature and air speed:

$$MRT = t_g + 0.24V^{0.5} (t_g - t_a)$$

[1] A more complete discussion of the nature, measurement, and distribution of radiation is included in Chap. 2.

where V is given in m/sec, and t_g and t_a, the Globe thermometer and air temperature, respectively, are in °C.

With a clothed individual, it becomes necessary to evaluate the radiation exchanges between (1) the individual and the clothing; (2) the clothing and the outside environment; and (3) the exposed skin surfaces and the outside environment. Because of the unknown temperatures involved and the complicated geometry of the body and clothing surfaces, it is essentially impossible to make a detailed evaluation of all the elements of the radiation exchange in this situation.

Givoni (1969) has suggested a much more complex relation by which the heat load as a result of solar radiation can be determined. It involves the inclusion of many empirically derived coefficients and is included here merely to show the nature of the factors that must be considered and how they might vary for an individual under different terrain, posture, and clothing conditions.

$$R = I_N \, K_{pe} \, K_{cl} \, [1 - a \, (V^{0.2} - 0.88)]$$

where R = solar radiation, kg cal/hr

I_N = normal solar intensity, kg cal/hr

K_{pe} = coefficient that varies with posture and terrain. Equals 0.386 for an individual sitting with his back to the sun in a desert; 0.379 for the same individual in a forest; 0.306 for an individual standing with his back to the sun in a desert; and 0.266 for the same individual in a forest.

K_{cl}, a = clothing coefficients. K_{cl} equals 1.0 for a seminude individual; 0.5 for one clothed in light summer clothing; 0.4 for one in industrial coveralls. a equals 0.35 for a seminude individual; 0.52 for one lightly clothed or with industrial coveralls.

V = wind speed, m/sec

Clothing seriously interferes with the transmission of short-wave radiation to the skin. While the amount of transmission depends on the nature of the interspersed material and the wavelengths of radiation, it is generally considered that no more than 5 percent of the short-wave radiation will be transmitted through average street clothing. Depending on its color and texture, clothing will also reflect a portion of the incident radiation. The proportion of radiation neither transmitted nor reflected will be absorbed by the clothing, changed into sensible heat, and used to raise the temperature of the clothing. This sensible heat will, of course, be transferred by conduction inward to the body or by conduction, convection, and radiation outward to be lost to the atmosphere. The net effect of clothing will, therefore, be great at first in protecting against a radiant heat load. This effect will be reduced in time as the clothing assembly gains heat.

Burton and Edholm (1955) have shown that any type of heating is more efficient if the source of the heat is located inside the clothing insulation layer surrounding the body. This insulation layer can consist of both a layer or layers of clothing as well as a layer of still air near the skin. The relative thickness of the still layer of air depends on the presence and nature of any overlying clothes layer and the wind velocity. These authors showed that

$$H_{eff} = \frac{H I_{ext}}{I_{total}}$$

where H_{eff} = effective heating at the skin

H = heat generated at some layer within the insulation

$\dfrac{I_{ext}}{I_{total}} =$ ratio of insulation outside heat source to total amount of insulation

Fabrics which permit penetration of scattered radiation into the deeper layers of the material will be heated more effectively by the radiation. Pratt (1952), using both light- and dark-colored pile fabrics of the same characteristics, determined that while the light-colored fabric resulted in more reflection outward (higher albedo), it also resulted in more scattering of radiation inward before it was absorbed by the material. This made the heating efficiency of the radiation more effective for the light-colored fabric because the layer at which energy is converted into sensible heat is further within the insulating material. This effect can be significant. Woodcock et al. (1955) reported one case in which the warming effect of radiation absorption within the clothing assembly was equivalent to a rise in ambient temperature of 18.3°C in a 1.8 m/sec wind. Woodcock (1964) has pointed out that many Arctic animals have lighter fur on their backs than on their undersides. Often, darker fur is found under the light-colored fur to absorb the radiation that penetrates inside the insulating layer. He states that while fabric color is important in terms of reflection, the texture of the outer surface is probably of greater importance in determining the place within the fabric at which absorption of energy occurs.

Conduction-Convection (C)

Heat transfer through a medium without the physical transfer of material is called "conduction," while heat transfer involving the physical transfer of a liquid or gas is known as "convection." In a fluid medium, such as air, both conduction and convection will occur at the same time so that it is difficult, if not impossible, to measure the effect of each of these processes separately.

Heat exchange by conduction-convection actually depends upon several factors:

1   Internal transfer of heat between the deep body tissues and the surface by thermal conduction and by blood circulation through the body and skin. The rate varies considerably with species and circumstances.

2   Surface transfer of heat. The rate of transfer depends on the surface transmittance coefficient (rate of transfer per unit area per unit difference in

temperature per unit time) and is markedly affected by the rate of air movement.

3   External transfer or the rate at which heat is removed from the immediate surroundings in contact with the body. This is primarily a function of convection or air movement.

The character of the surface, especially its roughness and extent of hair covering, influences the heat-transfer process. The greater the surface roughness and the amount of hair covering, the smaller will be the surface transmittance coefficient and the external transfer of heat.

Heat is transferred from one body to another by conduction at a rate proportional to the temperature difference between the bodies. In the case of a nude individual, the temperature difference is between the skin surface and the surrounding air. Gagge, Burton, and Bazett (1941) introduced the term I, the insulating value of clothing. They found that the insulating value of various layers of clothing is additive so that the total insulation value of a clothing assembly can be thought of as the sum of the insulating value of layers 1, 2, 3, ... (thus $I_1 + I_2 + I_3 + \cdots = I$). A new unit, the "clo," was introduced by these scientists in an effort to make the concept of thermal insulation more understandable. One "clo" unit was made equivalent to the insulating value of a clothing layer that permits 1 kg cal/m^2 hr of heat flow with a temperature difference of 0.18°C across the fabric. More important for the layman, 1 clo unit corresponds closely with the amount of insulation provided by an ordinary wool business suit. While the numerical definition may be strange, it is relatively easy to grasp the concept of the insulation value of a business suit and even to have a feeling for the insulation value equivalent to two or three business suits (2 or 3 clo). Clo units will be discussed more fully in a later section (p. 232).

The relation for the heat loss by conduction from the general body surface may, thus, be written as

$$H = \frac{5.55 \, (t_s - t_a)}{I_a + I_{cl}}$$

where H = rate of heat loss, kg cal/m^2 hr
       t_s, t_a = skin surface and air temperatures, respectively, °C
       I_a, I_{cl} = thermal resistivity of ambient air and clothing, clo units

The thermal resistivity of the air varies approximately as the inverse square root of the rate of air movement.

Often the radiation and conduction-convection terms are combined into one term $(R + C)$ which represents the dry heat exchange with the environment. The 1972 ASHRAE Handbook expresses this dry heat loss as

$$R + C = h_r(\bar{t}_{cl} - \bar{t}_r) + \bar{h}_c(\bar{t}_{cl} - t_a)$$

where h_r, h_c = linear radiation and mean convection heat-transfer coefficients applicable in the actual environment, respectively

$\bar{t}_{cl}$ = mean temperature of clothing surface, °C

t_r = mean radiant temperature of environment, °C

t_a = ambient air temperature, °C

This expression can be simplified by introducing the operative temperature (t_0), the average of the mean radiant and ambient air temperatures weighted by their respective heat-transfer coefficients, as follows:

$$R + C = h(\bar{t}_s - t_0)F_{cl}$$

where $h = h_r + h_c$

$\bar{t}_s$ = average skin temperature, °C

$F_{cl} = h_{cl}/(h + h_{cl}) = I_a/(I_a + I_{cl})$

Conduction-convection heat exchange also occurs from the respiratory tract. The relation may be written as

$$H_l = 0.000154q(t_l - t_a)$$

where H_l = net rate of conduction-convection loss from respiratory tract, kg cal/m^2 hr

q = volume of air respired, liter/hr (approximately 10 times total metabolic rate M)

t_l, t_a = temperature of lungs and ambient air, respectively, °C

The use of clothing clearly interferes with the removal of heat by convection from the body. If all air movement could be eliminated beneath the clothing assembly (perfectly windproof clothing), then heat would only be transferred from the body by radiation, conduction, and evaporation and the insulation efficiency of the clothing assembly would be greatly increased. However, wind will penetrate most windproof clothing to some degree and reduce its insulation efficiency. Activity by the individual sets up convective currents within the clothing, which will transfer heat to or away from the body and hence reduce its insulation effectiveness.

The use of two clothing layers with an air space between them eliminates much of the need to find the most wind-resistant materials. Fonseca et al. (1959) have shown that if the wind does penetrate the outer layer of clothing it will much more readily pass around the body to the downwind side through the air space and move out through the outer clothing layer again without ever penetrating the inner clothing layer. This is significant for even if it were possible, it would not be desirable to have completely windproof clothing. Some permeability to air is necessary. With two-layered clothing, more permeable fabrics can be used without losing the insulation effectiveness of the relatively still air layer near the inner clothing layer.

Evaporation-Convection (E)

Evaporation is the process by which liquid water is converted to vapor with a resulting loss of heat from the evaporating surface. The evaporation of perspiration is a major avenue by which the body exchanges heat with its surroundings. Since evaporation from a surface will almost always lead to the formation of convective currents, it is difficult to consider the exact effects of each process separately.

As long as moisture is present, evaporation-convection is dependent on (1) the temperature of the evaporating surface, (2) the relative humidity of the air in contact with the skin, and (3) the rate of environmental air movement. Since evaporation occurs from the skin as well as from the surface of the respiratory tract, both these surface temperatures need to be considered. The former is usually somewhat below body temperature while the latter is at or near body temperature. It is possible to consider the air in contact with the surface to be at the skin temperature. Skin temperature varies only slightly in the warm environments in which most evaporation occurs, so that absolute humidity rather than relative humidity can be considered to be the important governing value.

The efficiency of cooling as a result of sweat evaporation depends on the source of the heat for vaporization. Heat for evaporation in the respiratory tract or in the pores in the skin can be assumed to come almost entirely from the body and so it is highly effective in cooling the body. Evaporation from the body surface, especially when the rate of sweat production is high in relation to evaporation (resulting in a thicker layer of sweat on the body), will be less effective in cooling the body since some of the heat of vaporization will come from the air. The physical efficiency of sweat evaporation may be defined as the ratio of heat removed from the body by evaporation to the potential cooling of the evaporation (Givoni, 1969, p. 27). Givoni and Belding (1962) have emphasized that the ratio of the evaporated sweat (e) to the maximum evaporative capacity of the air (E_{max}) determines the rate and place of evaporation. As this ratio increases, the surface of evaporation occurs farther from the skin surface, and cooling efficiency decreases. Belding (1967) has suggested use of the relation

$$E_{max} = kV^{0.6}(42 - VP_a)$$

to determine the maximum evaporative capacity of the air in kg cal/hr man. The vapor pressure of the air (VP_a) is given in mm Hg and air speed V is in m/sec. The coefficient k varies from 2.0 for seminude individuals to 1.33 for lightly clothed man. The value 42 is the vapor pressure at the wet skin surface (in mm Hg).

Evaporation-convection heat loss to the surroundings may occur from the respiratory tract and from the general body surface. For the loss from the respiratory tract, the expression may be written

$$E_l = K_{el} q(p_l - p_a)$$

where E_l = rate of heat loss by evaporation-convection from respiratory tract, kg cal/m² hr

K_{el} = constant (approximate value 0.00031)

q = volume of respired air, liter/hr

p_l, p_a = water vapor pressure in respiratory tract and ambient air, respectively, mm Hg

The relation for the evaporation-convection loss from the general body surface may be written as

$$E_s = \frac{K_{es}\,(p_s - p_a)}{r_a + r_{cl}}\quad W$$

where E_s = rate of heat loss by evaporation-convection from body surface, kg cal/m² hr

K_{es} = constant (approximate value 5)

p_s, p_a = water vapor pressure at skin and air temperature, respectively, mm Hg

r_a, r_{cl} = resistivity of ambient air and clothing to outward passage of water vapor (equivalent cm of still air)

W = proportional wetness of skin (varies from 0.05 to 1.0; assumed to be 0.1 for individuals not exercising vigorously)

The expression $r_a + r_{cl}$ is of the same order of magnitude as the expression $I_a + I_{cl}$. For purposes of rough calculations these terms are often considered equal.

The 1972 ASHRAE Handbook has provided a slightly different expression for the total evaporative heat loss from the human body as a function of the rate of weight loss from the body ($\Delta w/\Delta t$), where w is in grams and t is in minutes:

$$E = \frac{60\lambda\,\Delta w}{A_D\,\Delta t}$$

where E = total evaporative heat loss from body, W/m²

λ = heat of vaporization of water, 0.7 W hr/g

A_D = Du Bois surface area of body, m² (for average size man 5'10", 170 lb, A_D = 1.8 m²)

E can be divided into two parts, as before: the evaporative heat loss due to respiration E_{res} and the evaporative heat loss from the skin surface E_s. The heat loss due to respiration is proportional to the rate of air movement through the lungs, which is, of course, related to the metabolic rate. Fanger (1967) has developed the following relation for E_{res} (W/m²):

$$E_{res} = 0.0023M(44 - P_{dp})$$

where P_{dp} = saturated vapor pressure, mm Hg, at dew point temperature

M = metabolic rate

The evaporative heat loss from the skin can be determined by subtracting E_{res} from the total evaporative heat loss E. If the entire skin surface is assumed to be wet, it can be shown that

$$E_{s\,(max)} = \frac{2.2\,h_c(P_s - P_{dp})}{1 + 0.143h_c\,I_{clo}}$$

where P_s = saturated vapor pressure, mm Hg, at skin temperature
   I_{clo} = insulation of clothing, clo units
   h_c = convective heat-transfer coefficient, W/m^2 °C

For a nude individual, the denominator becomes unity.

Evaporation is an effective way of losing heat since more than 0.5 kg cal of heat is lost for each gram of moisture evaporated. This loss can occur, however, only if moisture is constantly provided. A healthy nude individual will be able to lose enough heat through the evaporation of sweat to keep the body in thermal equilibrium under all but the most severe heat-stress conditions. However, the introduction of clothing interferes with the rate of cooling by evaporation in several ways. First, it reduces the wind effect on the body itself, thus lessening the opportunities for the moist air near the surface of the skin to be replaced by drier outside air. Second, clothing will absorb the moisture from the skin, thus increasing the amount of evaporation from the clothing assembly. While this cools the clothing, in turn, having an indirect effect on the cooling of the body, it is not as effective as direct evaporation from the skin itself.

The rate of cooling by evaporation in the presence of clothing is determined by the humidity or vapor pressure and the resistance of the clothing assembly. For each combination of humidity and clothing, there is a maximum evaporative heat-transfer possible. If the needed heat loss by evaporation is below this maximum, then the actual amount of sweating provides the limiting value of cooling. This is generally true in tropical areas where clothing assemblies offer little resistance and sweating can usually bring into adjustment the other factors of heat gain and loss. In cold climates, however, the heavy clothing normally worn imposes a severe limit on the heat loss by evaporation. It is quite possible that activity will force the individual so clothed to produce copious amounts of sweat which are absorbed into the clothing assembly, providing very little cooling to the body itself. This moisture may freeze in the clothing assembly as it approaches the cold outside environment, resulting in difficulty in movement and some loss in thermal insulation due to the presence of the ice itself. More significantly, the later evaporation of the moisture from the wet clothing when activity is reduced will result in what has been called an "after-exercise chill." The heat loss through this evaporation will cool the clothing and, in turn, the body at a time when little heat is being produced. It may result in a significant cold stress.

As previously indicated, the clo unit was introduced by Gagge, Burton, and Bazett (1941) to define the insulation value of clothing that would allow a heat flow of 1 kg cal/m^2 hr with a temperature gradient of 0.18°C across the fabric. An ordinary wool business suit has an insulation value of approximately 1 clo. Another unit, the thermal ohm (T-Ω), has also been used by some scientists. The T-Ω (in °C m^2/W) is the thermal resistance of the clothing material (with the temperature difference expressed in °C) that permits heat energy to pass through at a rate of 1 W/m^2. Thus 1 clo = 0.155 T-Ω or 1 T-Ω = 6.45 clo units. The advantage of the T-Ω unit is that it specifies the amount of electric heating needed. Another quantity, the Tog unit, has also been defined. This unit is just one-tenth the size of the T-Ω unit. Thus, 1 Tog = 0.645 clo (equivalent in insulation to light summer clothing) or 10 Tog = 1 T-Ω = 6.45 clo insulation units. The clo unit still seems the most easily understandable.

It is generally accepted that a normal nude individual is comfortable under light wind conditions at 86°F (30°C). With a normal business suit (1 clo) this same individual will be comfortable at 70°F (21°C). Roughly speaking, therefore, a drop in ambient temperature of 16°F (9°C) requires one additional clo unit of insulation. This would mean that at 54°F (12°C), 2 clo units would be needed and at 38°F (3°C), 3 clo units would provide comfort. Continuing to 5 clo units of insulation, approximately the maximum that can now be attained with any clothing assembly that still permits mobility, this would indicate that a resting individual could be comfortable at 6°F (-14°C). Below this value some additional heat production by the individual would be necessary for comfort. Figure 8-1 provides the general relation between ambient temperatures and insulation required for comfort for different activities. The values are not precise because of the many sources of uncertainty resulting from such things as intensity of the activity, cut of clothes, and varying thickness of clothing over different parts of the body.

Various investigators have measured the insulation value of different fabrics and combinations of fabrics both in the laboratory under controlled conditions and in nature. In still air, practically all fabrics such as wool, and some types of cotton or fur, give an insulation value of approximately 4 clo units per inch (25 mm) of thickness. It is the still air trapped within the fabric that seems to provide the insulation rather than the nature of the fabric itself. Burton (1943) determined the insulation value of fine steel wool as opposed to ordinary wool and found only a 12 percent reduction for the steel wool. Very little heat is evidently conducted along the fibers themselves so that the nature of the fibers in the assembly is relatively inconsequential. The ability of the material to retain or hold air in a relatively still condition, however, is of great significance.

In spite of this finding, most people have experienced differences in warmth between different clothing materials even though their thickness may be the same. This difference seems, in part at least, to be related to problems

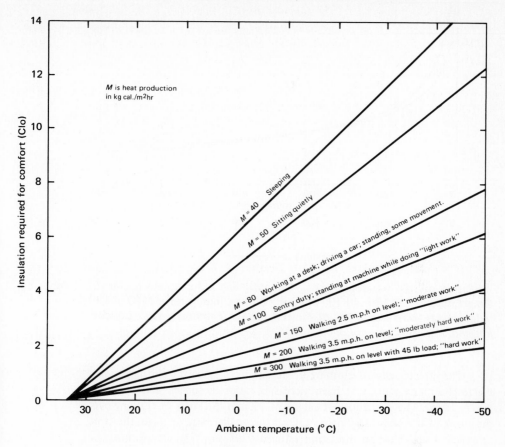

Prediction of total insulation required for prolonged comfort at various   FIG. 8-1
activities in the shade as a function of environmental temperature. (*With
permission of the W. B. Saunders Company, Philadelphia, from Belding,
1949, p. 353.*)

of air movement through the fabric or to the influence of moisture in the
fabric under varying conditions of wind or fabric compaction.

Cotton and wool react differently to moisture. Cotton becomes limp
and is easily compressed when moist, while the wool fibers do not change
their ability to hold their original thickness and to serve as a trap for air.
Cotton fibers, packed more tightly together, will provide a more satisfactory
barrier to air movement than will wool. Siple (1944) points out that in a
strong breeze a thin cotton windbreaker may be more effective and provide
more insulation than a woolen sweater of greater thickness which the breeze
can penetrate. The sweater would be the better insulator in still-air
conditions.

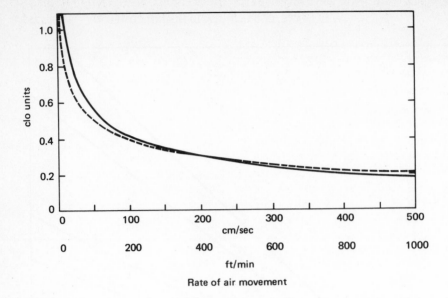

FIG. 8-2

Insulation of still air around men. The solid line is for an environmental temperature of -40°C (-40°F), while the broken line is for a 25°C (77°F) temperature. (*With permission of Edward Arnold (Publishers) Ltd., London, England, from Burton and Edholm, 1955, p. 51.*)

    The effectiveness of clothing insulation decreases as the clothing covers the smaller body parts that have increasing ratios of surface area to radius (Siple, 1944). The insulation efficiency of 12 mm thickness of clothing over the trunk of the body is 99 percent. Over the smaller head or thigh, the same thickness would have an insulation efficiency of only 95 percent, over the hands or feet it would decrease to 80 percent, and finally over the fingers its efficiency would become only 70 percent. With greater thicknesses of clothing these differences are magnified. As more insulation is added to the finger, there is a greater area through which heat is transferred and from which it can be radiated or removed by convection. Increasing clothing thickness, therefore, will soon result in the condition where the small addition to effective insulation is outweighed by the increased weight and thermodynamic problems, as well as loss of ability to move.

    In the atmosphere directly over a surface, there is a layer, of varying thickness, of relatively still air that will provide an additional layer of insulation. The thickness of this layer of relatively still air and, thus, its effective insulation is a function of wind speed, being equivalent to about 0.8 clo for nearly still air, to 0.4 clo with wind speed of 1 m/sec, and 0.19 clo with a 5 m/sec speed (see Fig. 8-2).

CLOTHING PROBLEMS

Clothing should permit regulation of the heat exchange between the individual and his environment. Thus it must have the dual role of reducing heat loss in cold conditions as well as preventing heat gain (or interfering with heat loss) under hot conditions. There is a strong tendency for the uninitiated to dress too warmly for activity under cold conditions, to wait too long before removing clothing layers, and to wait too long before adding additional layers after activity has stopped.

The insulation value of clothing may be varied through the use of loose clothing. With activity, the clothes will billow, allowing free exchange of air with the outside and promoting the removal of any heated air. This will cool the individual and prevent the build-up of sweat in the clothing assembly under cold conditions. Under conditions of rest, the loose-fitting clothing may be pulled more closely around the individual. The trapped air will serve as effective insulation; the lack of movement will keep the air confined and prevent its free exchange with the outside.

Such clothing will have some disadvantages under windy conditions for the looseness of the fit will allow some penetration of wind between clothing layers or between the clothing and the skin. If this air is cold, and if activity is minimal, such ventilation could result in significant cooling. This disadvantage is evidently not too serious, for this type of clothing assembly has been used successfully for centuries by the Eskimo under very severe cold conditions and the Arabs under very hot conditions.

MOISTURE IN CLOTHING

Most basic texts report that an increase in moisture in clothing will reduce its insulation value since water is a better heat conductor than air. Subjective reports of the extreme feeling of cold that accompanies a damp, overcast, windy day might suggest that the argument is valid. However, study of the problem shows that water drains out of the air spaces in clothing; the remaining moisture is not sufficient to reduce the insulating properties of clothing to the degree necessary to explain the measured heat losses. Woodcock and Dee (1950) have shown that the observed heat loss could be explained by assuming no change in the dry insulation properties of the clothing (even though the clothing is wet) and by attributing the additional heat loss wholly to evaporation from the wet clothing. Tests on nude individuals fail to indicate any real differences between damp cloudy and dry cloudy days, although the feeling of increased cold is verified by surveys of clothed individuals.

Woodcock (1964) discusses in some detail the question of absorbent and nonabsorbent clothing. He points out that, at low temperatures, moisture produced by the body through exercise will be absorbed into absorbent clothing but it will also condense onto nonabsorbent clothing. Later evaporation of this moisture will remove heat and produce cooling of both the clothing and the individual. Thus it makes little difference which type of clothing is selected. However, he points out, the relative absorbency of clothing might make some difference depending on the type of activity of the wearer. When an individual goes from a warm, dry environment out into a cold, damp environment, moisture is absorbed into his clothing. Heat of absorption is produced, warming his clothes and providing slightly greater initial protection against the cold. The individual will, of course, lose this moisture (and some heat) as the moisture evaporates when he returns indoors, but this is less important since he is again in a warm, dry environment. Absorptive clothing smooths out, at least partially, the temperature variation between indoors and outdoors. For the individual who must remain for long periods in a cold environment but who might be changing his activity and sweat production, Woodcock (1964) feels that nonabsorbent clothing is probably best since the after-exercise cooling due to evaporation will be less. Under cold conditions there is a great advantage in keeping the layers of clothing near the skin dry to reduce evaporative heat loss. Not only does this argue for proper ventilation to prevent sweating and, hence, moisture accumulation in the clothing, but it also suggests why an undergarment such as the fishnet or string vest is so effective. This garment is almost entirely holes which cannot retain moisture. The inmost layer of clothing and air is largely dry, only the net or string segments being wet. The air trapped within the mesh also acts as an effective insulating layer.

PROBLEMS OF THE EXTREMITIES

Head, hands, and feet all create special problems for individuals exposed to cold conditions. Because they are extremities, blood flow reaching them is often restricted due to vasoconstriction of the peripheral blood vessels so that the temperature of these areas can fall well below body temperature. Fingers are almost perfect cylinders with large surface areas for the volume contained, thus making them good heat exchangers. They are, however, difficult to protect from the cold, for when covered with several layers of cloth and insulation they become almost inoperable as fingers; the same can be said for the foot. Actually only about 6 mm of insulation can be placed around fingers if they are to maintain any mobility. This results in about 1 clo of insulation, hardly adequate for protection under extremely cold conditions. Thus gloves with fingers in them must be replaced with mittens in which fingers can be combined for added warmth and increased protection.

There are special problems with feet especially if a cold, wet environment is encountered. Rubber boots or some other water-barrier layer will keep out the moisture; the same layer will prevent moisture evaporated from the foot from getting out. Sweat absorbed by the insulation inside the boot will be trapped there until the boot is removed. Under cold conditions this can lead to the itching, burning, swollen foot problem of trench foot even though environmental moisture has not entered. Insulation sealed between two vapor-barrier layers in boots will prevent moisture from reaching it from either inside or outside and eliminate most of the trench foot problem. New developments in water-repellent materials that will exclude outside moisture but allow water vapor from inside to pass through offer real promise for future improvements.

When the outside temperature is well below freezing, there is no need to be concerned with the vapor barrier to prevent entry of environmental moisture. Under these conditions insulation against the snow or ice and extremely low temperatures is the main concern.

Possibly the greatest heat loss is through the head area. Burton and his associates have suggested that at 60°F (15°C) about one-third of the body's heat production can be lost from an unprotected head. At 5°F (-15°C), approximately three-fourths of the body's heat production is lost from the head, while at -40°F (-40°C) the unprotected head can lose all the heat an inactive man produces. Insulated hats with flaps down over the ears and possibly a covering over the face are definitely required at very cold temperatures to prevent undue heat loss and possible frostbite of cheeks, nose, or ear areas.

The breathing of extremely cold air will increase heat loss from the body appreciably. For example, at a temperature of -40°C Liopo (1966) has estimated that about 20 percent of the total amount of heat lost from a resting body will be through normal respiration. Under limiting conditions of cold, it is desirable to reduce this avenue of heat loss by respiratory preheating techniques. Mixing the incoming cold dry air with some exhaled air not only will warm the air taken into the lungs but will also increase its moisture content. This reduces the heat and moisture loss from the lungs and conserves bodily heat. Various face or nose masks are available to permit some degree of respiratory preheating.

CLOTHING DESIGNED FOR TROPICAL AND ARCTIC CONDITIONS
By way of summary, it is worthwhile to list briefly the desirable characteristics of clothing designed for wear in the Arctic and Tropics. Since either wet or dry conditions (wet or frozen in the Arctic) may exist in these areas, it is necessary to consider four different climatic combinations. Table 8-3 provides basic design characteristics that should be incorporated in any clothing assembly for wear under the postulated climates. Conditions of wear, or activity, and the nature of the nonclimatic external environment might suggest slightly different clothing characteristics.

Other considerations in addition to climate enter into clothing design and selection. For example, there is the need to provide clothing for protection against environmental hazards whether they be mechanical (rocks, jagged ice, burning sand, etc.) or plant and animals (thorns, snakes, etc.). Certain pieces of clothing, such as leggings and heavy boots, are needed in all climatic areas to provide such protection. Special uniforms and helmets are also required for the protection of soldiers regardless of climatic conditions. There is usually a need for pockets for carrying essential items regardless of the climate. Belts or suspenders must be used if trousers are to be worn. Openings must be supplied to permit the carrying out of normal bodily functions without having to remove too many garments.

There are also psychological aspects of clothing which often override other considerations. These include such things as style, fit, and smartness. If custom dictates a certain style or fit, they are often accepted even though the wearer becomes less comfortable within a particular climatic environment. Economic considerations must also be evaluated. The most desirable clothing assembly for protection against an inhospitable environment may be too costly for general use. A less expensive assembly may be accepted even with the loss of climatic comfort. The clothing assembly accepted will certainly be, in part, dictated by the cost and availability of the material or assembly. While all these considerations enter into the final decision as to what clothing is selected, only the climatic aspect is of immediate concern to the applied climatologist.

HUMAN COMFORT: COMFORT INDICES

One major preoccupation of "human climatology" is to understand the effect of a given set of climatic conditions, "the climate," on the comfort of the individual. To do this, it is first necessary to have some index or expression of comfort. What is comfort, when is the individual in equilibrium with his environment, and when is environmental stress minimized? These are most difficult questions to answer because of the number of variables involved, the difficulty of precise measurement, the importance of physiological as well as psychological factors, and the effect of secondary reactions.

A wide range of external and internal factors influences comfort. Comfortable conditions for one may not necessarily be comfortable for another. What is comfortable at one time to an individual may not be comfortable at a later time to the same individual. Many studies of "human comfort" have been undertaken. Reactions to different combinations of

TABLE 8-3

Clothing design characteristics for extreme climatic areas

HOT, DRY CONDITIONS	HOT, WET CONDITIONS
Close weave	Open weave
Light color to reflect radiation	Color immaterial
Moderate thickness	As thin as possible
Body well covered, head covering	Minimum coverage
Loose fit	Well-cut; good fit
Some underclothes desirable	Minimum underclothing

COLD, DRY (FROZEN) CONDITIONS	COLD, WET (UNFROZEN) CONDITIONS
Layered clothing, small air spaces	Layered clothing, small air spaces
External vapor barrier unimportant	External vapor barrier necessary
Absorptive layer next to skin	Absorptive layer next to skin
Head covering	Head covering
Respiratory preheating desirable	Respiratory preheating unnecessary
Fairly loose fit, bellows action or with vents to prevent overheating	Fairly loose fit, bellows action or with vents to prevent overheating
Face mask to prevent frostbite	Face mask unnecessary
Light colored with darker layer beneath to permit absorption of radiation within clothing	Light colored with darker layer beneath less important

climatic conditions by a large number of human subjects of different ages and sexes, dressed in various clothing assemblages, and exposed both indoors and outside, have been determined. All investigators realize the difficulties in such experimentation, yet the general agreement in results suggests that the concept of a range of conditions within which the majority of human subjects feel in comfortable adjustment to their climatic environment is valid and useful.

Body temperature is not a useful index of comfort since the body seeks to prevent any change in body temperature. Sweat loss is sometimes used as an index but it tends to depend on the nature of the environmental stress rather than on the total thermal effect of the stress. It does have certain advantages, however, in that it effectively combines the influence of different environmental factors and different work levels. It has been used in several expressions seeking to assess environmental influences. Skin temperature has also been used by some as a comfort index although it is less reliable than most other indices because of the variability of skin temperature over the body, problems due to degree of acclimatization, and racial variations in skin temperatures. Personal expressions of comfort or discomfort have also been used as indices in spite of their clear limitations. Since the actual effect of a given set of environmental conditions on an individual is the factor of comfort sought, it is almost mandatory that human reactions be sampled although individual variations make the results imprecise and somewhat unreliable.

Early work in the field of human comfort sought to establish the thermodynamic relations between bodily heat production or heat gain from external sources and the rate at which the environment can accept heat from the body. The specific goal was to establish an environmental optimum and to determine quantitatively areas of increasing stress on both sides of the optimum. The early work leading to the development of "effective temperature," "resultant temperature," "standard operative temperature," and "thermal acceptance ratio" are all examples of this approach.

The effective temperature index (see Fig. 8-3) was first developed by Houghten and Yaglou (1923) by sampling the instantaneous thermal sensations of human subjects moving between rooms maintained under different environmental conditions. Effective temperature is defined as the temperature of a still (average velocity 0.2 m/sec), saturated atmosphere which has the same general effect upon comfort as the atmosphere under investigation. Those particular combinations of temperature, humidity, and wind velocity conditions that produce the same thermal sensation on an individual are said to have the same effective temperature.

The effective temperature concept has been subjected to considerable testing and criticism over the years. One significant question relates to its method of experimental evaluation. As human subjects pass from wet environments to dry ones, any absorbed moisture evaporates and gives an instantaneous sensation of cooling. Similarly, in going from a warm, dry test

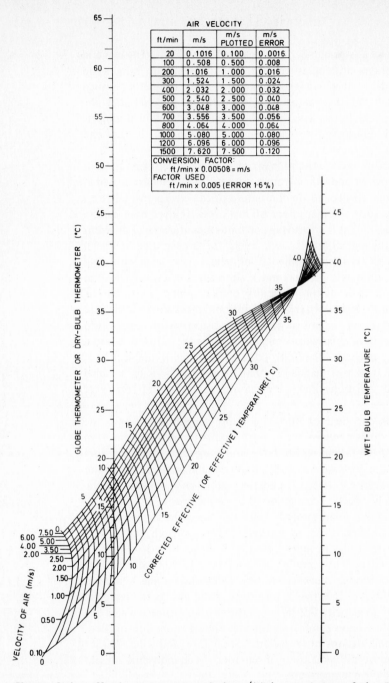

AIR VELOCITY

ft/min	m/s	m/s PLOTTED	m/s ERROR
20	0.1016	0.100	0.0016
100	0.508	0.500	0.008
200	1.016	1.000	0.016
300	1.524	1.500	0.024
400	2.032	2.000	0.032
500	2.540	2.500	0.040
600	3.048	3.000	0.048
700	3.556	3.500	0.056
800	4.064	4.000	0.064
1000	5.080	5.000	0.080
1200	6.096	6.000	0.096
1500	7.620	7.500	0.120

CONVERSION FACTOR:
ft/min x 0.00508 = m/s
FACTOR USED
ft/min x 0.005 (ERROR 1·6%)

Chart of the effective temperature index. (*With permission of the British Medical Assn., and the* Brit. J. Ind. Med., *from Ellis, Smith, and Walters,* 1972, p. 371.)   FIG. 8-3

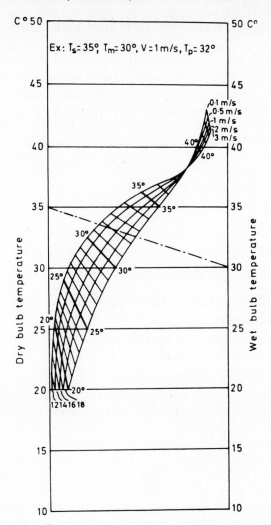

Chart of the resultant temperature index. (*With permission of* Chaleur Ind.,   FIG. 8-4
from Missenard, 1948, *p.* 171.)

condition to the cool, wet conditions, moisture is absorbed and the heat of absorption provides an instantaneous sensation of warmth. Thus the nomograms tend to overestimate humidity effects since they refer to instantaneous thermal sensations.

A number of modifications to eliminate certain of its deficiencies have been proposed. Missenard (1948) developed his "resultant temperature index" (Fig. 8-4) in part to eliminate the "instantaneous sensation" weakness of the effective temperature approach. He used a longer exposure time of the subject to the given climatic conditions so that thermal equilibrium between the individual and his environment is attained. Wet and dry bulb temperatures and wind speed are required in order to evaluate the resultant temperature.

Givoni (1963), in evaluating this index, found (1) a small over-estimation of the humidity effect at temperatures below 30°C; (2) an underestimation of the cooling effect of wind speeds at low index values. The resultant temperature still agreed better with available physiological responses than did the effective temperature.

During World War II, the Climatology Branch of the Office of the Quartermaster General developed the "thermal acceptance ratio" (R_a), the ratio of the rate at which the atmosphere will accept heat to the rate the body is producing heat.

$$R_a = \frac{H_a}{M}$$

where H_a = thermal acceptance, kg cal/m² hr
$\qquad M$ = metabolic rate of heat production, kg cal/m² hr

As defined by the Quartermaster Corps,

$$H_a = E_k(44.8 - p_a) + C_k(97 - T_a) + 2.6(97 - T_a)$$

where E_k = evaporative constant
$\qquad p_a$ = vapor pressure of the air, mm Hg
$\qquad C_k$ = convection constant
$\qquad T_a$ = air temperature, °F

The expression for H_a is limited by the fact that the radiation heat load is not considered and that evaporation and wind speed factors are only included by means of constants.

Another development during World War II was the so-called "predicted four hours sweat rate" index (P_4SR) by McArdles and his associates at the Royal Naval Establishment (1947). This index provides a value of the sweat rate resulting from 4-hr exposure to different values of metabolic heat production, clothing, wind speed, and dry and wet bulb temperatures. It can be evaluated by means of a nomogram (Fig. 8-5).

The determination of sweat rate is actually carried out in steps. The first step is to determine the 4-hr sweat rate for resting individuals in a seminude condition in a given climatic environment (dry and wet bulb temperature and wind speed). The next step involves correcting this sweat rate as the mean radiant temperature (Globe thermometer temperature), metabolic rate, or clothing level vary significantly from those postulated for the basic rate. Two types of corrections are applied. First, if the mean radiant temperature differs from the dry bulb temperature, if the metabolic rate exceeds the resting metabolic rate, or if the weight of the clothing exceeds 600 g, corrections are applied by adding to the wet bulb temperature. Second, the corrected basic 4-hr rate is further adjusted by the addition of values based on the metabolic rate and degree of clothing of the subject. The P_4SR index has proved to be quite reliable and useful, and the values of

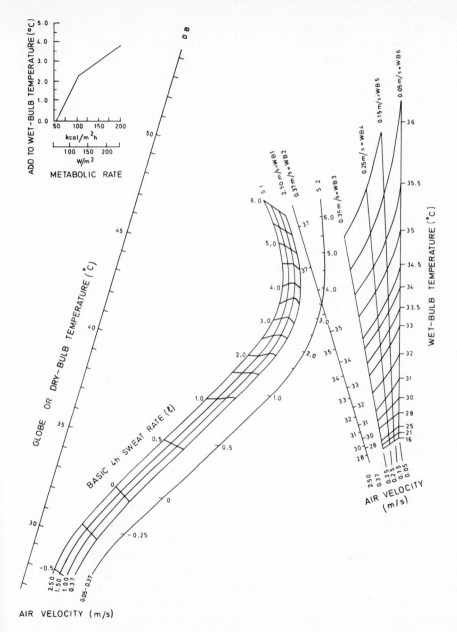

Chart of the P_4SR index (4-hr sweating rate). (*With permission of the British* FIG. 8-5 *Medical Assn., and the* Brit. J. Ind. Med., *from Ellis, Smith, and Walters,* 1972, *p.* 373.)

sweat rate are in good agreement with measured values within the limits of clothing and metabolic activity for which it was derived.

One final index, the index of thermal stress of Givoni, is too involved to be described in detail here, but it does point to the current direction being taken by investigators in seeking to express the relation of an individual with his variable environment. Givoni (1963) attempts to develop a biophysical model including all avenues of heat gain and heat loss by the body, as well as various levels of work and clothing. He assumes that if thermal equilibrium is possible the rate of sweat production should be a measure of the amount of cooling necessary to maintain this balance.

The basic relation is given by

$$S = \frac{(M - W) \pm C \pm R}{f}$$

where S = required sweat rate, kg cal/hr
   M = metabolic rate, kg cal/hr
   W = work performed, kg cal/hr
   C = convective heat exchange, kg cal/hr
   R = radiant heat exchange, kg cal/hr
   f = a dimensionless number indicating cooling efficiency of sweating
   $1/f = e^{0.6(E/E_{max} - 0.12)}$*

Substituting in the basic relation, Givoni obtains the relation

$$S = \{ M - 0.2(M - 100) \pm \alpha V^{0.3} (t_a - 35) + I_N \cdot K_{pe} K_{cl} [1 - a(V^{0.2} - 0.88)] \} / f$$

where α = a clothing coefficient
   t_a = air temperature, °C
   I_N = normal solar intensity
   K_{pe} = a coefficient depending on posture and terrain
   K_{cl}, a = coefficients depending on clothing
   V = wind speed

The formula would be quite unworkable without many nomograms to permit evaluation of various terms from available meteorologic and clothing data. As it is, the formula, while including all the variables known to influence environmental stress, still has so many empirically derived constants and functions that it lacks mathematical elegance. It provides reasonable values of sweat rate when compared with measured values under different environmental, activity, and clothing conditions. So many of the terms are difficult to measure or not regularly available that it is limited in applicability.

* $E = (M - W) \pm C \pm R$ in kg cal/hr and $E_{max} = pV^{0.3} (42 - VP_a)$

where p = a clothing coefficient
   V = airspeed, m/sec
   42 = vapor pressure of skin at 35°C, mm Hg
   VP_a = ambient vapor pressure, mm Hg

COMFORT DIAGRAMS

Many investigators have sought diagrams, classifications, or simple tempera-ture-humidity indices to evaluate the degree of stress under different conditions without having to include all the terms given in the previous relations. As the expressions and diagrams become simpler and utilize fewer variables, their reliability becomes more suspect. Because of the role that some of these indices play in our lives (for example, the *THI* index of the Weather Bureau appears regularly in many papers and in some areas is utilized to determine working hours), it is desirable to review them here.

The discomfort index (*DI*), or the same index now known as the temperature-humidity index (*THI*), was developed to provide an easily evaluated index describing the degree of discomfort of various combinations of temperature and humidity. Various formulations of these indices are available, providing slightly different results. The differences are not significant since the index merely tries to identify a zone in which the majority of individuals feel comfortable or uncomfortable. The U.S. National Weather Service uses either of the following formulations for the DI:

$$DI = 0.4 \, (T + T_d) + 15 \quad \text{or} \quad DI = T - 0.55 \, (1 - 0.01 \, \text{rh}) \, (T - 58)$$

In both cases, T = air temperature, $^\circ$F
T_d = dew point temperature, $^\circ$F
rh = relative humidity of air, percent

Applying the relations, it was found that below an index value of 70 essentially no discomfort was experienced, but that as the index increased above 70 an increasing proportion of the individuals experienced discomfort. Fifty percent of the people tested were uncomfortable at an index value of 75, while above 80 most individuals were experiencing some sort of discomfort. The value of 85 on the discomfort index (or *THI*) is used by some government offices to dismiss workers because of the acute discomfort being experienced by most individuals. These values, of course, have been developed by testing a middle-latitude United States population and may not necessarily apply elsewhere.

Another index, the wet bulb Globe temperature (*WBGT*) is a temperature that can actually be measured (unlike the effective temperature or the discomfort index) by covering a Globe thermometer with a moistened muslin wick. Since the blackened globe may be 15 cm in diameter, it is not a simple task to keep the wick moist at all times. The *WBGT* is usually approximated by means of the simple relation

$$WBGT = 0.2t_g + 0.1t_a + 0.7t_w$$

where t_g = dry globe temperature, $^\circ$C
t_a = dry bulb air temperature, $^\circ$C
t_w = wet bulb temperature, $^\circ$C

The U.S. Marine Corps has used this index to guide their outdoor activities under hot conditions. They find that with a *WBGT* reading above 29.5°C outdoor drill should be limited to just a few hours a day, while if the *WBGT* is over 31°C all outdoor training should be suspended. Such preventative measures greatly reduce problems due to heat stress.

The reaction of a human to temperature and humidity conditions is strongly influenced by wind speed. With higher wind speeds, more warm moist air near the skin surface is removed, evaporation is promoted, and the individual feels cooler. Under cold conditions, the higher wind speeds remove the heated layer of air around the body. The increased heat loss to the environment produces a "windchill" effect.

Siple and Passel (1945) defined windchill as "the measure of the quantity of heat which the atmosphere is capable of absorbing within an hour from an exposed surface one meter square." They provided a formula for calculating the total cooling power of the atmosphere in shade and without regard for evaporation (in kg cal/m² hr) as follows:

$$K = (\sqrt{100v} + 10.45 - v)\,(33 - t_a)$$

where v = wind speed, m/sec

t_a = air temperature, °C

Often the windchill equivalent temperature is determined instead. This temperature is the temperature under a light wind condition (defined as 5 mph or 2.2 m/sec) that is equal in cooling power to a given combination of actual temperature and wind speed. Table 8-4 lists the windchill equivalent temperatures for a wide range of cold air temperatures and wind speeds based on an evaluation of the Siple and Passel formula. The windchill index has been widely used and studied. It has many shortcomings, but it still provides a rough measure of the heat loss due to wind speed. It is limited by the fact that (1) it applies only to the heat loss from a bare skin surface, (2) it does not consider heat loss due to respiration, and (3) the formula cannot be used for wind speeds greater than 20 m/sec. The skin surface temperature has been assumed to be 33°C.

Any short-wave solar radiation is assumed to reduce the heat loss or windchill effect. Bright sun in middle-latitude conditions might add as much as 100 to 200 kg cal/m² hr to the individual and thus reduce the values of heat loss by that amount if the diagram is applied to nonshade conditions.

Steadman (1971) has broadened the usefulness of the windchill index by providing a basis for considering the influence of various wind and temperature combinations on clothed individuals. He assumes that the individual is wearing appropriate winter clothing, arguing that since under very cold conditions nearly all the body is clothed, clothing must be as important a factor in environmental protection as wind or temperature factors.

Steadman considers a typical adult with a clothed surface area of approximately 1.7 m². The face (3 percent of the body surface) is bare, while the hands and feet (12 percent of the body surface) are covered with about 7 mm of good mitten material or 2 cm of leather sole. The remainder of the body surface has sufficient clothing insulation to maintain thermal equilibrium. Body temperature is assumed to equal 37°C, while clothed skin is assumed to achieve a 33°C temperature and the hands, feet, and face have a 30°C temperature. Outdoor activity (walking at 1.3 m/sec) which will result in the production of 162 kg cal/m² hr of heat is assumed. Steadman further considers

TABLE 8-4

Windchill equivalent temperatures (°F) from Siple's formula*

TEMP., °F	CALM	WIND SPEED, MPH							
		5	10	15	20	25	30	35	40
32	66	32	22	16	11	7	5	3	2
30	64	30	20	13	8	5	3	1	-1
28	63	28	18	11	6	3	0	-2	-4
26	62	26	16	9	3	0	-3	-5	-7
24	61	24	13	6	0	-3	-6	-8	-10
22	60	22	11	3	-2	-6	-9	-11	-13
20	59	20	8	0	-5	-9	-12	-14	-16
18	58	18	6	-2	-8	-12	-15	-17	-19
16	57	16	4	-5	-10	-15	-18	-20	-22
14	56	14	1	-8	-13	-18	-21	-23	-25
12	56	12	-1	-10	-16	-21	-24	-26	-28
10	55	10	-4	-12	-19	-24	-27	-29	-31
8	54	8	-6	-15	-21	-26	-30	-32	-34
6	53	6	-8	-18	-24	-29	-33	-35	-37
4	52	4	-11	-20	-27	-32	-35	-38	-40
2	51	2	-13	-23	-30	-35	-38	-41	-43
0	50	0	-15	-26	-32	-37	-41	-44	-46
-2	49	-2	-17	-28	-34	-40	-44	-47	-49
-4	48	-4	-20	-30	-37	-43	-47	-50	-51
-6	47	-6	-23	-32	-40	-45	-50	-53	-54
-8	46	-8	-25	-35	-43	-48	-53	-56	-57
-10	45	-10	-27	-38	-46	-51	-56	-59	-61
-12	45	-12	-29	-40	-48	-54	-59	-62	-64
-14	44	-14	-32	-42	-51	-56	-62	-65	-67
-16	43	-16	-35	-45	-54	-59	-64		
-18	42	-18	-37	-48	-56	-62	-67		
-20	41	-20	-39	-51	-59	-65	-70		
-22	40	-22	-42	-53	-62				
-24	39	-24	-44	-56	-65				
-26	38	-26	-46	-58					
-28	37	-28	-48	-61					
-30	37	-30	-50	-64					
-32	36	-32	-52	-67					
-34	35	-34	-54						
-36	34	-36	-57						
-38	33	-38	-59						
-40	32	-40	-62						

* *With permission of American Meteorological Society, from Steadman, 1971, table 1, p. 675.*

the significant heat loss through respiration in his development. He estimates it to equal 20 percent of the total body heat loss under cold, dry conditions. In addition, he adjusts the measured wind speed to represent the slower values existing at the height of an individual rather than at the height of most standard anemometers.

Substituting the data for effective wind speed for active individuals, for convective and radiative heat loss, for respiratory heat loss from the lungs, and for the resistance of clothing into a computer program, Steadman obtains values of the required clothing thickness (in mm) for thermal equilibrium as well as the windchill. The windchill equivalent temperature (Table 8-5) can be obtained from this information for any selected combination of wind and temperature conditions. Comparisons of the Steadman (Table 8-5) and Siple and Passel (Table 8-4) values of windchill equivalent temperatures show the great differences, especially at low wind speeds, introduced by the various assumptions of Steadman. The Siple and Passel values appear unduly responsive to low wind speeds. Steadman questions whether one would expect at $0°F$, for example, that a change from a 5 mph wind speed to calm conditions would be equivalent to a temperature change of some $50°F$.

Steadman has provided a diagram of windchill isopleths for uncovered surfaces at $30°C$ (Fig. 8-6) as well as a graph of the thickness of clothing (in mm) needed to insulate all the body surface except the hands, feet, and face (Fig. 8-7). Figure 8-7 shows that under very severe conditions, no thickness of clothing will really be able to maintain thermal equilibrium under the postulated conditions of light activity. Only increased activity or use of outside heating sources (e.g., electric heaters) can provide for equilibrium under extreme conditions.

In spite of its previously discussed limitations (p. 239), the effective temperature (ET) developed by Houghten and Yaglou in 1923 for ASHRAE has been one of the most widely used thermal indices. As originally derived, it assumed that subjects were wearing 1 clo unit insulation under sedentary conditions. The numerical values of effective temperature were identified where the heat-stress lines intersected the 100 percent relative humidity line. A new effective temperature scale ET^* has now been identified (ASHRAE, 1972) in which temperature values along the 50 percent relative humidity line have been used for the numerical ET^* index, since these humidities represent conditions that are found more commonly in nature.

Based on the results of studies on over 1,600 college-age students at Kansas State University, exposed under different temperatures and relative humidities for varying time periods and with low wind speeds (< 0.17 m/sec), ASHRAE has recently produced a new climatic comfort chart (Fig. 8-8), the information on both comfort zones and effective temperatures being plotted on the ASHRAE psychrometric diagram. Two comfort zones are shown: the shaded area (ASHRAE study) being for individuals with 0.8 to 1.0 clo value clothing and engaged in slightly more than sedentary activity, while the diamond-shaped area (based on Kansas State University data) is for individuals with 0.6 to 0.8 clo value clothing and sedentary activity. These activity and clothing differences account for the slightly different comfort zones on the chart. The area of overlap provides optimum design conditions for comfort: an $ET^* = 76°F$ or $24.5°C$ (this also represents the dry bulb air temperature or mean radiant temperature); relative humidity 40 to 60 percent; and airspeed less than 0.23 m/sec.

The dashed diagonal lines on Fig. 8-8 are lines of constant wetness resulting from regulatory sweating (after Gagge and his associates, 1971).

TABLE 8-5

Windchill equivalent temperatures (°F) based on clothing requirements*

TEMP., °F	CALM	WIND SPEED, MPH							
		5	10	15	20	25	30	35	40
32	34	32	27	24	21	17	14	12	10
30	32	30	25	21	18	15	12	10	7
28	30	28	23	19	15	12	9	6	4
26	28	26	21	17	13	9	6	3	1
24	26	24	19	14	10	7	3	0	-3
22	25	22	16	12	8	4	1	-3	-6
20	23	20	14	9	5	1	-2	-6	-9
18	21	18	12	7	2	-2	-5	-9	-13
16	19	16	10	5	0	-4	-8	-12	-17
14	17	14	8	2	-3	-7	-12	-16	-20
12	15	12	6	0	-5	-10	-15	-19	-24
10	13	10	4	-2	-8	-13	-18	-23	-28
8	11	8	1	-5	-11	-16	-21	-26	-32
6	9	6	-1	-7	-13	-19	-24	-30	-36
4	7	4	-3	-10	-16	-22	-28	-34	-40
2	5	2	-5	-12	-19	-25	-31	-38	-44
0	3	0	-7	-15	-22	-28	-35	-42	-49
-2	1	-2	-10	-17	-25	-31	-39	-46	-54
-4	-1	-4	-12	-20	-28	-35	-42	-50	-58
-6	-3	-6	-14	-22	-30	-38	-46	-54	-63
-8	-4	-8	-16	-25	-33	-41	-50	-59	-67
-10	-6	-10	-19	-28	-36	-45	-54	-63	
-12	-8	-12	-21	-30	-39	-48	-58	-68	
-14	-10	-14	-23	-33	-42	-51	-62		
-16	-12	-16	-26	-36	-45	-55	-66		
-18	-14	-18	-28	-38	-49	-59			
-20	-16	-20	-30	-41	-52	-63			
-22	-18	-22	-32	-44	-55	-66			
-24	-20	-24	-35	-47	-58				
-26	-22	-26	-37	-49	-62				
-28	-24	-28	-39	-52	-65				
-30	-26	-30	-42	-55	-68				
-32	-27	-32	-44	-58					
-34	-29	-34	-47	-61					
-36	-31	-36	-49	-64					
-38	-33	-38	-51	-67					
-40	-35	-40	-54	-69					

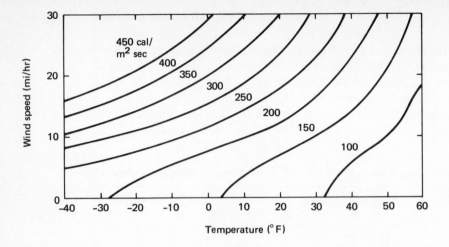

Windchill isopleths shown as a function of wind speed and air temperature. FIG. 8-6
(*With permission of the American Meteorological Society, from Steadman,*
1971, fig. 3.)

These lines have been drawn at 5°F dry bulb temperature intervals on the 50
percent relative humidity line. Not only do these intersections represent the
new effective temperature (ET^*) scale but the lines of constant wetness can
serve to delineate broad comfort zones.

Most attempts to provide simple comfort diagrams have been limited by
failure to obtain sufficient agreement among test individuals to give validity
to the concept of a comfort zone. The American Society of Heating,
Ventilating and Air Conditioning Engineers, in reviewing the work of many
others, noted that several different relations between effective temperature
and percentage of individuals indicating comfort had been obtained.
Differences in sex and age show up—women generally preferring slightly
warmer conditions for comfort than men, and older individuals slightly
warmer conditions than younger people. Humidity seems to affect the
comfort of men more than women. Individuals from warm areas appear to
prefer slightly warmer conditions than those from colder areas. Olgyay used
this type of information in the preparation of a comfort zone diagram,
although he clearly pointed out that with the wide range of reactions to
various conditions it would not be possible to define a zone with real
boundaries. He visualized conditions varying gradually outward from the
center of the comfort zone—from thermal neutrality, to slight degrees of
stress, to real discomfort (Olgyay, 1963, p. 18).

Figure 8-9 is a copy of Olgyay's bioclimatic chart for atmospheric
comfort derived for inhabitants of the midlatitudes of the United States. The
"desirable comfort zone" lies between 70 and 82°F (21 and 28°C)
temperature and 30 and 65 percent relative humidity, although two so-called
practical comfort zones were added to the ends of the desirable zone by
including higher and lower relative humidity values where no thermal stress

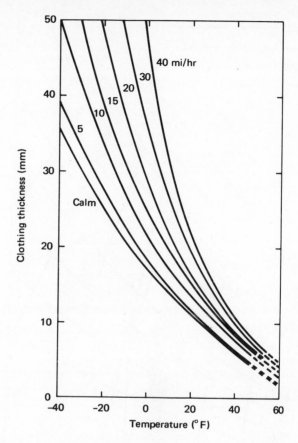

Thickness of clothing (mm) required to insulate 85 percent of the body's surface. Data from Table 8-5. (*With permission of the American Meteorological Society, from Steadman, 1971, fig. 2.*)   FIG. 8-7

was found (about 18 to 77 percent, summertime values). A winter comfort zone is also shown just slightly lower on the dry bulb scale than the summer zone. Air temperatures in the 80s or high 70s are not comfortable if the relative humidity is above about 55 percent.

The bioclimatic chart includes several series of lines to permit accounting for different wind velocities, humidities, and radiation conditions. Since air movement cools the body, the upper comfort limit for temperature can be raised by increasing wind speed. This rise is quite marked for low wind speeds, but as the wind exceeds about 1.5 m/sec the rate of rise of the upper comfort limit slows appreciably.

Many individuals are affected by changes in the moisture content of the air. Some of the adverse effects of increasing vapor pressure can be counterbalanced by increasing wind velocity. Siple, for example, has

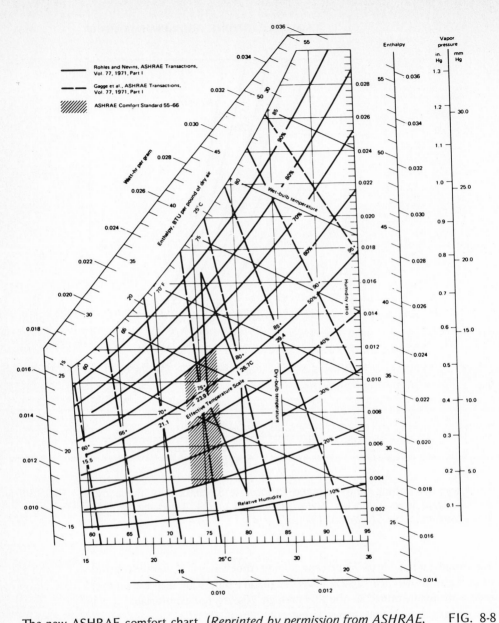

The new ASHRAE comfort chart. (*Reprinted by permission from ASHRAE,* 1972, *"Handbook of Fundamentals," fig.* 17, *p.* 137.)   FIG. 8-8

suggested that within reason 1 mm Hg (1.3 mb) of increased vapor pressure can be balanced by 1 mph (0.4 m/sec) of increased wind speed. Evaporation of water will remove heat from the air. It thus will lower dry bulb temperatures and increase the limit of the comfort zone in the high-temperature, low relative humidity portion of the chart.

Heating due to radiation is important in the low-temperature portion of the diagram. Increased radiation (outdoor conditions only) can counterbalance

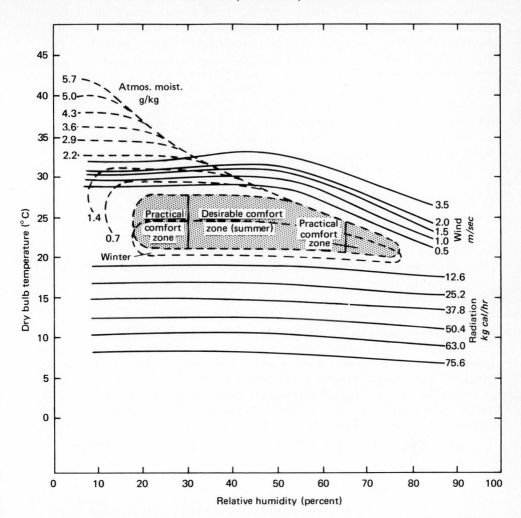

Bioclimatic chart for United States temperate zone inhabitants. (*From   FIG. 8-9
Housing Research*, 1954, *Housing and Home Finance Agency; investigations
conducted by V. Olgyay.*)

a decrease in dry bulb temperature, up to a point, and so lower the bottom
limit of the comfort zone. In this case, it is assumed that an increase in solar
radiation of 12.6 kg cal will balance a dry bulb temperature decrease of
2.2°C.

The bioclimatic chart, while greatly oversimplified, is immediately
useful to anyone who seeks to apply climatic information to a practical
human problem in clothing, architecture, or health. Each set of dry bulb
temperatures and relative humidity values locates a point on the diagram.
From the particular location, one can immediately see whether the condition

is one of general comfort and, if not, what corrective measures might produce a feeling of comfort on the part of most individuals. For example, if the plotted temperature and humidity point falls above the upper comfort zone line, comfort can be restored by increasing wind velocity up to a point. However, under very dry conditions (say 33°C and 20 percent relative humidity) comfort cannot be restored by wind alone but only by adding moisture to the air. In this case, an addition of about 2.2 g of moisture/kg of air should produce a feeling of comfort. With a wind of 3.5 m/sec only 0.3 to 0.4 g of moisture/kg of air need be added to produce a feeling of comfort.

Above the line marked "shading line," comfort can only be achieved in the shade, while below that line, as temperatures decrease, exposure to certain amounts of radiation is desirable for the restoration of comfort.

CLASSIFICATION BASED ON HUMAN COMFORT
Some investigators have attempted to provide classifications of climates in terms of human comfort. Most existing classifications express climate in relation to plant growth or vegetation zonation, although climate and its effect on human health and comfort is certainly as relevant an element to classify. There is, of course, great difficulty in trying to delimit areally such a quantity as comfort.

Terjung (1966) has attempted a bioclimatic classification by using monthly climatic data that he has first separated into daytime and nighttime conditions since both influence comfort but in somewhat different ways. He begins with a standard comfort chart on which he outlines 11 regions on the basis of the wet bulb and effective temperatures. These are listed in Table 8-6.

Recognizing that wind speed is not considered in the comfort chart, Terjung next makes use of the windchill relation of Siple and Passel mentioned previously. Because the windchill relation considers only shade conditions and since the occurrence of solar radiation tends to offset cooling

TABLE 8-6

Boundaries and designations of 11 comfort zones*

ZONE	UPPER (OR WARM) BOUNDARY, °F	DESIGNATION
−6	−40°F ET	Ultra cold
−5	−4°F ET	Extremely cold
−4	+14°F ET	Very cold
−3	+35°F ET	Cold
−2	60°F ET (56°F WB when rh above 70%)	Keen
−1	64°F ET (60°F WB when rh above 70%)	Cool
0	72°F ET (68°F WB when rh above 70%)	Comfortable
1	78°F ET (74°F WB when rh above 70%)	Warm
2a	74°F WB up to rh of 70%	Hot
2b	86°F ET	Oppressive
3	above 86°F ET	Extremely hot

* With permission of Association of American Geographers, after Terjung, 1966, p. 148.

due to wind, Terjung makes a twofold adjustment. He obtains the percentage of actual sunshine for a particular latitude and month. He then converts this to hours by multiplying by the number of hours of daylight at the particular place and into energy by multiplying this last figure by the average incoming solar radiation (assumed to be 200 kg cal/m^2 hr) to obtain a value for the quantity of heat that can be gained from solar radiation. This value is deducted from the heat loss resulting from windchill as found on the standard windchill chart. The difference is then divided by the number of hours in the day to obtain a value of heat loss or gain per hour. An example of the computation process for San Francisco, California, is included in Table 8-7.

The computations result in two sets of indices: the first from the application of the temperature and relative humidity data to the standard comfort chart and the second from the evaluation of the radiation and windchill effects. These two indices each consist of two parts, one for daytime conditions and the other for nighttime conditions. The index derived from the comfort chart is made up of the zone numbers of the 11 comfort zones given in Table 8-6 (for example, $-1/-2$ for day/night). The index derived from the radiation and windchill computations is made up of the actual values of heat gain or loss (for example, $-270/-550$ kg cal/hr).

Terjung divided the various possible combinations of indices into letter classes, which have also been identified by means of descriptive phrases to make the letter combinations more meaningful to those unfamiliar with the indices themselves. Terjung considers that the comfort index is more important than the radiation and windchill index since the latter would not be particularly significant under shade or overcast conditions with light winds. However, neither index should be considered independently since both influence the heat exchange between man and his environment and thus affect human comfort.

Terjung's approach is preliminary. Little is really known about the factors that influence human comfort and the complex physiological, psychological, and cultural factors within man that influence perception of comfort. His efforts are to be commended in view of the difficulties in applying climatic data to such a problem. It is to be hoped that he, or others following his lead, will continue to develop this important aspect of applied climatology.

A CASE STUDY IN APPLIED CLIMATOLOGY: CLIMATE, CULTURE, AND THE NEW WORLD MIGRATION OF PRIMITIVE MAN

Knowledge of the relation between climate, heat loss, and clothing has been utilized to determine something of the antiquity of man in the Western Hemisphere (Mather, 1954c). Early man is generally believed to have moved eastward and northward through Siberia and across the Bering Strait. While this general path of migration may be assumed to be reasonable, it is necessary to consider possible times for migration in view of general climatic conditions and the cultural equipment of the possible migrants.

TABLE 8-7

Examples of application of comfort indices for selected stations*

San Francisco, Calif., Lat. 37° 47′ N; elevation 52 ft; July (about 14.5 hr of possible sunshine)

DAILY MEAN MAX. TEMP.	MIN. TEMP.	DAILY MEAN RH LOW	HIGH	MEAN HOURLY WIND SPEED	PERCENT POSSIBLE SUNSHINE
64.4°F	53.4°F	75%	92%	11.3 mi/hr	64

Comfort index

Day: apply 64.4°F and 75% rh, result: −1
Night: apply 53.4°F and 92% rh, result: −2

Combined: −1/−2 (or: C_2)

Wind effect index

Day: (1) Apply 64.4°F and 11.3 mi/hr. Result: −400 (in. kg cal/m² hr)
    (2) Multiply −400 by 14.5 hr. Result: −5,800
    (3) Take percentage of actual sunshine (64%) of 14.5 hr. Result: 9.4 hr (rounded off to 65%)
    (4) Multiply 9.4 by 200 (kg cal/m² hr). Result: +1,880
    (5) Find difference between −5,800 and +1,880. Result: −3,920
    (6) Divide −3,920 by 14.5. Result: −270 (or −*b*)

Night: Apply 53.4°F and 11.3 mi/hr. Result: −550 (or −*c*)

Combined: −270/−550 (or −b_2)

Physiological climate:

Symbol: C_2-b_2 (from −1/−2, −270/−550)
Term: "Cool day with pleasant wind effects, keen night with cool windchill."

Yuma, Ariz., Lat. 32° 40′ N; elevation 199 ft; July (about 14 hr of possible sunshine)

DAILY MEAN MAX. TEMP.	MIN. TEMP.	DAILY MEAN RH LOW	HIGH	MEAN HOURLY WIND SPEED	PERCENT POSSIBLE SUNSHINE
107.7°F	81.5°F	22%	49%	9.8 mi/hr	91

Comfort index

Day: Apply 107.7°F and 22% rh. Result: +3
Night: Apply 81.5°F and 49% rh. Result: +1

Combined: +3/+1 (or: EH_3)

Wind effect index

Day: (1) Apply 107.7°F and 9.8 mi/hr. Result: off nomogram
    (2) Take percentage of actual sunshine (91%) of 14 hr. Result: 12.6 hr (rounded off to 90%)
    (3) Multiply 12.6 by 200 (kg cal/m² hr). Result: +2,520 (in. kg cal/m² hr)
    (4) Divide +2,520 by 14 hr. Result: +180 (or *c*)

Night: Apply 81.5°F and 9.8 mi/hr. Result: −100 (or −*a*)

Combined: +180/−100 (or c_1)

Physiological climate:

Symbol: $EH_3 c_1$ (from +3/+1, +180/−100)
Term: "Extremely hot day with very discomforting heat addition to the skin, warm night with warm wind effects."

---

* *With permission of Association of American Geographers, from Terjung, 1966, p. 160.*

Early migrants to this hemisphere by way of Alaska almost certainly spent more than 1 year in moving from temperate latitudes of Asia to comparable latitudes in North America. By the shortest route over a Bering Strait land bridge, this distance is more than 3,000 mi (4,800 km). Since the early migrants to this hemisphere probably had no idea of where they were going or of the character of the region around them, it must be assumed that the paths they followed during migration would be far from direct. Unless they were following a food source or being pursued, the necessity for continued, almost daily, movement would be lacking and several generations of migrants might develop during the passage through the sub-Arctic region. Therefore, they may have faced the rigors of many Arctic and sub-Arctic winters.

At present, the region around Tierra del Fuego, in southern South America, is the home of the poorly clad Alacaluf, Yahgan, and Ona peoples. Before contact with Europeans occurred, the Alacaluf and Yahgan peoples used to wear only small capes or mantles of seal, sea-otter, or fox skin. The capes, covering only the shoulders and chest, were worn with the fur outward. Frequently, they were left off entirely. No head coverings were used, and the people went barefooted unless on a long journey, when moccasins of sealskin were worn. The children were generally naked.

The Ona tribe, living in the interior hill region of the island chain, possessed slightly more in the way of clothing. Both men and women wore long capes of guanaco or fox skin which reached from the shoulders to the ankles. The capes could be tied across the chest. Leggings of guanaco skin were used in heavy snow. The Ona wore moccasins made from the skin of the guanaco's legs. These were tight fitting and relatively water resistant. The men wore triangular peaked fur hats, while the women usually wore under the cape another fur garment, with the hair inward, which reached from the armpit to the knee. This garment was tied at the waist.

For people with the culture and clothing of the Alacaluf or Yahgan, the freezing point might well be the lower temperature limit for possible survival. Under cool or cold, wet conditions the individual and his clothing become wet. In drying, water evaporates and rapid cooling occurs. In the absence of much wind, it may be more efficient to remove wet clothing and to allow the rapidly dried skin to serve as a protection against the cold. With temperatures below freezing, wet conditions would not exist and the use of clothing would always be beneficial. However, with temperature conditions slightly above freezing, lack of any clothing would probably prove to be more efficient than the possession of a large amount of damp clothing. Thus the clothing habits of the Alacaluf and Yahgan peoples are probably well adapted to their environment, but they would not be efficient at all if the temperature were to drop below the freezing point. There are no nearly naked races of men existing today in regions with mean monthly temperatures below freezing.

The insulation value of the clothing possessed by the Alacaluf and Yahgan peoples of Tierra del Fuego must be almost negligible. Even the Ona should not be expected to have clothing of more than 1 or at the most 1.5 clo insulation value. By contrast, the clothing of the Eskimo would appear to have an insulation value of 3 to 4 clo units, although the method of wearing the clothes and their fit would determine, to a great degree, their insulation value. These values are, of course, only subjective estimates based on an interpretation of the descriptions of the clothing. It is not possible to assign clo values to each individual piece of clothing, for it is the integrated effect of the entire clothing assembly which is of importance in insulation determinations.

The Heat-Exchange Relation between Mand and His Environment

The heat-exchange relationship between man and his environment has been discussed in detail in earlier sections of this chapter. From discussions of metabolic energy production and the various activities of the Alacaluf and Yahgan peoples, their average energy production may be assumed to be about 260 kg cal/m² hr (p. 221). Since 10 to 15 percent of this energy production goes to perform external work while the rest is produced as heat, the total heat production by the Yahgan and Alacaluf peoples as a result of their daily activities might be about 230 kg cal/m² hr.

To compute the radiation loss from individuals along the Bering Strait migration route, several simplifying approximations may be made. First, all radiation terms concerned with short-wave solar radiation may be disregarded. The most critical conditions for primitive man would be the winter period in the most northerly section of the Bering Strait migration route. Since the latitudes involved are 60 to 65°N, the amount of solar radiation received in the winter will be small enough to be negligible in the final computations. Second, for purposes of calculation, man in sub-Arctic regions may be thought of as a vertical plane of surface area 1.75 m² exposed to radiation from two horizontal infinitely extending planes—one, the terrain under him, and the other, the sky over him. A vertical plane will receive one-half the radiation received by a horizontal plane facing the reflecting terrain and sky (Blum, 1945).

Using a mean Bering Strait air temperature of -10°C (approximates blackbody temperature of surroundings), a mean temperature at the prevailing cloud level of -10°C (approximates blackbody temperature of sky), and the assumptions outlined in the previous paragraph, the radiation heat exchange between man and his surroundings (R) may be calculated from the expression

$$R = \frac{K_r}{2} [(T_s^4 - T_g^4) + (T_s^4 - T_a^4)]$$

where K_r = a constant (4.93×10^{-8})

T_s = temperature of exposed clothing and skin surface, assumed to be 300 K

T_g, T_a = mean blackbody temperatures of surrounding terrain objects and sky, respectively, both assumed to be 263 K

The radiation heat loss is found to be 232 kg cal/m² hr. The body loses as much heat to the surroundings by radiation as it produces through metabolism.

Since the two terms, metabolic heat production and radiation loss, essentially balance one another under the postulated conditions, the net loss of heat from the individual to the environment depends upon the magnitude of the conduction-convection and evaporation-convection heat exchanges. Expressions for determining the rate of conduction-convection and evaporation-convection heat loss from humans under varying conditions have been provided in earlier sections (pp. 226, 229). From these relations, charts showing the rate of heat loss from the body by conduction-convection and evaporation-convection heat exchange for different ambient temperatures, clothing insulation values, and wind speeds have been prepared. For the purposes of calculation, the clothing insulation was varied from 0.5 to 4 clo units. Ambient temperatures were allowed to range from 15 to $-20°C$ (59 to $-4°F$), while the wind speeds were 2, 6, and 12 mph.

The charts, one for each of the wind speeds, are presented in Fig. 8-10. They represent the net rate of heat loss from the body as a result of evaporation-convection and conduction-convection heat exchange for different ambient temperatures and clothing insulation values. In our particular case the charts may also be used to express the total heat-exchange relation for man since the metabolic heat production equals the net rate of heat loss by radiation. When this equality is not found, the radiation heat-exchange term must also be included in the overall heat-exchange relationship.

To determine the heat loss from individuals with a culture similar to that of the Alacaluf the previously postulated temperature, wind, and clothing insulation values may be used to enter the charts.

Assuming a wind of 12 mph, frequently experienced in Arctic locations, the chart of Fig. 8-10, prepared for a 12-mph wind speed, can be used. The clothing of the Alacaluf has a maximum insulation value of 0.5 clo. Following down the 0.5 clo line until the postulated Bering Strait ambient air temperature of 14°F ($-10°C$) is reached, a rate of heat loss of 300 kg cal/m² hr is found. The amount of heat that may be safely lost from the body over a short period of time depends, of course, on the individual. Belding (1949) indicates that up to 80 kg cal/m² may be removed from the body without danger. Greater losses than 80 kg cal/m² will bring discomfort and finally death. The maximum amount of heat loss before death occurs is not known, but it might well be of the order of 100 to 200 kg cal/m². Thus the tolerance time for people such as the Alacaluf exposed to moderate Bering Strait climatic conditions would appear to be $\frac{1}{2}$ hr or less. This is the amount of time that they could remain outdoors in search of food and fuel. Any realistic account of the food- and fuel-gathering processes of such native peoples reveals the need to be outdoors for several hours almost every day. Stockpiling or storage of food is not practiced by the Fuegians or many other such peoples.

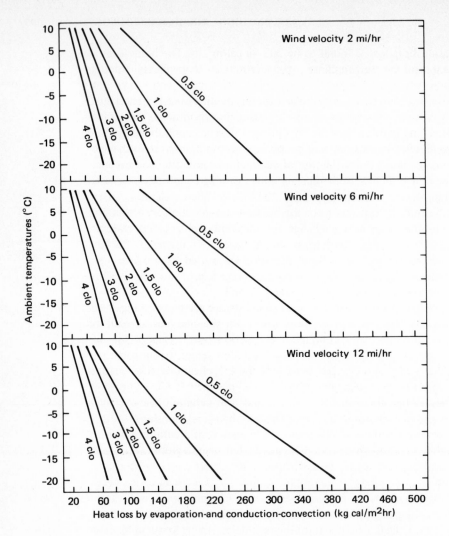

Heat loss from a human by means of evaporation- and conduction-convection at different air temperatures, wind speeds, and degrees of clothing. (*With permission of the* Southwestern J. Anthro., *from Mather, 1954c, p. 318.*)   FIG. 8-10

     The survival time in the Bering Strait region for tribes with a cultural development similar to that possessed by the Ona can also be determined. Accepting a clothing insulation of 1.5 clo units and with climatic conditions as previously postulated, heat loss will be about 120 kg cal/m² hr. On the basis of the previous discussion of tolerance times, individuals so clothed would be able to survive outside in the Bering Strait region somewhat over an hour. The Eskimo, or any migrating races with similar cultures, would be able to survive generally for unlimited periods of time outdoors under the postulated Bering Strait climatic conditions.

Discussion

The development of the heat-exchange relation and its application to the survival of poorly clothed individuals in sub-Arctic climatic conditions is not exact. Certain assumptions and generalizations have been made to achieve quantitative results. The results, therefore, indicate only roughly the survival chances of people in cold climates.

In developing the heat-exchange relation, several approximations were made so that the radiation term could be evaluated. To disregard all short-wave solar radiation is permissible for only a short period during the winter. The effect of solar radiation would be to reduce the net radiative heat loss from the body and thus reduce the amount of clothing insulation required for the maintenance of thermal equilibrium. In evaluating the heat-exchange relationship, average metabolic energy figures, mainly determined on white subjects in midlatitudes, were used. The reduction of the insulation value of clothing, due to bellows action resulting from activity, could not be quantitatively included in the calculations. However, for clothing insulation of 0.5 clo or less, bellows action should be of little importance. The body may be brought more into thermal equilibrium with its environment by reducing the amount of heat stored within the body. Permitting the skin temperature to drop reduces the amount of clothing insulation required to maintain thermal equilibrium, for with a smaller temperature gradient between the skin and the environment less heat transport from the body will occur. A skin temperature of 80.6°F (27°C) has been assumed in the previous calculations but it is only approximate and it may, for any individual, be as much as 6 or 7°F (3.3 to 3.9°C) in error.

Clearly, the survival times of various groups in the Bering Strait climates are only approximate. However, in the case of people with cultures similar to those of the Alacaluf, Yahgan, or Ona, the extremely short tolerance times to the rather moderate climatic conditions that were postulated would indicate that even if errors of as much as 100 percent were present in the calculations, the conclusions would be unchanged.

The above findings lead to several interesting conclusions. If one assumes that the present Arctic climate is warmer than any since the last glacial period, then the Arctic region would be effectively closed to the migration of nearly naked people since before that time. The position of the Fuegians and other migrants with similar cultures in the Western Hemisphere poses an intriguing problem. It would seem that they must have either entered this hemisphere possessing a culture different from what they now have, more adequate clothing and better housing, in order to survive the cold Arctic conditions of the last 15,000 to 25,000 years, and have subsequently lost that cultural complex, or that they entered this hemisphere during a mild climatic period since or perhaps before the last great glacier advance. The selection of the proper alternative in this case must await the interpretation of more detailed archaeological work.

HISTORY OF MEDICAL CLIMATOLOGY

Weather and climate influence human health. Man has been aware of a possible connection almost since the dawn of time and well before the actual relationships were understood. While Hippocrates (approximately 460 to 377 B.C.) probably made the first major contribution to our early thinking on this subject, it is clear that he followed a tradition of healers who, through experience and observation, recognized the seasonal nature of some illnesses and correctly identified certain complaints and sicknesses as weather dependent. Medical climatology is, thus, an old field, although it has never received the emphasis that it deserves. Applied climatologists have been slow to move into the field; even from the first it has been the physician, not the climatologist, who has broadened himself to include climatology within his own investigations.

In recent years, there has been a growing understanding of the relation of climate to human well-being. A number of careful studies of mortality and morbidity in relation to climatic variables have appeared. The climatic characteristics of recognized health resorts are undergoing study, and the influence of certain less well-known factors such as atmospheric electricity and noise are under investigation. Some climate-controlled chambers have been constructed that are large enough to be utilized by groups of hospital patients for testing purposes.

Little of real scientific value in medical climatology was possible, of course, until the widespread use of basic meteorologic instruments made available the needed climatic information. There was a companion need for reliable statistics on mortality and morbidity without which little more than suggestions or inferences could be made. John Locke and Dr. Charles Goodall in England, in the latter part of the seventeenth century, made an effort to obtain worldwide information on both weather and morbidity in order to undertake more broadly based weather-disease studies. They sent questionnaires to physicians and scientists in other parts of the world, asking for information on mortality statistics, the seasonal pattern of temperature and other available climatic parameters, as well as information on the seasonal distribution of the diseases prevalent in each area. Very few data were forthcoming as a result of this early effort.

In the Western Hemisphere, small-scale studies of weather and health were begun by a few physicians in the American colonies during the period before 1770, but most were limited to Hippocratic-type generalizations due to the lack of quantitative data (Cassedy, 1969). A physician, Cadwallader Colden, probably made use of the first meteorologic instruments in America. He settled in the New York area about 1716 and seems to have had both a thermometer and a barometer in use from about that time. In 1720, he wrote a short note entitled "Account of the Climate and Diseases of New York" which followed the Hippocratic tradition.

Colden noted the great range of weather experienced in America and suggested that this would lead to considerable differences in the weather-health relationship in the various parts of the colonies, although he could still recognize seasonal similarities in diseases in places far removed. For example, he pointed out that while spring, or the period of breaking up of hard winters, might come later in some areas than others, it still resulted in an increase in "pleurisies and inflammatory fevers." He felt that the summer months were the most difficult of the year healthwise, with a greater number of individuals sick or dying than at other times. The fall was extremely healthful with mild, dry weather and fairly clear skies in the middle Atlantic colonies, at least. He associated rheumatic pains and some pleurisies with the cold weather of winter. Colden also accepted one of the then current ideas of climatic change—namely, that as a result of deforestation and the opening up of the land, the climate was improving. As a result, he reasoned that the inhabitants were becoming healthier.

Dr. John Lining of Charleston, South Carolina, compiled one of the earliest meteorologic records. Beginning in 1737, his observations ran for nearly 20 years. While interested in the climatic record itself, Lining was also eager to apply the observations to the better understanding of seasonal trends in epidemic diseases. Two other Charleston physicians took up the task of relating weather and health after Lining. Lionel Chambers, once a medical associate of Lining, published an "Account of the Weather and Diseases of South Carolina" in 1776, a statewide survey of the seasonal trends in climate and diseases. While mortality statistics were few and many of his generalizations were based on just 10 years of weather data in Charleston (1750–1759), his early attempt to survey a broader region deserves mention. About the same time, George Milligen wrote "A short description of the Province of South Carolina with an account of the air, weather, and diseases at Charles-Town" (written in 1763 but published in 1770). He relied extensively on the weather data of Lining, while only suggesting general weather-health relationships because of the paucity of mortality and morbidity data.

These early and, in large measure, nonquantitative studies set the stage for later advances. The lack of real data on weather and especially on diseases and mortality was, at once, apparent; the repeated calls for more quantitative data resulted in their gradual accumulation. Two hundred years later we are still asking for more and better data in order to refine those weather-health relationships already identified. Our present progress in this area of applied climatology owes a real debt to the early investigators who recognized the problems and sought answers even without sufficient factual data.

PHYSIOLOGICAL CLIMATOLOGY

Principal Thermoregulatory Responses[1]
The human body can stand only a small deviation in its temperature from normal. Prolonged exposure to stressful conditions may ultimately result in a significant change in body temperature and possibly death. The body, however, has a remarkable capacity for adjusting to thermal stress by producing more heat or by increasing heat losses to the environment without undue bodily stress. These physiological responses to thermal stress need to be understood for when they are weakened or blocked by poor health or disease the body becomes seriously affected by the stressful conditions. The principal thermoregulatory responses can be listed as

1   Changes in blood flow to the outer body tissues

2   Water shifts in the blood

3   Shivering and sweating

4   Changes in respiratory activity

5   Changes in posture and behavior

1   *Changes in blood flow* As air temperatures decrease and the body seeks to minimize heat loss to the surroundings, the arterioles that supply blood to the skin area become increasingly constricted, which reduces the convection of heat from the deeper body areas to the skin. By allowing the extremities as well as the general skin surface to come to a lower temperature,

[1]   The material in this and the next three sections has been adapted, with permission, from a series of notes provided by Dr. D. H. K. Lee, Associate Director for Scientific Information and Communication, National Institute of Environmental Health Sciences, Research Triangle Park, North Carolina, as well as from Lee, 1964.

constriction of arterioles results in a smaller temperature gradient between the skin and the surrounding air. The extent of vasoconstriction of the blood vessels is, of course, limited; some feel that it has already reached its maximum amount in a nude individual exposed to an air temperature of about 70°F (21°C). Some degree of dilation of the blood vessels feeding the skin area is also possible under hot conditions. This increases the heat transport to the surface and helps promote its loss to the surrounding environment.

2   *Water shifts in the blood* Sudden exposure to hot or cold conditions results in rather rapid water additions to or withdrawals from the bloodstream. These changes appear to have two purposes: (*a*) to influence thermal conductivity and assist in heat regulation and (*b*) to compensate for changes in the capacity of the circulation system. There is some evidence to indicate that as much as 5 to 7 percent change in the water content of the blood occurs within the first 30 min of exposure to the hot or cold conditions.

3   *Shivering and sweating* "Goose pimples" or "gooseflesh" is a familiar response to cold conditions. While it is of little importance to man, it could serve a purpose in a furred or feathered animal by increasing the depth of the insulating layer. Shivering, an involuntary and automatic contraction of the muscles in response to cold conditions, is more effective in man. Considerable heat can be produced for short periods of time through shivering; five times the basal metabolic rate can be produced for an hour or so while twice the basal rate may be produced through shivering for much longer periods of time.

Sweat glands, under the control of the nervous system, manufacture sweat from materials in the bloodstream. The number of active sweat glands and their rate of production vary over the body. The amount of sweat produced by the body should provide sufficient evaporative cooling to balance deficiencies in the other avenues of heat loss. There is an effective limit, of course, for once the body is completely covered with sweat (100 percent wetted surface), increased production results in no additional cooling; the total wetted surface area cannot be increased. The capacity of the body to produce sweat can be varied somewhat. Rates of 1,500 cc/hr can be maintained for short periods, while rates of 1,000 cc/hr (1 liter/hr) can be maintained for long periods if sufficient water is present in the body.

4   *Changes in respiratory activity* Man does not change his respiratory activity appreciably to increase heat loss in hot conditions or to conserve heat loss under cold conditions. Panting, with the tongue exposed, is common among many animals in hot environments. In man, only a slight increase in rate and depth of respiration is noted under repeated exposures to hot environments.

5   *Changes in posture and behavior* Animals show a fairly marked tendency to stretch out under warm climatic conditions in order to encourage

heat loss from as much of the body as possible and to curl up or tuck extremities under them to conserve heat in cold conditions. Man, too, will undertake somewhat similar changes in position, especially in sleep, in order to adjust his heat loss to the environmental conditions. Increased physical activity usually accompanies exposure to cold in response to the need to increase heat production. Whether any lack of activity under hot conditions can be directly linked with the conditions of the thermal environment has to be more carefully evaluated. Certainly the ingenuity with which man has attacked the problems of adjustment to both hot and cold conditions through clothing, shelter, and mechanical devices suggests that climatic stress encourages man to modify his behavior pattern in such a way as to bring himself more into equilibrium with his environment.

Secondary Stresses

As the body undergoes the foregoing thermoregulatory changes, a number of other internal changes occur which are not necessarily thermoregulatory themselves but which can result in bodily stress. These include dehydration, changes in blood flow, changes in body heat, feelings of comfort and discomfort, changes in food intake, and changes in work capacity.

Water loss by sweating can amount to 7 liters/day. While there may be some water saving due to reduced urinary activity, the principal water supply must come from increased intake. A considerable volume of fluids is lost from body tissues; muscle, skin, and many glands lose appreciable amounts of water in proportion to their weight. Without enough salt, water ingested from outside is not retained by the body and hence will not help stem dehydration or improve the efficiency of blood circulation.

Under hot conditions, blood flow is changed markedly. Arterioles serving the viscera, the alimentary canal, possibly the kidneys, and the muscles carry less blood although there is an increase in blood volume to balance the increase in total capacity of the circulation system. As long as the ratio of volume of blood to capacity of the circulatory system (volume/capacity ratio) remains unity, an efficient circulation system exists. However, if dehydration causes this ratio to become less than unity, then some areas will not receive all the blood they need. Usually those areas to which blood must be pumped against gravity, such as the head, will be the first to feel this effect, with the resultant failure of certain mental functions. Lack of energy, headaches, and nausea are all symptoms of a lack of blood or oxygen reaching the brain. The function of the cerebral cortex is disturbed; mental alertness is reduced followed by a lack of judgment, irritability, and finally loss of consciousness as circulatory conditions become more difficult with increasing heat stress.

One of the earliest and, unfortunately, most unreliable reactions of the body to changes in the thermal environment involves human feelings of comfort or discomfort. These reactions are influenced by many psychological factors which themselves can change with time and other stresses within the individual. Human sensations of comfort or discomfort are often not reproducible from one time to another. These sensations seem to be more strongly influenced by relative differences than by absolute values, by gradients of temperature, and especially by percent of skin wetted, since this is a particularly annoying condition. Despite these problems, careful study of the reactions of different individuals under controlled test conditions, with proper statistical treatment of the results, can provide useful information unavailable in any other way.

Reactions to heat also involve loss of appetite. This is partly a biological adaptation since the extra heat production by food alimentation is unnecessary. There appears to be some relation between decreased oxygen flow to the brain and a lowered rate of contraction of the gastric muscles, although the rate of secretion is not greatly affected. Cold conditions result in an increased appetite since carbohydrates are utilized more rapidly in the life processes under colder conditions.

Under hot conditions, the circulatory system must supply increased amounts of blood to the surface. Less oxygen reaches the cerebral cortex; the muscles receive a less adequate supply of blood. Metabolic products accumulate and fatigue or anoxia cause a decrease in mental alertness and judgment. The result will be a limited capacity for physical work. Unacclimatized individuals in tropical environments are likely to show this reaction for a few days, but under shade conditions in the Tropics a healthy, acclimatized worker should not be greatly affected. Under moderately cold conditions, the capacity to work seems to be limited only by the heavy clothes necessary for thermal equilibrium. As air temperatures fall well below zero and shivering develops, work performance decreases markedly; an increasing effort must be directed toward maintaining body temperatures. Figure 9-1 shows a typical distribution of labor efficiency in relation to thermal stress.

Factors Affecting Human Reaction to Thermal Conditions

Age, nutritional level, degree of hydration, exercise, amount of sleep, mental attitude, general health, and acclimatization all influence an individual's reaction to stressful thermal conditions. Children over 1 year in age actually withstand heat stress more easily than do adults (possibly due to better general health, better mental attitude, or better physiological adaptation as a result of greater surface area in proportion to mass). Older people often find the additional heat stress too much for a body possibly already in poor balance due to other factors; the death rate under hot conditions is higher than at other times. In general, hot conditions are more associated with circulatory failures, while cold conditions are more often associated with respiratory problems.

Caloric intake appears to be strongly correlated (negatively) with air temperature. Based on many surveys with troops under both hot and cold conditions, the caloric intake selected by the individuals themselves was found to decrease almost linearly as air temperature rises. There is little

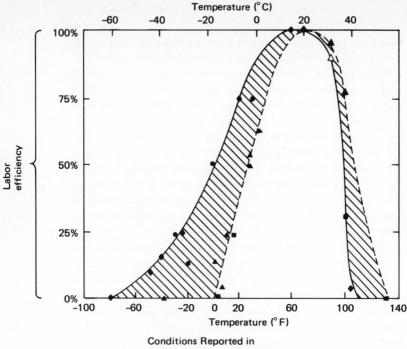

Conditions Reported in
♦ Industrial construction
□ Automotive industry
▲ Airframe industry
△ Locomotive and car building
● Petroleum exploration and recovery
■ Coke chemicals

Labor efficiency at various temperatures. Labor efficiency estimates made
independently by consultants in industries named. Compiled by L. E. Doyle,
University of Illinois. (*With permission of the American Meteorological
Society, from Russell, 1957, fig. 2, p. 5.*)   FIG. 9-1

agreement concerning the influence of thermal stress on the proportion of
protein in the diet.

Any bodily condition which affects the degree of hydration of the
body becomes a matter of vital importance under hot conditions. Thus any
condition that increases water output or withdraws water from the
circulatory system will influence the volume/capacity ratio and seriously
jeopardize the ability of the body to maintain its heat balance. As mentioned
previously, water will not be retained by the body unless there is sufficient
salt. The average diet contains 10 to 15 g of salt in excess of requirements;
the body can also store appreciable amounts. Heavy sweating, especially by
unacclimatized individuals, however, can exhaust these reserves so that
additional salt intake is desirable for proper hydration.

Lack of sleep under conditions of marked thermal stress rapidly
produces a decrease in performance and ability. While this is actually true

under all temperature conditions, sleep loss under stressful temperature conditions appears to magnify the difficulties which result.

The lowered metabolic rate during sleep reduces heat production. It eases the problem of maintaining a heat balance under hot conditions while it increases the problem under limiting cold conditions. Heat loss actually seems to be increased during sleep, which makes the problem of maintaining a proper balance under cold conditions even more difficult.

Mental attitude is a most significant aspect of any adjustment to stress. Individuals with good mental attitudes feel better in carrying out their assignments under stressful conditions and actually will be better adjusted to the environmental conditions than will others who seem to have better physical or physiological reasons for success. The reason for this importance of mental attitude is not clear.

General health is obviously connected with reaction to stress since most of the body's systems become involved in one way or another with any adaptation to stress. Therefore, any disturbance of the general bodily functions through impaired health decreases the body's ability to adjust to stress.

PHYSIOLOGICAL FAILURES UNDER CLIMATE STRESS

Failure, in this case, does not necessarily mean death but rather the point at which the individual is essentially powerless to improve his situation without outside help. There are both acute and chronic physiological failures due to heat stress.

Acute Physiological Failure under Heat Stress

1   *Heat cramps:* spasms in the voluntary muscles due to a reduction in the sodium chloride concentration in the blood below a certain critical level. The spasms will disappear with the ingestion of salt and the rise in the salt content of the blood.

2   *Dehydration:* results in a reduction of blood volume and thus produces heat exhaustion. Continued dehydration may result in other disturbances of cell functions and their deterioration. As dehydration continues, nervous irritability, depression, loss of muscular efficiency, acid accumulation in the tissues, uremia, and finally death may follow. Increased water ingestion with some addition of chlorides, glucose, and possibly alkalies will alleviate the situation.

3   *Heat exhaustion:* state of collapse resulting from lack of sufficient blood supply to the cerebral cortex. Heat exhaustion results from an inadequate heart output which, in turn, results from a fall in the volume/capacity ratio below unity. This latter factor may develop from (*a*) increasing vasodilation of the circulatory system, (*b*) decreasing blood volume or dehydration, or (*c*) lowered cardiac efficiency due to disease or lack of physical fitness or training. Heat exhaustion is possibly the most common

crisis in hot environments, being almost universally experienced to some degree. Fatigue, headache, nausea, fainting, below-normal temperatures are all common symptoms. Recovery normally follows rest in a cool place with the feet raised to help the blood flow to the brain.

4   *Heatstroke (including sunstroke):* results from failure of the heat-regulating mechanism to maintain a proper balance as the body temperature rises. The critical body temperature for man is 106°F (41°C). Above this point, deterioration sets in—the higher the body temperature, the more rapid the deterioration. Damage to the nervous system and other tissues, leading finally to death, results from too long duration at such high temperatures.

The lack of thermal balance with the environment and, hence, heatstroke may follow from (a) reduced heat loss through inhibition of sweating, inadequate peripheral circulation; (b) increased heat gain through exposure to intense sun or high temperatures; (c) increased heat production due to exercise, fever or other diseases; or (d) interference with the heat-regulating center due to injury or infection. Relief follows from cooling the body as rapidly as possible by wrapping it in wet sheets, increasing air movement, or even packing it in ice. The body temperature must not, of course, be lowered too fast.

Adolph (1947), in a classic monograph on the physiology of man in the desert, answered many earlier questions concerning water requirements under different levels of work both in the sun and in the shade. He also studied in some detail the effect of clothing and housing on the well-being of individuals exposed to desert conditions. Based in part on his work, the U.S. Quartermaster Corps produced a diagram giving the suggested daily water supply for men at rest in the shade, as well as under conditions of moderate and hard work in the sun (Fig. 9-2). At 40°C (104°F) nearly 10 liters/day of water is required for resting individuals, while at hard work in the sun the water requirements rises to about 17 liters/day. Adolph could suggest no practical method for reducing the strong feeling of thirst or the general body deterioration that accompanies lack of water. One interesting result of his work, however, was a series of maps showing survival time without water for various desert areas of the world (Fig. 9-3 for the southwestern United States is one example). The map is an excellent example of applied physiological climatology.

Chronic Physiological Failure under Heat Stress

Three long-term physiological failures under hot conditions can be noted. No similar chronic physiological failures to cold have been identified. Of these three chronic conditions, only one, heat exhaustion, has an acute counter-part.

1   *Chronic heat exhaustion* An individual whose circulatory system is barely able to meet the severe demands for heat loss in hot environments may suffer chronic heat exhaustion. Any heat overload through undue exposure to solar radiation, extra workload, or mental or physical stress can result in an

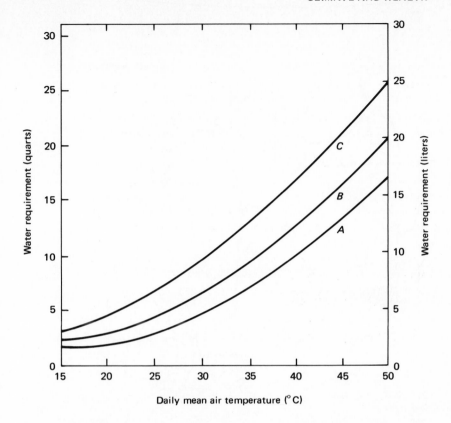

Recommendations for daily water requirements for men at rest in the shade (A), performing moderate work in the sun (B), and hard work in the sun (C). (U.S. Army Quartermaster Corps.)

FIG. 9-2

inability of the cardiovascular system to pump enough blood to the brain area; fatigue, nausea, occasional fainting will result. An individual's overall efficiency will be impaired by his inability to meet short-period overload conditions.

2   *Renal stones* There seems to be a higher incidence of renal stones in tropical dwellers, possibly due to the higher concentration of less soluble mineral salts and urates in the urine.

3   *Tropical dermatosis* Certain skin problems develop under hot conditions due to more frequent sweating, use of certain soaps, friction between wet clothing and the skin, and possibly the take-up of water by some of the superficial flaky cells. "Prickly heat" is a common problem although not particularly severe. In some cases, skin irritation, blocking of the sweat glands with plugs of hard material, actual rupture of the sweat ducts under continued pressure from accumulating sweat, and finally permanent blocking

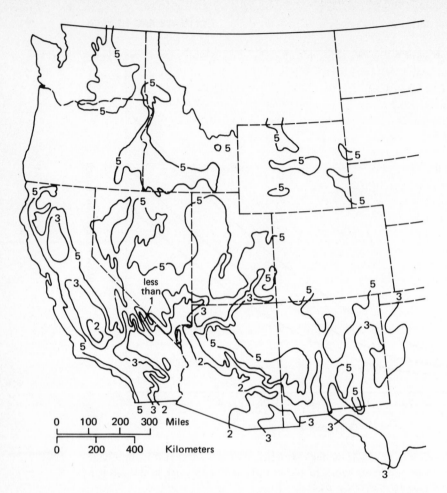

Survival time in days, without water in July in the southwestern United States. If men rest in the shade by day and walk only at night, they should be able to continue walking the number of days indicated by the lines. Not walking at all greatly increases survival time. (*Reprinted with permission of the author, from Adolph, 1947, p. 275.*)   FIG. 9-3

of the glands can develop. Lotions and powders do not help but often further increase the problem. Application of fatty substances and more effective cooling of the skin with a reduction in sweating should lead to improvement in the skin problems.

Sunburn, blistering, and even skin cancers are all associated skin problems for those exposed for periods of time to intense solar radiation. With exposure to those wavelengths of radiation that have the most pronounced effect on the skin (from 0.29 to 0.32 μm), the following sequence of events will occur: immediate pigment darkening, erythema (reddening), followed by delayed pigment darkening (Arndt et al., 1969). Skin color may change within a few minutes after exposure due to the reversible oxidation of the melanin already in existence in the pigmented skin. Skin reddening begins within 2 to 5 hr of exposure, reaching a peak about 18 to 25 hr after exposure, and then fades. Klarmann (1949) reports that if the exposure is five times greater than that necessary to produce a barely perceptible reddening, a painful burn will occur. An exposure 10 times greater than that necessary for perceptible reddening will result in a blistering burn.

Delayed darkening, resulting from new pigment formation, will occur whether the immediate pigment darkening occurred or not. This delayed darkening occurs some 2 to 3 days after exposure and reaches a maximum 4 days after exposure (Arndt et al., 1969). The radiation most responsible for reddening as well as for delayed darkening has a wavelength of 0.297 μm, although if sufficient radiation in the range 0.32 to 0.65 μm is available, both reddening and delayed darkening can occur (Pathak et al., 1962). Window glass, while screening wavelengths less than 0.32 μm, will not therefore protect completely against all the effects of solar radiation.

With exposure to sunlight, there appears to be a thickening of the stratum corneum. It is entirely possible that the ability of this layer to absorb some solar radiation provides a measure of protection in the event of prolonged exposure to insolation (Blum, 1959) although absorption by the melanin appears to be the main defense against the ultraviolet wavelengths (Kligman, 1964).

Both acute and chronic effects of exposure to ultraviolet light appear to be increased by heat. Animals maintained at higher air temperatures developed more skin tumors under ultraviolet radiation than those at lower temperatures and had a decreased life span. Augmenting the injurious effect of ultraviolet radiation by heat may also occur in man (Freeman and Knox, 1965).

One other problem, resulting from the more intense glare under tropical conditions (especially in arid regions without much shade), does seem to bother some newcomers to the Tropics. Usually the problem of "glare asthenopia," as it is called, goes away in a few days, but for some individuals it may persist, causing a general feeling of discomfort and headaches to continue.

Acute Physiological Failure under Cold Stress

When the body is subjected to cold conditions for a period of time, three areas of possible physiological failure have been identified.

1   *Hypothermia:* essentially cold stroke which results from the inability of the body to maintain a proper heat balance with the cold surroundings. It can be brought about by (a) increased heat loss due to low surrounding temperatures or increased convection from the body, due to inadequate clothing insulation; or (b) decreased heat production due to sleep, lack of activity, low food intake, or lowered body temperatures.

If the deep body temperature drops much below 95 to 96°F (35 to 36°C), the heat-regulating mechanism tends to break down and most humans become fairly inactive. There have been examples of individuals surviving with rectal temperatures below 80°F (26.7°C) but it is not common. With such low body temperatures, death usually results from heart failure or irreversible failure of the heat-regulating mechanism. To counteract hypothermia, rapid rewarming is most effective at first in order to prevent further loss of body temperature. Later, slower rewarming is preferable to permit adjustment of the circulatory system. Heating in warm or hot water is best for rewarming a large area while radiant heating is effective for more restricted areas if care is taken to prevent local overheating.

2   *Frostbite:* results from local freezing of body tissue. Not all tissues freeze at the same temperatures since factors of air movement, blood supply to the area, curvature, and distance from the central body core all combine with air or mean radiant temperature to determine the possibility of frostbite. Extremities are particularly vulnerable due to distance from the central body core.

As the tissue freezes, it becomes hard and either blue or white in color, depending on the condition of the capillary blood vessels. Damage to the tissue results usually from slowing of the blood flow to the area, clumping of red cells, or the escape of fluids through the walls. With thawing, the tissues often swell as fluid moves from the blood vessels into the tissues, causing intense pain.

3   *Trench foot:* may result if body tissues are kept wet and cold for a period of time even without freezing. Symptoms similar to frostbite are likely to occur with swelling, gangrene, or possibly death of the tissue. The foot is particularly susceptible, hence the name. If the foot can be kept warm and dry by means of various insulating and water-resistant layers, the condition should not develop.

There is some question about possible damage to the respiratory tract from breathing very cold air. There have been examples of individuals coughing blood under quite cold conditions, but animal experiments with air down to −50°C have not produced local respiratory damage. While it might be desirable to reduce the large heat loss from the respiratory tract that occurs under such conditions through some form of respiratory preheating, there is no proof that the very cold conditions result in direct damage to the respiratory tract.

ACCLIMATIZATION
Acclimatization leads to a modification of certain bodily functions to permit better adjustment to stressful climatic conditions. Often, acclimatization has been given credit for the excellent performance of some groups under stressful conditions, when better mental attitude or possibly other factors deserve the credit. While there is some degree of acclimatization under fairly severe climatic conditions, it may not be great and it often occurs rapidly

(within a few days or weeks) so that it can scarcely be used to explain many of the recorded group reactions to stressful conditions. The well-attested, increased lung capacity of natives living high in the Andes must, of course, develop slowly over time as a result of long exposure to the lowered oxygen intake and air pressures at those high elevations. Adjustments to heat and cold, however, do not normally take so long.

The ability to survive at high elevations with reduced atmospheric pressure results from the capacity of the body to adjust physiologically to the low level of oxygen at those high elevations. Oxygen cannot be stored within the body so that it must be obtained continually from the atmosphere and transported to body tissues. Hemoglobin, an iron-containing protein in the red-blood corpuscles, plays a basic role in this transfer system. The amount of oxyhemoglobin formed depends upon the partial pressure of oxygen (Edholm, 1966). When this partial pressure is 100 mm Hg (133 mb) or higher, pressures normally found in the lungs at sea level (partial pressure of oxygen in the atmosphere is about 150 mm Hg [200 mb]), the hemoglobin is essentially saturated with oxygen. The percent saturation of hemoglobin decreases as the partial pressure decreases but not in a linear relation (see Fig. 9-4). As elevations increase and pressures decrease, the percent saturation of the hemoglobin with oxygen decreases rapidly. At 18,000 ft (5,500 m), atmospheric pressure is about half that at sea level; the hemoglobin is about 50 percent saturated, and the volume of oxygen delivered to the tissues is greatly reduced. If the brain is not provided with sufficient oxygen for only a few seconds, consciousness is impaired—this can occur with a partial pressure of oxygen in the brain of less than 40 mm Hg (53 mb).

A man from lower altitudes can adapt to live at 5,500 m but some strain on the heart may develop. One of the changes that occurs in an individual moving to high elevations is the increased formation of hemoglobin to enable more oxygen to be carried to the brain and tissues. While this partially compensates for the reduced oxygen saturation of the hemoglobin itself, it is important to recognize that not only the volume of oxygen but also the pressure of the oxygen is important in the oxygen-transfer process in the lungs. Deep and rapid breathing is an adaptive change that tends to reduce the gradient in the partial pressure of the oxygen between the atmosphere and the lungs. Individuals adapted to high-altitude living have larger chests and lung capacities than lowland inhabitants. The increased hemoglobin content does increase the viscosity of the blood, making the heart work harder in pumping blood. This, of course, increases the possibility of heart damage (Edholm, 1966, p. 343).

Little real evidence exists of any appreciable acclimatization to cold conditions. There is, of course, a learning factor so that individuals increase their ability to conserve heat, but this is not acclimatization. Man in a cold environment must maintain a nearly constant body temperature. Unable to make many bodily adjustments, he has succeeded in his effort to maintain a constant body temperature through clothing and shelter.

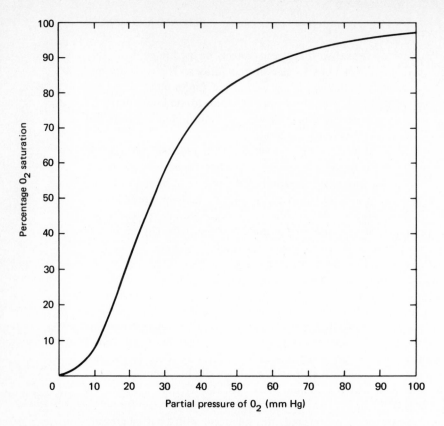

The dissociation curve of oxyhemoglobin. At a partial pressure of oxygen of 100 mm Hg, hemoglobin is about 97 percent saturated. (*With permission of the Royal Meteorological Society, from Edholm,* 1966, *p.* 342.)

FIG. 9-4

Man is much more able to show rapid adjustments to hot conditions than to cold conditions. One significant adjustment occurs in the case of the rate of sweating. A newcomer to the Tropics will not sweat as much as natives. As a result he feels more uncomfortable; with a smaller rate of heat loss, his body temperature rises and he may even suffer heat collapse. If he is exposed to the hot conditions gradually, the newcomer very rapidly increases his rate of sweating, which increases evaporative cooling and keeps the rise in body temperature within acceptable limits.

There is little real evidence of any reduction in basal metabolic rate as a result of exposure to hot conditions although, clearly, there is a reduction in free metabolism. There may also be a greater degree of muscular relaxation during periods of inactivity which will lower heat production slightly. A slower pace, avoidance of unnecessary effort and exertion, a more relaxed approach are all part of the way of life under hot conditions. They serve to adjust heat production to the stress of the environment but they are really not part of acclimatization—rather they might be considered acculturization.

The salt concentration in the sweat of unacclimatized individuals is greater and more variable than is the salt content in the sweat of native tropical dwellers. Some workers have also suggested a 5 to 10 percent increase in blood volume with acclimatization to the Tropics although these results need to be checked further due to the lack of precision in sampling methods. When individuals in the Tropics are pushed to the point of heat exhaustion, the rise in the pulse rate does not seem to be as great in acclimatized as in unacclimatized individuals. This may result, in part, from a slightly lower body temperature by those native to the area as well as a reaction to experience, which has taught that such an increase in pulse rate is largely useless as long as there is not a satisfactory volume/capacity ratio.

It must be understood in all these discussions that individual variations can be significant and can mask any of the foregoing small adjustments due to age, acclimatization, health, or mental attitude. There are great variations in individual reactions to both heat and cold. While some of these differences are actually due to physical or physiological disabilities, there does appear to be a wide variation in reactions among individuals who are otherwise fairly healthy and of normal appearance and reactions. It is these variations that make the definition of average comfort zones so difficult.

AIR POLLUTION AND HEALTH

There is a growing literature on the relation between air pollution and human health. Since air pollution is primarily a meteorological phenomenon, such relations are within the purview of the applied climatologist working with the proper health officials. Exceptional situations such as the December 1952 air pollution episode in London in which thousands of lives were lost, or the less severe examples of Donora, Pennsylvania, in 1948 or the Meuse Valley in 1930, are well documented pollution-weather relationships. More significant and more difficult to investigate are such questions as "how harmful to human health are the ordinary levels of air pollution found in many urban environments," or "is continued exposure to low levels of air pollution as harmful as short-term exposure to very high levels of pollution?" The unknown effects of prolonged exposure to low levels of pollution are particularly frustrating; we need the information to evaluate our current environmental crisis, but sufficient reliable and definitive data are just not available.

One recent study by Hodgson (1970), of Cornell Medical College, has provided some answers concerning the effect of normal day-to-day variations in air pollution. Hodgson analyzed, by multiple regression techniques, the monthly figures of mortality, air pollution levels, and temperatures for the 31-month period of November 1962 through May 1965 for New York City. He divided his mortality data into two age groups, 65 and over, and 64 and under, in order to study age effects as well. Hodgson felt that the 31-month period would be short enough to eliminate dietary or socioeconomic changes in the large New York City population. He also considered that other factors (such as heart attacks resulting from overexertion or excitement) should have only slight influence on such a large population and should be relatively stable except for clearly climatic situations such as the increased heart attacks during heavy winter snowfalls.

Hodgson considered both gaseous and particulate air pollution. He used sulfur dioxide measurements to represent gaseous pollutants. Particulate matter was included in the investigation since there is an indication that the combined effect of sulfur compounds and particulates produces more severe stresses than does either one of these pollutants alone. Air temperature was also studied because of the recognized stressful effects of extreme temperatures independent of air pollution conditions. The study did not consider the cumulative or long-range effects of pollution or temperature stress but rather only the short-run or essentially day-to-day relations between mortality and selected meteorologic variables.

Hodgson determined the average daily mortality per month due to respiratory conditions, heart conditions, and all other causes. He also determined the average daily concentration per month of particulate matter and sulfur dioxide, as well as the average daily number of degree-days per month.

A highly significant relation was found between mortality from respiratory and heart conditions and both the concentration of particulate matter (in coh units) and degree-days. Sulfur dioxide concentration did not seem to be a highly significant variable. Hodgson concluded that nearly three-fourths of the variation in mortality from heart and respiratory diseases could be explained by variations in air pollution and air temperature. Air pollution appeared to be more significant than temperature.

Even more important was the finding that a change in 1 coh unit in average daily concentration of particulate matter resulted in nearly a 9 percent change in average daily mortality from respiratory and heart diseases. Concentrations may actually change from day to day by 5 coh units or more with a 2-coh unit change from month to month entirely reasonable. This would lead to monthly changes in mortality of 15 to 20 percent. The essentially linear relation found led Hodgson to conclude that changes in low levels of air pollution were as significant and resulted in the same percentage change in mortality as did changes in high levels of pollution. Statistical tests using the data separated by age showed little difference in susceptibility to environmental conditions for those under 65 as opposed to those over 65. In fact, contrary to expectation, mortality in those under 65 seemed to be more highly related to pollution conditions than in those over 65.

In another recent study of air pollution and health, Lave and Seskin (1970) tried to evaluate the economic costs of the ill health resulting from air pollution. Using available health statistics, the authors formulated a series of multiple regression coefficients relating mortality and morbidity from bronchitis, lung cancer, other cancers, and pneumonia, to air pollution. The health data were separated into various population groups (males, females, country–city dwellers, smokers, nonsmokers, etc.) From the regression coefficients, the authors concluded that death from bronchitis could be reduced anywhere from 25 to 50 percent by a reduction of the level of air pollution to the *lowest* level now found within the areas being studied. Similarly, death from lung cancer could be reduced by 11 to 44 percent by such a reduction in air pollution levels. Hagstrom and his associates (1967) reported the death rate from cancer to be 25 percent higher in polluted areas than in areas of relatively clean air in a study of the Nashville, Tennessee, area, a figure quite comparable to that reported by Lave and Seskin. These authors felt that their figures justified the conclusion that a reduction of air pollution to the lowest levels currently existing within each area would reduce mortality and morbidity rates for heart disease by 10 to 15 percent.

Absenteeism can also be related to air pollution episodes. For example, Cornwall and Raffle (1961), in a study of the work habits of bus drivers, found that 20 to 35 percent of the absences attributed to sickness could really be associated with air pollution problems. Dohan and Taylor (1960) obtained a correlation coefficient of 0.96 between SO_3 concentration in the air and absences of more than a week due to respiratory diseases for female employees of RCA.

Summing these and other results, Lave and Seskin conclude that decreasing particulate pollution by 10 percent would lower the total death rate by 0.5 percent while decreasing the minimum concentration of sulfates by 10 percent would result in a decrease in the total death rate by 0.4 percent. Considering the present-day amounts spent on each illness, they suggest possible savings that could be made by reducing air pollution to the lowest levels currently found in the areas under study (Table 9-1).

The figures are only approximate, and they do not include information on other illnesses related to air pollution or the costs involved in man-days of work lost. They do, however, suggest something of the significant economic loss resulting from air pollution and emphasize the pressing need to reduce high values of pollution to levels at least as low as those now found within different local urban areas. The study has attempted to provide realistic figures in that the values achieved are those that would result from a change not to absolutely clean air but to air still containing certain low levels of pollution currently found in nature, a much more feasible goal of pollution abatement.

TABLE 9-1

Estimated dollar savings by reducing air pollution to natural background levels*

Bronchitis	$250–$ 500 million
Lung cancer	$ 33 million
Respiratory diseases	$1222 million
Cardiovascular mortality	$ 468 million
Other cancers	$ 390 million

* *Lave and Seskin, 1970, copyright 1970, by the American Association for the Advancement of Science.*

The preceding section has emphasized the effect of air pollution on health and mortality. There is an equally serious environmental stress from heat, especially in the big urban heat islands, that must be considered in any evaluation of environmental problems. While there is a growing recognition of the problem of thermal stress, many reasons exist why it is difficult to obtain a clear picture of the actual effect of thermal stress. First, detailed statistics on mortality or morbidity due to heat stress are not available. Ellis (1972) has pointed out that heat illness and death in a healthy community in peacetime can almost always be prevented. Listing the cause of death as due to heat may result in some inquiry, liability for damages, criticism, family distress, or other problems. Thus, it is often more convenient to list other factors, undoubtedly aggravated by the heat stress, as the actual cause of death. This makes it difficult to obtain more than just rough estimates of heat deaths; additional analyses are necessary in order to obtain an understanding of the seriousness of the thermal stress problem.

In an analysis of 5 years with marked heat waves, 1952 to 1955 and 1966, Ellis concluded that excess deaths resulting from heat-aggravated or heat-related illness were at least 10 times greater than the actual number of deaths listed as due to excessive heat in the General Mortality Tables. Thus the effect of a stressful climatic condition can be seen more in the excess of deaths from all causes rather than from deaths just due to the specific climatic condition because of errors in death certification.

Another reason for the lack of clearly defined data involves the complex interrelationships among many factors of stress possibly masking the effect of any one individual factor. Thus air pollution episodes at the time of heat stress might make it difficult to determine the contribution of each stress to the actual mortality rates.

Buechley and his associates (1972) have achieved the following relation to describe the relation between temperature and mortality in the eastern United States:

$$TMR = \text{cycle} + 10_e^{0.2\,(F^1 - 90)}$$

where TMR = temperature-specific mortality ratio forecast
  cycle = expected mortality ratio for the date
  e = base of Naperian logarithms
  F^1 = previous day's maximum temperature, °F

Evaluation of this relation shows that with maximum temperatures at 95°F (35°C) excess total mortality will be 27 percent of the year's average daily mortality, while at 100°F (37.8°C) it will rise to 75 percent of the yearly average daily mortality. At 105°F (40.6°C) excess total mortality will be some 200 percent of the average daily mortality, while at 110°F (43.3°C) the excess jumps to 546 percent of the yearly average.

By considering the distribution of deaths both within urban and surrounding areas, these investigators were able to suggest the possible additional effects that stress created by the urban heat island might have on death rate. Analyzing data for the very hot first week of July 1966 in the New York City area, these authors achieved the following relation among the temperature-specific mortality ratio forecast (TMR) and various characteristics of the local community.

TMR = -99.11 + 2.73 min. temp. + 26.24 log density -1.55 X % of incomes over \$10,000 + 2.13$e^{0.2}$ (max $F-90$)

Substituting data for suburban areas around New York City into that relation provided a forecast of about 160 percent over expected mortality for July 4, 1966, while a value of 260 percent over expected mortality is found for the core New York City areas. The difference, some 150 to 200 extra deaths in the core city area, can be attributed to the urban heat island as well as to the city living conditions. Henschel et al. (1968) concluded that during the hot July of 1966 the death rate in St. Louis due to heat within the city limits was some 5.5 times greater than in the suburban area around the city proper. Clarke (1972), in analyzing the effects of urban structure on heat deaths, pointed out that the lack of adequate relief from high temperatures at night during heat waves may be the most significant climatic difference in heat stress between the city and suburban areas. He also cautioned that many nonthermal factors need to be considered in the analysis: more affluent suburban dwellers might have better housing, a higher percentage of air conditioned houses, more knowledge of proper health precautions in hot weather, more medical care available, different physical activity, and more adequate clothing to survive the stressful conditions. Thus care must be taken in drawing conclusions from available tabulated data and simple regression coefficients.

In a comprehensive study of death from heat illness and heat-aggravated illness in the United States, Ellis (1972) suggests that excessive heat may be more lethal than high levels of air pollution. Gover (1938) concluded that high death rates due to heat followed by a day several successive days of extreme temperature. Excess death rates during a second extreme temperature period during a year are quite small when compared with the excess death rates from the first hot period for the summer.

Ellis went on to conclude that infant (less than 1 year old) deaths from acute heat illness were very much greater than for any other age group below 50 years. This he attributed to the less stable thermoregulatory control in infants and the greater sweat loss in proportion to body weight than for adults. Healthy older people are capable of adjusting to the stressful hot conditions, although as age increases, the ability to adjust to limiting conditions appears to decrease. Ellis indicates that the higher mortality rates due to heat illness in those over 70 years is probably not due to lack of acclimatization but rather to the increased demands made on the less than perfect cardiovascular and cerebrovascular systems.

Ellis devotes some time to considering the effects of our present air conditioned society. While he recognizes air conditioning as a first line of defense against heat, he points out that those accustomed to air conditioning are not acclimatized to stressful conditions and may suffer the most if power shortages eliminate this artificial control of high temperatures and humidities.

MORTALITY, MORBIDITY, AND THE SYNOPTIC WEATHER PATTERN

Study of the body's reaction to individual weather elements (such as in a controlled climate chamber) provides some understanding of the direct physiologic responses to these stressful conditions. But, it is not enough to consider a body's reaction to an isolated drop in temperature, for example. In nature, the drop in temperature is often associated with the passage of a cold front and the arrival of a mass of cold polar air replacing a warm, moist, tropical air mass. This change is often accompanied by precipitation, changing cloudiness, a wind shift, and a change in pressure. While the influence of each of these individual items can be studied by itself, it is clear that it is the combined influence of all the factors acting in concert that is of greatest significance to the body's reaction. We need to consider the whole complex of meteorological elements as they are instantaneously grouped to form what we call weather and seek to relate this to human morbidity and comfort sensations.

One important analysis technique in applied climatology involves relating the phenomenon under study to the general synoptic weather situation. As described by Jacobs (1947), the analysis technique involves locating the phenomenon under study in its proper position with respect to idealized fronts, air masses, and circulation conditions as they might occur on an average weather map. Thus, in studying the influence of weather on aircraft accidents, each accident would be plotted on a hypothetical weather map in relation to standardized pressure systems and fronts (that is, 200 km ahead of a warm front, 30 km behind a strong cold front, in the warm air wedge between a frontal wave, etc.). The juxtaposition of the accidents and idealized weather map phenomena provides a revealing insight into possible weather influences on accidents. Such an analysis technique has also provided new insights into the influence of weather on mortality and morbidity.

Let us consider a rather simplified weather pattern to relate to the available data of illness and physiologic stress. Weather maps are prepared by plotting data of pressure, temperature, winds, cloudiness, humidity, and other weather factors taken simultaneously at many different stations. Lines are then drawn on these maps to connect places that have the same atmospheric pressure. These isobaric lines reveal areas of the map in which the pressures are higher than others as well as areas where the pressures are lower than those around them. These high- and low-pressure areas have distinct circulation patterns, with the air spiraling around the high-pressure area in a clockwise direction in the Northern Hemisphere and moving slowly outward from the center and spiraling inward toward the center of a low-pressure area with a counterclockwise motion.

The air slowly moving outward from the center of the high at the earth's surface must be replaced, and this can only occur as a result of subsidence of air from higher elevations. As this air descends, it is heated adiabatically since it is moving into a region of higher pressure and work is done on it to compress it. Because its saturation vapor pressure increases with temperature, clouds or precipitation will probably not form in this air even if moisture is evaporated into it. The high-pressure area is generally a region of clear skies and fair weather.

The low-pressure area or cyclone at the earth's surface (not to be confused with the special intense form of cyclone more properly called a tornado) has air spiraling in toward the center. This air can only move out of the center by rising upward. As air ascends into regions of lower pressure it cools adiabatically. Cool air contains less moisture in vapor form than warm air. Thus if the air is near saturation, only a small amount of ascending motion is necessary to cause clouds to form and possibly to initiate precipitation. Consequently, greater amounts of cloudiness and precipitation are associated with low-pressure areas than with high-pressure areas.

As air spirals in toward the center of a low-pressure area, there is an opportunity for cold air from the north or northwest on the left side of the cyclone center (in the Northern Hemisphere) to come into contact with warmer air from the south or southeast on the right side of the center. Thus, along a roughly north-south (or northeast-southwest) line through the center of the low-pressure area, one can often identify the boundary between these two unlike masses of air. Such a boundary or dividing line is called a "front": a cold front if cold air is replacing warm air at the surface and a warm front if warm air is moving in to replace cold air at the surface. Since the warm air is lighter and less dense than the cold air, it will rise over the cold mass of air. In the process it cools adiabatically and may give up moisture in the form of clouds and precipitation. Thus fronts in low-pressure areas may contribute to the bad weather often found in those systems.

Low- and high-pressure areas are quite large in extent, often being many hundreds of miles in diameter. They move slowly across the country, usually with the speed and in the general direction of the upper-level airflow. Figure 9-5 is an example of a daily weather map adapted from a U.S. Weather Bureau map, showing a deep low-pressure area over western Kansas with moderate high-pressure centers located to the east and west over the New England states and over the Great Basin area of Utah and Nevada. A warm front extends eastward from the low center through central Kansas and Missouri. Further east a cold front is found through Tennessee and North Carolina. A fairly strong cold front runs south from the center of the low through the Panhandle of Texas and Oklahoma. The frontal system separates the warm, moist maritime tropical air from the Gulf of Mexico over the southeastern quadrant of the country from the colder, drier continental polar air covering the northern and western portion of the country. In Kansas and Missouri, the warm air is slowly pushing northward (a warm front), while east of this, along the same frontal surface, the cold air is pushing southward (a cold front).

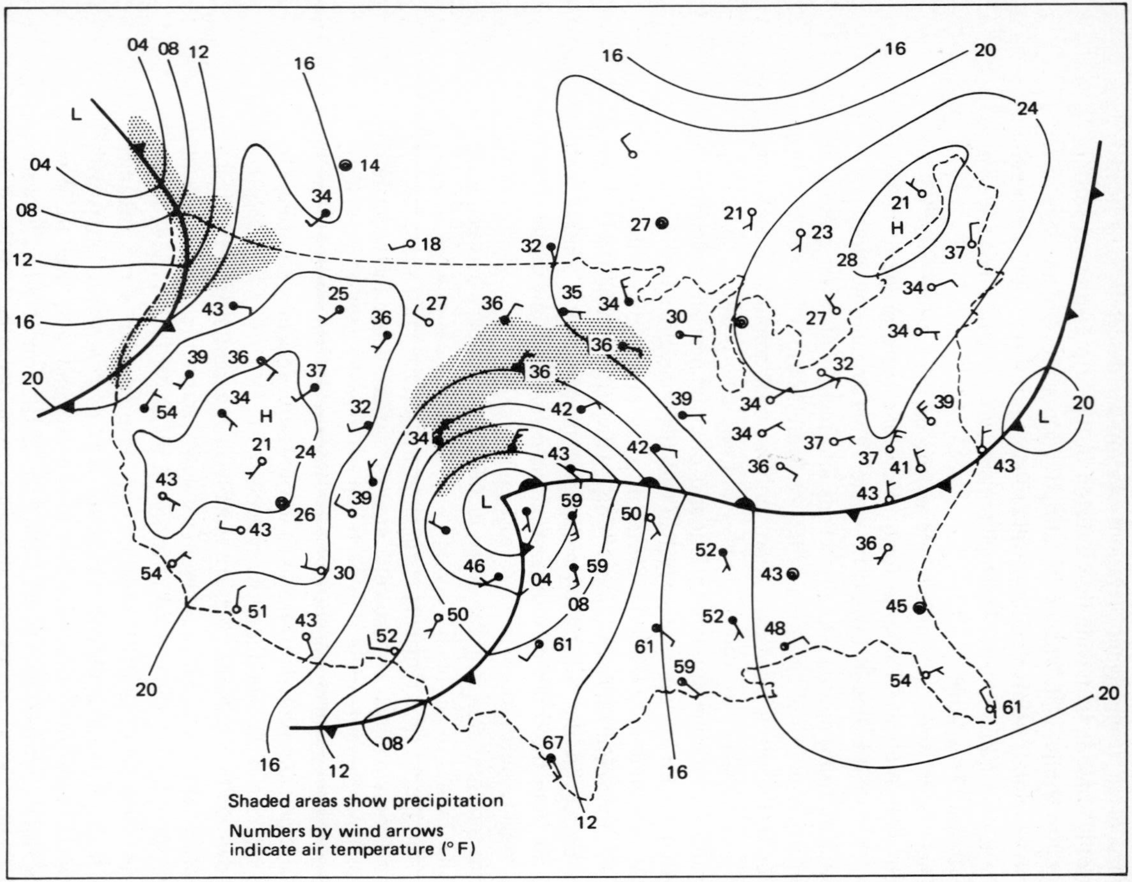

Weather map for Sunday, Nov. 8, 1970. *(Adapted from NOAA, the National Weather Service.)*   FIG. 9-5

It is fairly simple to describe the circulation and weather conditions accompanying idealized fronts and pressure systems. Let us consider a typical sequence of weather starting with a high-pressure area over our locality. In the center of the high the sky is generally clear, the air fairly calm and stable, and there are few if any clouds so that an appreciable amount of sunlight occurs, provided air pollution is not a factor. In our modern society, we find that stable high-pressure areas may lead to conditions of increased turbidity since pollutants will not diffuse rapidly but rather are concentrated near the ground by the descending air.

As the high moves slowly eastward we will find ourselves located on the left side of the high with generally southerly wind flow. This will be light and variable at first, but as the low-pressure area which follows the high approaches, the wind should pick up in strength and regularity. Airflow around the forward edge of a low-pressure area will be from the south or southeast, and continued flow from this direction in the Northern Hemisphere should result in bringing warm, moist air over our locality.

The leading edge of a mass of warm, moist air may be identified as a warm front. Far ahead or to the north of the warm front one would experience first a high cloud deck of cirrus or cirrostratus clouds that will, in time, lower to a complete layer of altostratus and finally to low stratus clouds with light continuous rain. When the warm front passes, the wind shifts rapidly from easterly to southerly, the precipitation stops, and the sky clears. Warm, muggy weather is often experienced after the front passes.

As the low-pressure area continues to move eastward, a cold mass of air will approach from the west. We will notice a build-up of towering cumulus clouds and possibly see a line of thunderstorms approaching. As the cold front passes, some heavy showers or thunderstorms may be expected, the wind will shift rapidly from southerly to west or northwest, the temperature will drop rapidly, and the warm, muggy air will be replaced by brisk, invigorating air from the north. Wind speed often picks up. The low-pressure area moves on eastward and is replaced by a high-pressure area. First the airflow is from the north around the forward side of the high, the skies clear, and the wind begins to die down. As the center of the high approaches, the wind becomes light and variable and the skies are almost clear.

This ideal sequence of weather, which occurs often in somewhat modified form in the midlatitudes of the Northern Hemisphere, is illustrated in Fig. 9-6. The weather that occurs during the passage of these various systems can be divided into a number of phases as indicated on the diagram. Phases 1 and 2 are associated with conditions in the high-pressure area, phase 3 with the southeasterly flow and rainy conditions ahead of a warm front, phase 4 with conditions in the increasingly strong warm, moist air from the south or southwest ahead of the approaching cold front, phase 5 in the cold front itself, and phase 6 in the strong northerly or northwesterly flow behind the cold front. Phase 1 would then be repeated again as the second high-pressure area arrives at the station.

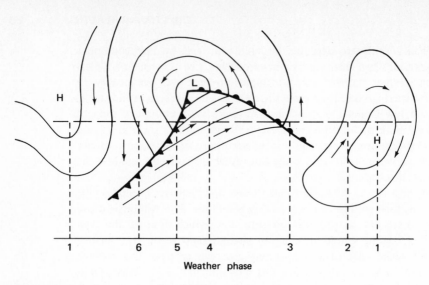

Weather phase

Cross section through idealized high- and low-pressure systems, illustrating   FIG. 9-6
different weather phases (see text).

To apply the synoptic climatology technique to the analysis of mortality and morbidity statistics, the figures on human illness and death can be related to one of the six listed weather phases. This has been done for a great many human reactions and illnesses with illuminating results.

As an example, Landsberg (1969b) reports on a German test of human reaction time and weather, based on some 20,000 visitors to a traffic exhibition during a 10-week period. While considerable individual differences were found in the time needed to push a button following a light change, when the reaction times were grouped by weather phase, a significant relation was found. Fast reaction times occurred with weather phases 1 and 2. Reaction times slowed in phase 3 and were slowest of all in weather phase 4. During phase 5, reaction times were still slow. The reaction times in weather phase 6 were somewhat faster, being about the average of the reaction times for all the previous five phases.

Weather phase 4 is associated with the warm, muggy weather in the warm sector of a developing frontal wave. The barometer is falling as the low-pressure area approaches, and the high humidities and warm air temperatures along with the scattered thunderstorm activity sometimes found ahead of the approaching cold front often result in considerable physiological and psychological stress for many humans. This may slow their reaction times.

Industrial accidents have also been related to idealized synoptic conditions. Weather phases 3 to 5 were associated with an above-average number of time-lost accidents, while phases 1, 2, and 6 had a below-average number of such accidents in one industrial plant with some 6,000 workers. While some accidents may be related to slower reaction times, other factors such as lack of carefulness or mental alertness are undoubtedly related to the weather phase also. Other studies have shown that accident rates seem to be related to temperature extremes. Temperatures over 24°C or below 12°C are both correlated to higher accident rates. Thus it is possible that any significant weather stress may result in a higher than normal accident rate due to slower reaction times, less concentration on the task at hand, and a subconscious concern for the return of more comfortable conditions, all of which prevent full attention being given to safety considerations.

Brezowsky (1964) has provided us with a thorough summary of weather and morbidity relationships, including a number of diagrams which correlate weather phase with deviations in percentage from the normal or expected value. Figure 9-7 shows the variation of nine different clinical complaints, averaged together, and expressed as deviations from normal during the weather sequence from a cold, clear high-pressure area through a warm, moist low-pressure area to the cold, dry conditions following the low center. The weather phases used by Brezowsky are slightly different from those described previously. Phases 1, 2, and 3 in the present case correspond to phases 1 and 2 of Fig. 9-6, while phases 5 and 6 correspond to the earlier phase 6. In spite of slight changes in nomenclature, there is a significant increase in morbidity as the weather changes from the cool, fair weather of the high-pressure area to the more oppressive warm, muggy weather associated with the warm sector of a cyclone. The return to cooler, drier conditions results in a return to below-normal morbidity statistics.

Plotting deviation in number of deaths per 6-hr interval in Munich against the phase or type of weather results in the same pattern of change as found for morbidity (Fig. 9-8). There is a strong negative deviation in number of deaths during the cooler and drier periods with a marked positive deviation in deaths during the more oppressive warmer, wetter periods. Four years of death statistics were used by Brezowsky in preparing this graph. Since high positive deviations were found with both warm, muggy weather and with cold, humid weather (and a slight positive relation was found in extremely dry weather), it appears that stressful conditions of any type, whether cold, warm, wet, or dry, can result in an increase in mortality rates.

One significant and well-reported relationship between weather and human reactions occurred with the hurricane of September 1938 in New England (Huntington, 1945). The freshman class at Massachusetts State College was taking standard psychological tests on the day before, the day of, and the day after the passage of the hurricane. Scores on the first and last day were nearly average, being slightly above on the first day and somewhat below on the third (actually the fourth day since the students helped clean up on the day after the hurricane). Scores received on the second day, the day of actual passage of the hurricane, were so high that the college average increased from the 75th percentile (long-term average) to the 95th percentile on that test. While the students were not experiencing any real stress that day, they were certainly aware of the stressful conditions outside. Evidently they were not distracted and they did not suffer from any lack of

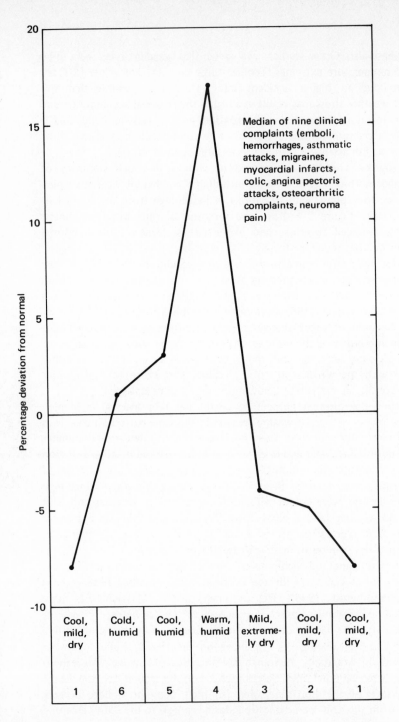

Variation with weather phase of nine clinical complaints. (*With permission of Elizabeth Licht, Publ., Physical Medicine Library, after Brezowsky,* 1964, *p.* 366.)   FIG. 9-7

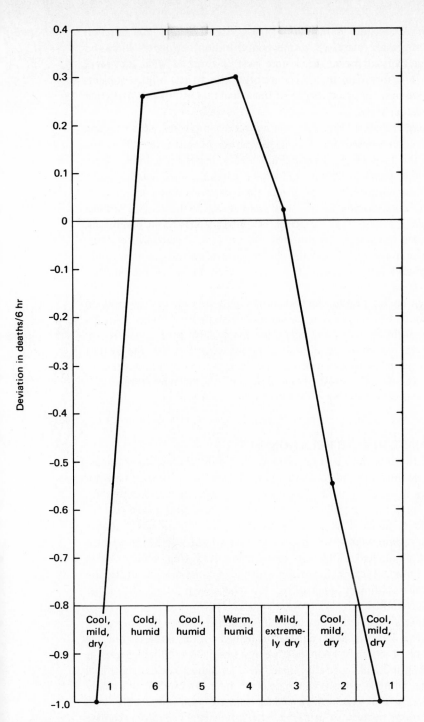

Relation of deaths to weather phases in Munich from 1954 to 1957. (*With permission of Elizabeth Licht, Publ., Physical Medicine Library, after Brezowsky, 1964, p. 377.*)   FIG. 9-8

concentration or decline in mental alertness. Something in the situation (Huntington suggested increased ozone concentration) evidently increased their mental activity or made them more alert or aware of what they were doing. While it is fortunate that such opportunities to test human subjects come rarely, we may never be certain of the actual reasons for this particular reaction to weather stress.

Landsberg (1969b) has also related scores on tests with climatic conditions. He considered the feelings of stress or comfort within the classroom at the time of the learning and testing experience. In this case, several carefully matched groups of school children were taught under different room conditions. For one group, the conditions were controlled by heating and air conditioning so that they were always in the zone of optimal comfort, while a second group experienced ordinary classroom conditions, heated as needed but not air conditioned. On the basis of standardized tests, it was found that the children in the climate-controlled classroom not only learned faster and more successfully but also made higher scores on their tests.

To check on the results, the two groups were reversed and the group in the climate-controlled room again learned better and scored higher than the group in the ordinary classroom. Since the major difference in the rooms resulted from the use of air conditioning, it is not surprising that the greatest differences in the test scores between the two groups came at a time when air conditioning was needed. Stress at the warm end of the temperature scale slowed the process of learning or resulted in less attention or care in lesson preparation.

OTHER CLIMATE-HEALTH RELATIONSHIPS

In a study of the seasonal variation of mortality in the United States, Japan, and England, Momiyama and Katayama (1966) show that Americans are not as affected by seasonal climatic differences as inhabitants of other nations. They show (Fig. 9-9) essentially no change in death rate through the year for American infants under 1 year of age and only a slight seasonal change (greater death rate in winter) for individuals over 70 years. In England, there is a slightly more marked winter maximum in the death rate, while in Japan there is a well-marked seasonal variation with the maximum of the death rate in winter. Similar, although less marked, differences were found when other age groups were studied.

The authors reason that seasonal climatic stress causes the changes in death rate. The lack of any change through the year results from the widespread use of efficient room-heating systems in the United States as opposed to England and especially Japan. With an almost constant artificial indoor climate, climatic stress is eliminated and the death rate remains nearly constant through the year. Where indoor climate varies markedly with less-efficient room heaters as in Japan, the influence of outdoor climate becomes more clearly defined. Their data not only suggest a seasonal climatic influence on deaths but also provide strong motivation for effective climate conditioning of one's home.

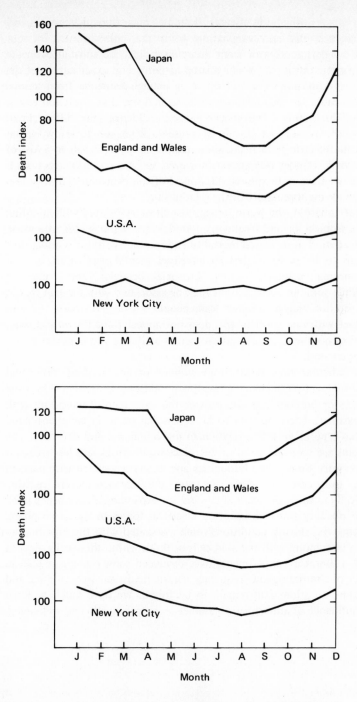

Seasonal fluctuations of total death index by countries for (top diagram)
children under 1 year (1958 to 1962) and for (bottom diagram) individuals
over 70 (1958 to 1962). (*With permission of the Meteorological Research
Institute, Tokyo, Japan, from Momiyama and Katayama, 1966, p. 282.*)

FIG. 9-9

There is a growing body of evidence that electromagnetic waves, as well as ions (charged particles originating from the bombardment of our atmosphere by cosmic rays or from decaying radioactive substances in the earth), have some influence on our mental activity. The exact nature of the relationship is still unknown but there is enough evidence from animal experiments, and a few short human tests, to indicate that a relation exists. Possibly the fluctuating electromagnetic field influences the human brain waves although we are not certain at present. What we do know is that changes in the electric field make some people sleepy or result in a loss of mental alertness. It does not appear, however, to be a universal reaction; it may be that we have not explored all aspects of the problem so that certain unknowns mask the true nature of the relationship.

Weather phase 4, the warm, muggy condition associated with an influx of maritime tropical air and frequently found in the warm sector of a wave, not only is a period of increased mortality and morbidity, but it is also a time which seems to be more related to increased mental depression, acts of suicide, abnormal conduct, rioting, antisocial behavior, and crimes of violence. While some of these actions may be associated with the fact that in city areas the hot, muggy weather leads people out on the streets and into closer contact with one another, this does not appear to be the only answer. The role of atmospheric conditions on mental attitudes and behavior is still not well understood.

Scars, whether they result from wounds or amputations, often will show a painful reaction to changing weather conditions. Deformed skin tissue such as corns or bunions can also provide the same painful sensations with changing weather. There appears to be a seasonal trend in weather-related pains of this type as well as a relation to individual wet and dry days. The seasonal pains are more frequent in wet or cloudy periods and less frequent during sunny, dry times. High humidities are closely associated with pains in scar tissues and corns. Whether the skin and scar tissues are reacting hygroscopically to the higher humidities and are expanding, thus causing pain, or whether possibly some unsuspected electrical forces in the atmosphere, accompanying the stormy conditions often associated with the high humidity, may be interacting with the skin cells or fluids within these cells remains to be seen. Laboratory experiments have produced some scar-tissue pain in the presence of alternating low-frequency electric fields. Answers to these and other puzzling questions still remain to be found. We have much to learn about the influence of atmospheric variables on human health and well-being.

CLIMATE 10
AND ARCHITECTURE

The migration of man into regions with climates less and less satisfactory to his needs for physical comfort has forced him to turn increasingly to clothing and shelter to modify the climatic environment. Man did not take long to understand the different properties of building materials and to adapt local materials with desirable properties in order to modify climatic extremes. In fact, it is quite instructive to study the housing of the early native peoples of an area; often, they have used local materials and their own local designs to achieve a marked degree of adjustment to their climatic environment. This has led to a wide variety of shelter types from the Arctic to the Tropics, from the humid areas to the arid—a variety that we no longer see as more reliance is placed on mechanical means (air conditioners, dehumidifiers, furnaces and space heaters, humidifiers, etc.) to create artificially a climate that satisfies our need for comfort.

REGIONAL VARIETY AND MECHANISTIC INVARIABILITY
Olgyay (1963) has discussed the regional differentiation of shelter types used by various native groups living under widely different climatic conditions in North America and eastern Siberia. The Eskimo in Alaska and northern Canada exist in an area of extreme cold and, thus, need shelters in which the heat produced by limited fuel supplies can be conserved and concentrated into a relatively small space. The Eskimo house is usually of a semi-subterranean nature, made of wood, whalebone, or in some cases ice and

snow. The wood or whalebone are used merely for the framework, the rest of the house being covered with sod, moss, or other available material. The igloo, being low and hemispherical in shape, offers little resistance to the wind and creates few eddies to increase turbulent heat exchange. The ice and snow shelter takes advantage of the excellent insulating properties of snow, while the icy interior effectively seals the structure from all air leaks. Entrances through tunnels opening downwind from the prevailing wind direction and at a lower elevation than the interior living quarters keep the cold air out of the shelter and trap within the shelter the air warmed by small whale or seal oil lamps and by human bodies. So effective is this type of structure that the Eskimo, especially the children, often use the minimum of clothing (or none at all) when inside, even though the exterior temperature is well below $-20°C$.

The tribes of northeastern Siberia, the Chukchee and the Koryak, are also well equipped to meet the rigorous winter weather of their homeland. They live in two-compartment houses consisting of a fur box, used for sleeping, placed inside a conical tent or a semi-subterranean sod house. The purpose of the fur box is to retain the heat generated by the presence of many bodies grouped together inside in sleep. The area between the sleeping compartment and the walls of the outside tent or house is used for storage and for general household chores during the day. The house, most especially the inner fur box, is heated with stone lamps. There is also some evidence that, in past times, some of the Chukchee used a sunken sod house with ribs of wood or whalebone and a subterranean entrance much as do the American Eskimo of today.

Along the Pacific Northwest coast, the Indians build large rectangular houses of split planks fastened to upright posts. Generally more than one family lives in each house. Sometimes the houses contain a central fire pit, in which case the higher regions along the walls are for sleeping (Martin et al., 1947). In some cases these plank houses are joined together, reducing the number of exposed surfaces; often, the shelter-within-a-shelter construction of the Chukchee and Koryak is employed. In areas with milder summers, the outer shelter can be removed for increased ventilation.

In the plateau region of the upper Columbia River, the Indians in pre-Columbian time built houses of a semi-subterranean nature, usually an earth lodge or pit-type house. The earth-lodge house was buried in the ground 4 to 6 ft and was 10 to 15 ft in diameter. It had a flat or conical roof and was covered with cedar planks, grass, mats, and brush. The smoke hole in the top of the roof was used in the wintertime for the entranceway.

Early Indians of central North America experienced fewer climatic extremes and, therefore, could achieve somewhat greater diversity in their housing. Typically, they adapted some form of wigwam, tepee, or wickiup, a fairly conical structure of poles covered with available skins and heated by a central fire. Entranceways were covered by flaps of skin which could be rolled or pulled back in warmer periods. These structures were quite portable, easily ventilated in warm periods, and fairly weatherproof in the colder periods of the year. Furs for added warmth were also used by individuals inside during cold periods.

Under the hot-dry climatic conditions found in the southwestern part of the United States, extreme climatic conditions again forced the native tribes to build with ingenuity. Excessive heat, hot, dry winds, and bright sun are the limiting climatic factors, and the resulting shelters were well designed to protect against these climatic factors. Adjoining structures were often built in order to expose fewer surfaces to the hot, dry winds and intense sun. Thick, adobe roofs and walls were employed because of their excellent insulating properties and their ability to slow the transfer of heat through the structure, so that the peak of heat load reaching the inside is delayed until after sunset when the outside air is much cooler. Windows were small and limited in number to prevent the entrance of hot, dry winds and radiation. Communal buildings were often oriented east-west to reduce the solar radiation loading in morning and afternoon hours. Such an orientation also increased solar heating in winter on the south wall and decreased it in summer due to the angle of the sun in the sky. Current scientific investigations have suggested few better natural ways to adjust to the limiting conditions imposed by such an environment.

Under hot, humid conditions found in the southeastern part of North America, the need is to provide shade from the high sun and to encourage evaporation of excessive moisture. This suggests open shelters and town sites to increase the possibility of free ventilation, with the shelters themselves in the shade of trees to lessen the impact of the solar heating. The Seminoles of Florida used large gable roofs which they covered with grass or thatch to protect against the sun and to provide shade over an extensive ground area. The shelters had no walls to interfere with the free flow of the wind. Floors were usually elevated to keep them dry and to permit ventilation underneath. The roof overhang was generally sufficient to protect the interior of the dwelling from rainfall, although rolled mats could be dropped on the windward side to prevent penetration of driving rainstorms.

In the last few decades, especially in the United States, we have paid less attention to the role of climate in architecture. With the increasing mobility of people, with their increasing affluence, and with the emphasis on status, styles developed in one area have been moved bodily to other regions for which they might not be climatically suited. Houses have been built without regard for climate, and materials have been utilized that accentuate rather than negate climatic extremes; we have created shelters that seem to be fighting rather than working with their environments to produce internal comfort.

Fitch (1950, p. 91) has pointed out that many designers and builders tend to think of the physical environment "as being either a meteorological vacuum or constant, or, at best, meteorologically symmetrical." Most of our city skyscrapers are designed alike on all four sides although it is clear that the north-facing side will experience far different conditions than the south-facing or the west-facing sides. A cooling or heating system must therefore be overdesigned if it is to maintain uniform thermal conditions inside, for the external heating or cooling load is clearly not symmetrical.

As we place less emphasis on the meteorological detail, we must place more reliance and, unfortunately, money on mechanical contrivances. These can usually maintain our buildings and houses in comfortable conditions but at a cost—in initial installation, in operating expenses, and in maintenance costs. Buildings designed to work with existing environmental conditions not only cost less to operate but may even provide more liveable conditions naturally.

What Fitch is calling for, and what the earlier builders all recognized as necessary, is a healthy regionalism in our architecture. It is up to the architect, working with the climatologist, to take account of the means and extremes of the climate, not only in the general region but also in the local or microarea, and to fashion a shelter which is tasteful, economically acceptable, and at the same time, climatically adjusted to take advantage of favorable local climatic conditions. The structure must first be designed for the macroclimatic conditions existing where it will actually be built. Thus the applied climatologist has the dual job of advising on the influences of both regional and microclimatology.

URBAN CLIMATES AND SITE SELECTION

There is not space in this brief chapter on climate and architecture for a detailed description of microclimatology or the complex variations in climates near the earth's surface. It should be clear already that those climates are of basic importance in the final comfort or discomfort of individuals living on the earth; these are the climatic conditions we seek to maximize either by judicious selection of site or by willful manipulation of the environment. Unfortunately man often changes these climates in the wrong direction, resulting in increased feelings of discomfort in many urban and even some rural settings. The following few paragraphs point out some of the more obvious microclimatic differences between sites as well as the local modifications that can be undertaken by man for his increased comfort; the reader who wishes more detail must refer to the basic work on micro-climatology by Rudolf Geiger (1965b).

Topography, vegetation, proximity to water bodies, and human modification of the surface conditions (such as by paving, changing the reflectivity or heat-transmission characteristics of the surface, or by modifying surface roughness) all result in differences in microclimates. Usually these changes will not affect the general climate of a region (except possibly in the case of a large metropolitan area), but they can and do result in large local changes in climatic conditions.

In recent years there has been a rapid increase in the number of urban climate studies (Peterson, 1969). Conditions are still so variable within an urban area, due to the effect of slopes, of roughness elements, of water or park areas, of different elevations, and of wind flows around obstacles, that it is extremely difficult, if not impossible, to generalize from one urban area to another. The same conclusion applies to a more rural area but to a lesser degree. It is still desirable to make microclimatic observations at each site, if at all possible, in order to understand the local regimes of heating and cooling, of wind speed and direction, of humidity, cloudiness, and pollution.

Mitchell (1962) has reviewed the temperature differences between cities and rural areas and discussed the existence of the so-called "heat island." He first tries to answer the question of whether these heat islands are merely topographic anomalies that would exist whether there was an urban complex or not. To answer this he cites several pieces of evidence suggesting that heat islands result from the presence of the city. These include the facts that (1) urban developments in quite diverse topographic environments all demonstrate typical heat islands, (2) heat islands can sometimes be shown to be weaker on Sundays than on weekdays, and (3) some heat islands have been shown to increase in intensity as the city grows. Mitchell has identified some of the contributing causes of the heat island. These include the absorption of solar radiation by buildings and paved areas having large heat-storage capacities as well as the virtual lack of any vegetation—trees or grass—that would utilize a portion of this energy in the process of evaporation. He suggests that the afternoon smoke and dust in the city may reduce the amount of energy received somewhat so that afternoon temperatures are comparable between city and rural areas. In the evening, the buildings and streets will cool down, but slowly, since they have absorbed a great deal of energy during the day and they will reradiate much of it back and forth between adjacent walls and other man-made surfaces. The city air cannot cool significantly, and even by morning it is still considerably warmer than the air in the surrounding countryside. Thus the urban heat-island effect is generally more clearly defined at night than in the daytime. This effect is noticeable in both winter and summer, although on the basis of limited data, it is more marked in the summer season. While the urban afternoon temperatures are essentially the same as those in the country, the nighttime temperatures average several degrees warmer in the city.

Winter conditions, while showing the same heat-island effect, may be the result of somewhat different causes. The heat produced by homes, offices, and factories contributes greatly to the atmospheric heat in the city. With turbulent mixing in the winter afternoon, this effect is not too great. In the relatively stable winter night, however, the nocturnal inversion, coupled with the smoke, water vapor, carbon dioxide, and other impurities in the urban atmosphere, will decrease outgoing radiation and result in warmer winter nighttime temperatures in the city than in the nearby country.

Landsberg (1950), in a survey of temperatures during an August afternoon and evening in Washington, D.C., identified four landscape features which exert strong influences on the microclimate of the city: (1) the buildings that so easily absorb solar radiation; (2) the river with its moderating influence on temperature; (3) the higher elevations with lower

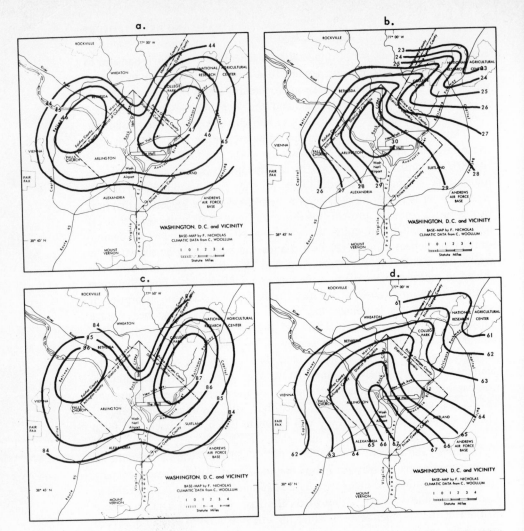

FIG. 10-1

Twenty-year mean maximum and minimum air temperature patterns for both summer and winter for Washington, D.C., and vicinity: (*a*) mean maximum temperatures (°F) for winter; (*b*) mean minimum temperatures (°F) for winter; (*c*) mean maximum temperatures (°F) for summer; (*d*) mean minimum temperatures (°F) for summer. Data for period 1946–1965. (*With permission, from F. W. Nicholas, 1971.*)

daytime temperatures, especially in summer; and (4) the forest with lower temperatures resulting from the shade and the increased evapotranspiration.

Nicholas (1971), working under the direction of Dr. John Lewis, University of Maryland, has prepared a very fine series of maps of the urban heat-island effect in terms of long-term average maximum and minimum temperatures for winter and summer periods (Fig. 10-1) as well as for individual winter and summer days (Fig. 10-2). While the long-term averages

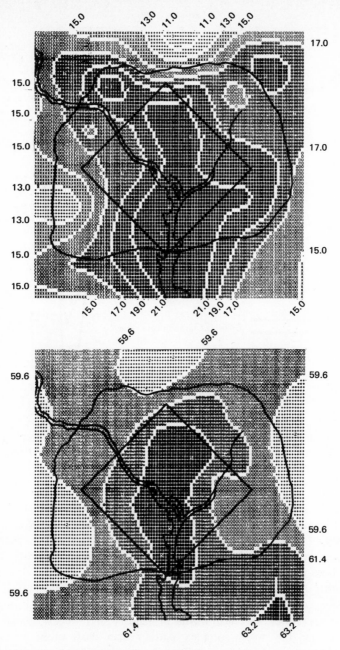

Minimum temperatures for Jan. 14, 1970 (top), and Aug. 13, 1969 (bottom), in Washington, D.C., and vicinity. Sky was clear, wind NW 5.5 knots, anticyclonic conditions on January 14; sky cover 2/10, wind S 3.5 knots, smoke and haze, anticyclonic conditions on August 13. (*Symap version 5 by Harvard University; climatic data by C. A. Woollum, USWB; see also Woollum and Canfield, 1968; computer printout by the University of Maryland C.S.C.; map analysis by F. W. Nicholas. With permission, from F. Nicholas, 1971.*)

FIG. 10-2

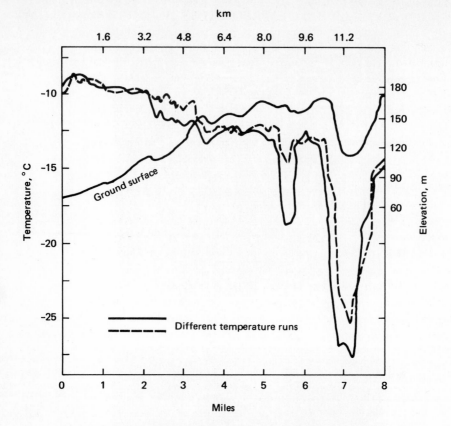

Results of temperature measurements made across Toronto on a winter night   FIG. 10-3
in 1935. (*With permission of the Royal Astronomical Society of Canada,
from Middleton and Millar,* 1936, *p.* 265.)

mask many of the local and short-period differences, they still reveal a
pronounced nighttime heating effect in the center city area (see also Woollum
and Canfield, 1968). Twin areas of maximum temperatures are found
generally outside the boundaries of the District of Columbia during the
daytime in both winter and summer. Temperature ranges across the city are
greater at night than in the daytime, emphasizing that the heat-island effect is
more pronounced in the evening. The distributions of Fig. 10-2, showing
minimum temperatures for clear nights in January and August, reveal a
somewhat similar but more complex pattern than those given by the
long-term averages, as would be expected. A somewhat larger variation in
temperature across the mapped area is found (14°F on the January night and
7°F on the August night).

    Temperature differences due to topography are even more clearly
revealed in a much-quoted study of the variation in temperature across
Toronto, Ontario, by Middleton and Millar (1936) (Fig. 10-3). The data were
taken on a clear winter evening. The warmest temperatures are near the lake,

which acts as a heat reservoir. There is a gradual drop in temperature with distance inland from the lake. A small river valley about 7 mi from the lake shore has become filled with cold air draining from the higher valley areas above, so that very cold temperatures (about $-27°C$) are found there as compared with $-8°C$ near the lake shore, a difference of $19°C$ within a 7-mi (11-km) distance. There is no reason to suspect that these temperature differences are unique in rolling areas where large lakes and river valleys occur in close proximity. Recognition by architects and home builders of the great temperature differences that can exist across suburban as well as urban areas can lead to more intelligent site selections or changes in landscaping or building construction on an already selected site.

The occurrence of cold air drainage into hollows or depressions in the evening, a phenomenon that has long been recognized by farmers and others concerned with frost pockets, can be utilized in several ways by builders and home owners. For example, Geiger (1965b) clearly shows how a raised embankment on a slope can dam up the cold air flowing slowly down the slope during the night and create a cold pool of air upslope from the embankment. Fitch (1950) suggests how this fact might be used by a builder whose home is on a slope. Construction of a wall or even planting a fairly impermeable hedge on the lower side of the property will effectively trap the thin layer of cool air that begins to move downslope shortly after sunset. The resulting cool pool of air may cover the yard area or even the lower portion of the house, depending on the degree of slope and the relative locations of the house and the barrier. In this situation, it is, of course, desirable to provide some way to release this cool air on those days in fall, winter, or spring when it is not desired. A gate in the wall or the use of deciduous vegetation is usually sufficient to eliminate the damming effect of the obstruction when it is not needed.

Forests or even a few rows of trees will greatly reduce wind speed. Zon (1941), on the basis of 10 years of wind records at a height of 4.6 m inside and outside a maple-beech-birch forest, showed that wind speed in the forest was less than 50 percent of the speed outside for 11 months of the year. In the summer months, with leaves providing the maximum shelter effect, wind speed in the woods dropped to about 25 percent of the speed found outside on the average.

Defined openings cut into forested areas for houses or a general thinning of a forested area without any well-defined openings result in quite different wind-flow conditions. The open cut will result in an eddy of wind circling downward along the downwind side of the opening, and then near ground level flowing against the general wind aloft (from the downwind to the upwind side) before rising along the upwind side of the cut. Thus the prevailing wind near the surface may be just opposite to the prevailing wind above the height of the trees themselves. In the case of the thinned forested area, winds should be in the same general direction all the way down to the ground although the wind speed will be greatly reduced within the thinned forest.

Architects and builders should be encouraged to utilize the climatic advantages of wooded tracts. A deciduous forest is more appropriate for summer coolness, while a coniferous forest is better for warmth in winter. Since the forest cannot be changed, the builder must choose which season is most severe and select his site accordingly. Clearings of any size result in conditions similar to those found in the open countryside.

Why is it in our modern housing developments we have so consistently bulldozed all the trees and vegetation from the site, installed houses with mechanical cooling devices and been willing to pay the higher costs for electricity, repairs, and maintenance when nature is willing to provide freely at least a portion of our comfort if we will but adapt to her comfort-producing devices? A few suburban developments, like the new town of Columbia being built between Washington, D.C., and Baltimore, Maryland, are carefully planned to protect some of nature's gifts. The area will eventually have 100,000 inhabitants in an area of 28 km^2 of rolling grassland, trees, recreational lakes, and golf courses. Every effort is being made to prevent the creation of the urban heat island and the other conditions which have been associated with the unchecked sprawl of so many of our cities. Yet even in this newly developing area, only partially complete at present, a measurable heat-island effect is beginning to develop. Landsberg (1969a) has reported on some preliminary observations (Table 10-1).

While the temperature differences noted on clear days are nowhere as great as found in Washington, San Francisco, or other large urban areas, already a 1.2°C difference in temperature has developed. As more of the area is suburbanized these temperature differences should become more marked. With careful attention to natural conditions and the maintenance of tree and grass areas between the built-up areas, these temperature differences can be minimized. The lack of temperature variations on cloudy days (Table 10-1) is expected, reaffirming previously understood principles.

Infrared surface temperature measurements from a helicopter flying over Columbia on a sunny morning show the same influence on buildings and paved areas (Table 10-2).

Because of the configuration of Columbia, it is likely that there will be a number of small heat islands interspersed with areas of lower temperature over grassed or wooded sections in place of the more massive heat islands that are now found in most urban complexes.

TABLE 10-1

Average summer midday air temperatures (°C) in the Columbia, Md., area*

EXPOSURE	CLEAR DAYS	CLOUDY DAYS
Lake shores	28.9	22.4
Woods and rural areas	29.5	22.5
Open residential areas	29.7	22.4
Business and dense residential areas	30.1	22.5

* *With permission from Landsberg, 1969a, table I, p. 6.*

SHELTERBELTS

Various investigators have looked at the distribution of wind speeds around shelterbelts or belts of trees. C. E. P. Brooks (1951) distinguishes four zones in the variations of wind speeds around a shelterbelt or windbreak (Fig. 10-4). The horizontal scale is given in units of the height of the break itself, while the vertical scale is in percent of wind speed to that of the free wind. The upwind zone (A) is a region of decreased wind speed before the break itself, resulting from a backing up of the wind flow as it approaches the fairly impenetrable windbreak. This zone might extend five to eight times the height of the break with an average speed of 70 to 80 percent of the free wind. Zone B, just behind the windbreak itself, is the zone with the greatest protection from the wind. The lowest wind speed is found at a distance from the break of about three to four times the height of the windbreak itself although this depends on the openness of the barrier. The more impenetrable the barrier, the shorter is the distance to the minimum wind speed, and the greater is the reduction in velocity. Zone C, found downwind from the barrier from six to twelve times the height of the barrier itself, is a zone of rather rapid recovery of the wind speed. In this zone, the wind may range from 40 to 80 percent of the free wind speed. This is a region of turbulent eddies, often with the wind circling downward after having risen over the barrier. Zone D is the region in which the wind speed returns gradually to what it was well upwind of the barrier. The wind speed achieves 100 percent of the free wind speed at distances about 25 to 30 heights of the barrier downwind.

These general figures are quite realistic for a wide variety of wind and shelterbelt conditions. Zon (1941) has estimated the downwind zone of protection to be 20 times the height of the tree barrier. Thus a tree belt 15 m high influences wind speed through a 300-m downwind zone. He emphasized the practical significance of this reduction in wind speed by means of tests on fuel consumption in farmhouses. Test houses around which the wind speed was reduced by 50 percent showed a 30 percent saving in fuel consumption.

Brooks (1951) feels that an ideal windbreak would be 6 to 9 m in width and made up of three rows of mixed coniferous and deciduous trees in order to maintain adequate density in both winter and summer. The belt

TABLE 10-2

Surface temperature values (°C) measured from
helicopter over Columbia, Md., on sunny morning*

ENVIRONMENT	T,°C	ENVIRONMENT	T,°C
Lakes	27.5	Open housing areas	32.2
Woods	27.5	Built-up spaces	34.7
Open spaces, fields	30.8	Parking lots & shopping center	36.0
Green surfaces in urbanized area	31.0		

* *With permission, from Landsberg, 1969a, table III, p. 8.*

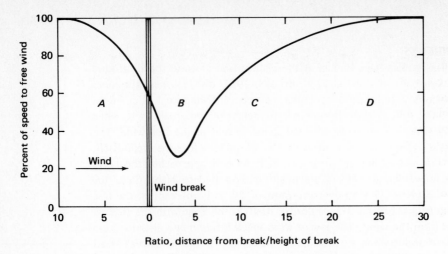

The response of the wind speed both up- and down-wind of a windbreak.   FIG. 10-4
(*With permission of the Philosophical Library, from Brooks*, 1951, *p.* 271.)

should be fairly uniform in height and of equal density from the ground to the top. In snow areas, shelterbelts should be at least 30 m from buildings and roads if possible so that they will not be in the zone of snowdrifts which accumulate behind the shelterbelts. Since shelterbelts reduce wind speed, they are generally employed to keep out cold winter winds. But they should not reduce the strength of cooling breezes in the summer. Thus the shelterbelt must be so oriented to block the prevailing winter wind but not to reduce the summer breeze significantly, a task for an applied climatologist.

INFLUENCE OF SURFACE MATERIALS

Surface materials, whether they be soils, grass, or pavement, differ in their heat conductivity and heat capacity. Air has very low values of both of these factors; thus the more air in the soil or surface material, the lower will be its ability to hold or conduct heat. When water replaces the air within the surface material, both heat conductivity and heat capacity are increased. At the same time, evaporation of the moisture will occur; large amounts of heat are required for evaporation and are transferred to the air as latent heat of vaporization. Thus we would expect that a given amount of radiant energy falling on the surface of the earth will result in greater heating in a dry desert than on the ocean shoreline although both areas might have the same sand. Temperatures will rise and fall more quickly over a dry sand than over a more moist loam or clay soil.

Consideration of the roles of heat capacity, heat conductivity, and evaporation leads us to the conclusion that moist, grassy surfaces have smaller diurnal and annual ranges of temperature than do drier sands or paved surfaces. Cultivation of a soil introduces more air spaces and hence increases the diurnal temperature range at the surface. If the surface temperature is near freezing, we can, therefore, expect more likelihood of frost in a plowed field than in an unplowed field with similar soil.

The color of soils will, of course, influence the amount of energy absorbed and the resulting rise in temperature of the soil. Black soils show a markedly higher temperature than brown or ash-colored soils.

Measurements have shown that bare soils are cooler by day and warmer at night than similar grass-covered plots. The effect of the grass cover increases with the length of the grass. Since the air above the surface responds to surface conditions, we would expect to find greatest nighttime cooling over a field of tall grass. Radiation fogs have often been seen to form earlier over tall grass than over short grass. In orchards, or around house sites therefore, short grass covers are preferable to tall grasses.

While paved surfaces clearly store up a great deal more heat and remain hot longer than unpaved or grass surfaces, it is a common architectural mistake to place paved surfaces close to houses and other buildings. In discussing the effect of shade on temperature, Landsberg (1950) has given us the following distribution of temperatures for a clear, sunny day (Table 10-3).

The unshaded surfaces all add appreciable heat to the air layer near the surface, while the shaded areas have temperatures quite close to the air temperature. Landsberg suggests that these temperature differences are conservative. Many series of observations show considerably larger differences, especially between grass or shade temperatures and those found over pavements or dark roofs.

Fitch (1950) has pointed out how remiss we have been in the juxtaposition of terraces and driveways with respect to picture windows or even regular windows. Large temperature differences can exist between a grassed area and a blacktopped drive or terrace or from a paved street across a tree-shaded lawn (Figs. 10-5, 10-6). Not only does a great temperature difference exist in just a few feet but the blacktop or other paved surface will radiate and reflect appreciable amounts of energy into the house. It will, thus,

TABLE 10-3

Variation in air temperatures over different surfaces on a sunny day*

SURFACE	TEMPERATURE
Air temperature in standard screen	25°C
Concrete walk, no shade	35°C
Dark slate roof, no shade	43°C
Short grass, no shade	31°C
Within leaves of oak tree	27°C
Soil in shade of oak tree	26°C

* From "Weather and the Building Industry," Proc. BRAB Conf. Rept. No. 1, 1950, Build. Res. Advis. Bd., NAS-NRC, Washington, p. 27, with permission.

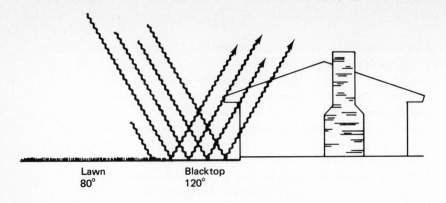

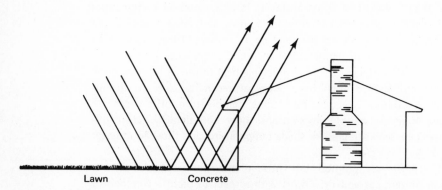

Radiant temperature (top, °F) and reflection differences (bottom) between   FIG. 10-5
lawn and blacktop terrace. ("*Weather and the Building Industry,*" *Proc.
BRAB Conf. Rept. No. 1, Building Research Advisory Board, National
Academy of Science–National Research Council,* 1950, *from Fitch, p. 95.*)

serve as an additional heating source which is unwanted in summer although
it may be desirable in winter.

To utilize a paved terrace to the fullest in both winter and summer,
Fitch suggests the need for movable awnings or partitions that can shade the
terrace in the summer or interrupt the sun's rays in the afternoon period
when they are coming from low western angles, yet which can be removed in
the winter so that the paved area can serve as a source of heat for the house
(Figs. 10-7, 10-8). He showed one set of measurements with a temperature
difference, in the sun, between a snow-covered lawn and a dark-surfaced
terrace shielded from the wind of 40°F (22°C) (between 30°F (- 1.1°C) over
the snow and 70°F (21.1°C) over the terrace) to emphasize the need to
utilize such areas in both winter and summer to improve the microclimate of
the home.

Terraces will also lose heat rapidly by long-wave radiation at night. To
conserve heat so that the terrace can be utilized longer in the evening, simple

Effective employment of sod and shade as elements of cooling (temperature   FIG. 10-6
in °F) ("*Weather and the Building Industry,*" *Proc. BRAB Conf. Rept. No. 1,*
Building Research Advisory Board, National Academy of Science–National
Research Council, 1950, from Fitch, p. 95.)

Overhead-door type of solar shading device. ("*Weather and the Building*   FIG. 10-7
Industry," *Proc. BRAB Conf. Rept. No. 1, Building Research Advisory*
Board, National Academy of Science–National Research Council, 1950, from
Fitch, p. 96.)

shading devices such as easily pulled canvas awnings or horizontal curtains
will reduce outward radiation and keep the terrace warmer during the early
evening hours.

Landsberg (1950) has also emphasized how the roof contributes to the
thermal balance of a house. To stress the need for (1) shade above the roof,
(2) insulation below the roof, and (3) ventilation within the attic area, he
provides an interesting set of measurements (Table 10-4).

With proper ventilation and insulation, the attic area can be kept quite
close to air temperature. It should not serve as a source of thermal discomfort
for the rest of the house.

The principle of cooling by evaporation is utilized by some builders to
cool roof areas. On flat-roofed buildings it is possible to cover the roof area

Sliding shutters for shading a west wall in summer. ("*Weather and the*   FIG. 10-8
Building Industry," *Proc. BRAB Conf. Rept. No.* 1, *Building Research
Advisory Board, National Academy of Science–National Research Council,*
1950, *from Fitch, p.* 96.)

with a thin layer of water which not only acts as a layer of insulation but also
keeps the roof cooler than it would be if dry as a result of the evaporation
from the surface. For most efficient operation, the water layer must be
maintained continuously on the roof, the roof must be watertight to ensure
no seepage into the roof or to the area below, and the stagnant water on the
roof must not become a breeding place for mosquitos or other insects.

Evaporative cooling is widely used in many hot, dry areas where "water
curtains," "water blinds," "swamp coolers," or various other devices are
employed. These are nothing more than an updating of the old technique of
cooling water in porous jars by letting the slow seepage through the container
evaporate on the surface. The evaporation of water slowly flowing down a
masonry wall, or a porous curtain, or even over some vegetation, will result in
cooler air temperatures in a small enclosed patio area or room. Small spraying
fountains can produce a similar result. The technique is most effective in hot,
dry areas where evaporation is rapid.

TABLE 10-4

Temperature observations above and beneath a roof surface*

LOCATION	TEMPERATURE
Outside air temperature, screen	31.1°C
Dark asbestos shingle roof, no shade	50.0°C
Temperature below insulated roof	42.2°C
Temperature in fan-ventilated attic below insulated roof	32.2°C

* *From "Weather and the Building Industry," Proc. BRAB Conf. Rept. No. 1,
1950, Build. Res. Advis. Bd., NAS-NRC, Washington, pp. 27–28, with permission.*

CLIMATIC APPROACH TO HOUSING DESIGN

Bioclimatic Approach to Architectural Regionalism
Olgyay (1963) uses the bioclimatic chart (see Fig. 8-9) as the basis for his climatic evaluation of housing design. Plotting monthly or preferably hourly values of temperatures and relative humidity for a particular area on the chart, one can identify quickly the problem periods of the year and the nature of the problems (length and severity of overheated or underheated period, too dry, too moist, etc.). The bioclimatic chart also provides information on possible modifications (more shade, more radiation, more wind, more or less humidity) in order to reduce the impact of an unfavorable climatic situation. This should suggest to the architect the type of design conditions to be incorporated into any building plans.

The generalized climatic data plotted on the bioclimatic chart provide only an overview of the climatic stresses in an area. For architectural design purposes more detail in terms of the influence of radiation conditions, the effect of various orientations, and the nature of possible shading and ventilation techniques are needed. The role of many of these important micro- or local effects is very clearly analyzed by Olgyay by means of a series of diagrams and charts.

Sun-Path Diagram
For a particular place, it is a relatively simple matter to develop a diagram which shows the position of the sun in the sky at any time of the year. Known generally as "sun-path diagrams," they are the projection onto a flat piece of paper of the path that the sun would follow across the sky on any given day. Figure 10-9 shows a typical sun-path diagram and its development. The observation point is in the center of the diagram, and the curved lines (the projections of the sun's path across the sky vault onto the flat paper) provide a means of locating the direction of the sun (but not the vertical angle) at any time and day of the year. For use in determining the shading characteristics of various buildings, overhangs, and exterior vegetation features, the sun-path diagram is modified so that the angle of the sun above the horizon can be included.

Shading Masks
Any solid object placed between the sun and the center of the diagram (point of observation) will cast a shadow on this point. The situation can also be reversed with the same effect; i.e., the light source can be placed at the point of observation and the shadow is then cast onto the sky vault. The areas of the sky vault covered by the shadow are then the portions of the sky from which no light can reach the point of observation as long as the solid or opaque object is present. If the sun itself moves through such an area of the sky vault, clearly the point of observation will then be in shade and receive no direct light from the sun. Figure 10-10a shows this situation (in this case with

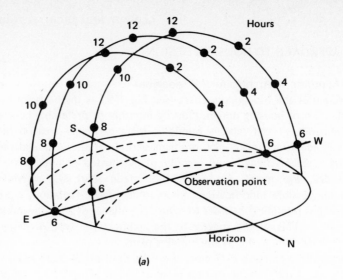

(a)

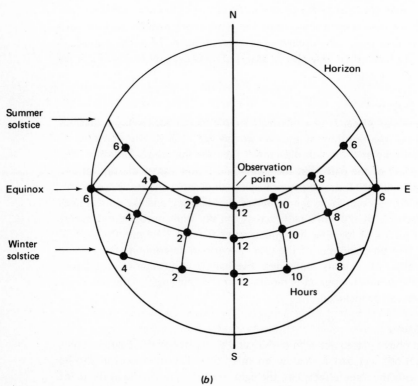

(b)

(a) Half-sphere of the imaginary sky vault with sun paths. The projections of the sun paths are shown in dotted lines on the horizon plane; (b) projected diagram of the sky vault. (*Prepared by V. Olgyay for* Housing Research, 1954, *p.* 48.)   FIG. 10-9

Situation:

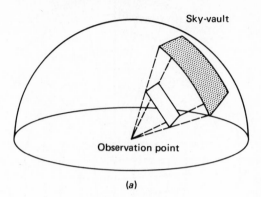

(a)

Diagram:

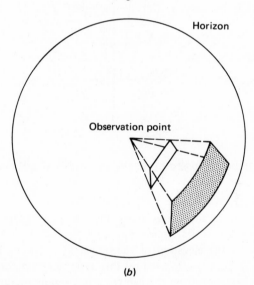

(b)

Diagram illustrating how light at point of observation will cast a shadow of FIG. 10-10 object on the sky vault or, conversely, the area from which observation point will be in shadow if sun rays were to come from that portion of the sky vault. (*Prepared by V. Olgyay for* Housing Research, 1954, *p.* 49.)

the light source at the point of observation), while Fig. 10-10*b* shows the same situation as it would appear with the rectangular shaded area in the sky vault projected onto the flat sun-path diagram.

Projecting the rectangular figure, or any figure, from the sky vault onto the sun-path diagram is most easily accomplished by means of a simple overlay protractor. On the upper half of the protractor (Fig. 10-11) are lines

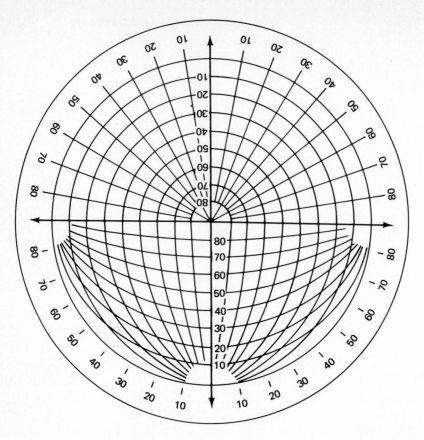

Mask protractor. (*Prepared by V. Olgyay for* Housing Research, 1954, *p.* 64.)   FIG. 10-11

radiating out from the center giving the bearing or azimuth angle from the center, while concentric circles drawn perpendicular to these lines provide information on the elevation of the object in question. Thus, if a tree (or other shading device) is located on this upper diagram correctly in terms of bearing and elevation from the center, it is a simple matter to transfer these coordinates onto the lower portion of the protractor, which shows profile lines parallel and at right angles to a horizontal line drawn on the upper portion of the diagram (Fig. 10-12).

Any geometric form, whether it be a line, a plane, or a solid object such as a building overhang, porch, tree, or other shading device, can thus be projected onto the lower portion of the protractor, and the area from which no direct sun's rays can reach the point of observation can be determined. Such projected shade diagrams are called "shading masks" and each geometric form will have its own distinctive shading mask. For example, a horizontal line parallel to a vertical wall will project as a profile line. A horizontal line perpendicular to the wall will be projected as a curved profile line perpendicular to the other profile line (Fig. 10-13).

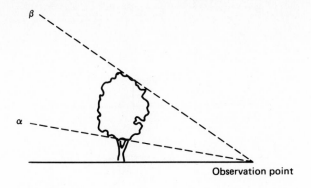

Observation point

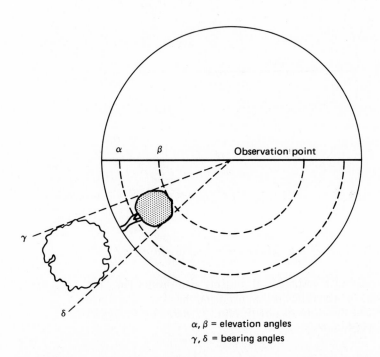

Observation point

α, β = elevation angles
γ, δ = bearing angles

Illustration of geometrical projection method with object plotted on diagram   FIG. 10-12
showing bearing and elevation angles. (*Prepared by V. Olgyay for* Housing
Research, 1954, *p.* 49.)

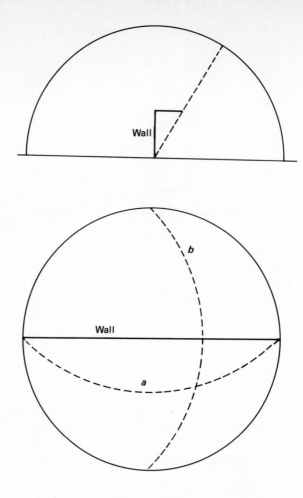

a = profile line, parallel to wall
b = profile line, normal to wall

A horizontal line parallel to a vertical wall projects as a profile line. The FIG. 10-13
profile angle is the angle between the horizon plain and the plain determined
by the horizontal line and the point of observation. (*Prepared by V. Olgyay
for* Housing Research, 1954, *p.* 49.)

      Most shading devices produce shading masks which can be simply
resolved into a series of three basic types: horizontal overhangs which
produce profile lines when projected; vertical walls or louvers which are
projected as radial lines, and eggcrate shapes which are a combination of the
horizontal and vertical forms previously discussed (Olgyay, 1954). Several of
the shading masks that can be developed from standard shading devices are
given in Fig. 10-14 from more elaborate diagrams by Olgyay. The small
diagrams illustrate the three basic types of shading devices as they might
appear in section, in plan view, and as a shading mask. The two lines marked

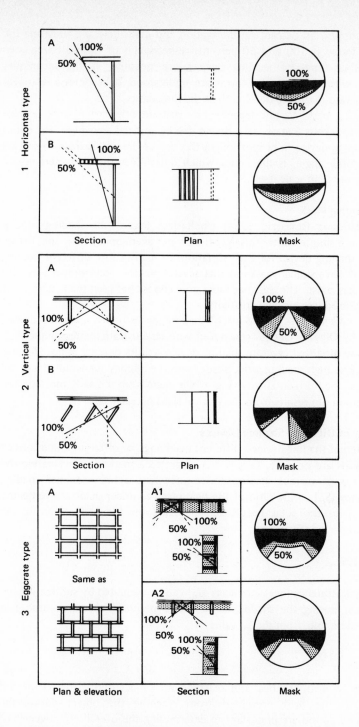

Basic types of shading devices in plan and section views and with accompanying shading masks. (*Prepared by V. Olgyay for* Housing Research, 1954, *pp.* 50–51.) FIG. 10-14

100 percent and 50 percent on the section and mask diagrams illustrate the situation found if 100 and 50 percent, respectively, of the wall area is in shade. Other types of shading devices are possible; if the techniques illustrated with these simpler forms are understood, it becomes a relatively straightforward matter to develop their shading masks.

Shading masks are not dependent on latitude, orientation, time, or scale of the device. Thus a mask once drawn can be utilized anywhere. The depth of the shading device in proportion to the size of the wall surface, expressed by the profile angle, is the factor which determines the characteristics of shading rather than scale itself.

Use of Shading Mask

Olgyay (Housing Research, 1954) emphasizes that if the shading need is known it is a simple matter to select the proper overlapping mask and, hence, the right shading device for each situation. In many cases, different shading devices will have similar masks so that several possible solutions to a shading problem may exist. The designer must now choose between them, and this is where he introduces his own creativity.

Several examples of the use of shading masks under actual conditions are given by Olgyay. One is reproduced here to illustrate the technique (Fig. 10-15). The modified eggcrate shading device is quite popular in hot climates because of its high shading ratio. Shade from the midday sun results from the presence of the horizontal overhang, while shade from the early morning and late afternoon sun results from the vertical slabs on each side.

Efficiency of Different Shading Devices

Since different shading devices, different overhangs, or different extensions of vertical walls are possible, Olgyay has suggested a method of evaluating the effectiveness of various shading devices. He first defines summer shading performance (S_p) as the ratio of the heat received under shade at overheated times (S_0) to the total heat load impinging on the surface during the overheated period (R_0):

$$S_p = \frac{S_0}{R_0} \, 100$$

The annual influence (heat efficiency H_e) can be evaluated by subtracting the energy loss during the shaded, underheated period (S_u), when heat gain would be desirable, from the S_0 values:

$$H_e = \frac{S_0 - S_u}{R_0} \, 100$$

Clearly, both the heat not reaching the object during the overheated period and the heat prevented from reaching the object during the underheated period are important in determining the overall effectiveness of the shading device. Suggesting that the shading at an overheated period is twice as important as loss of heat at an underheated period, Olgyay gives an equation for the shading effect ratio (S_e):

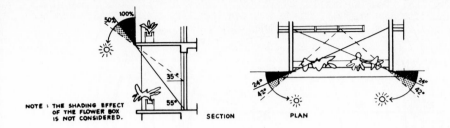

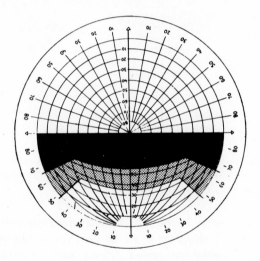

Detailed illustration of shading mask and plan and section views of fixed   FIG. 10-15
eggcrate shading device with horizontal slabs protruding farther than vertical
separations. Mask is a combination of symmetrical profile and radial lines.
Performs well on south walls in hot climates. (*Prepared by V. Olgyay for
Housing Research, 1954, pp. 58-59.*)

$$S_e = \frac{S_p + H_e}{2} = \frac{S_0 - S_u/2}{R_0} \, 100$$

Evaluation of this ratio for various shading devices, with different
shapes and orientations, provides a means of determining the annual
effectiveness of the device.

Olgyay has carried out the necessary calculations for the length of fixed
overhangs in the New York–New Jersey area and has developed a regional
shading chart (Fig. 10-16). He finds that for south exposure the optimum
overhang is 68° or 41 percent of the wall height. To the southwest and
southeast shorter overhangs are more effective on an annual basis, 26 and 22

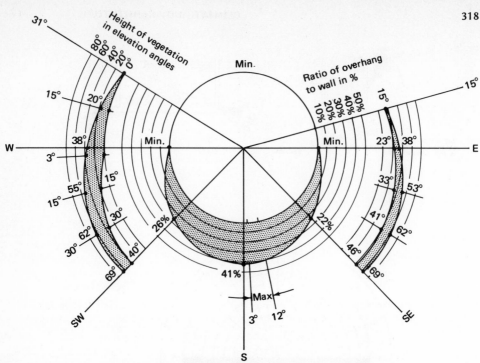

Regional shading chart giving recommended length of overhang and height of FIG. 10-16
vegetation for all orientations in the New York–New Jersey area. (*Prepared
by V. Olgyay for* Housing Research, 1954, *p.* 67.)

percent of the wall height, respectively. To the east and west directly, in these
latitudes, overhangs have a negative effect on an annual basis although some
shade is desirable during the overheated period. Olgyay suggests either
movable devices or the use of vegetation—preferably deciduous vegetation
whose leaves would be gone during the underheated period when solar
radiation is desirable in these latitudes. The regional shading chart provides
information on lengths of overhangs in all directions as well as on heights of
trees or other vegetation and their preferred orientation for most effective
shading. Such an approach does not suggest any thought of a "meteorological
vacuum" or "meteorologically symmetrical" conditions surrounding the house
but rather an awareness of the different local climates on each side of each
building.

SOL-AIR APPROACH TO BUILDING ORIENTATION
Olgyay (1963) recognizes that both radiation and temperature act together to
produce the level of heat experienced by a body. Thus orientation must
consider not only the radiation receipts but also the heat impact of the
diurnal temperatures.

From the bioclimatic chart, Olgyay first determines the extent of the
overheated and underheated periods. If cool climates predominate, the

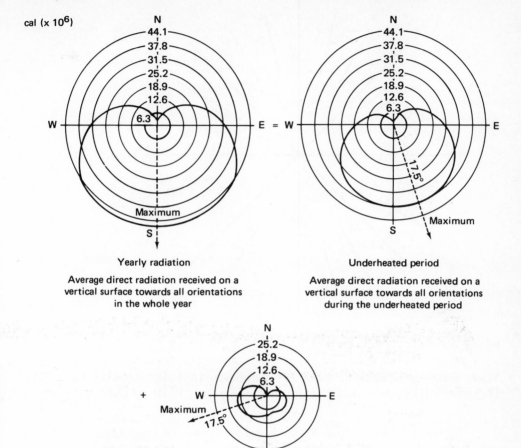

cal (x 10⁶)

Yearly radiation

Average direct radiation received on a
vertical surface towards all orientations
in the whole year

Underheated period

Average direct radiation received on a
vertical surface towards all orientations
during the underheated period

Overheated period

Average direct radiation received on a
vertical surface towards all orientations
during the overheated period

Average direct radiation received on a vertical surface toward all orientations   FIG. 10-17
in whole year as well as in underheated and overheated periods in New
York–New Jersey area. (*Prepared by V. Olgyay for* Housing Research, 1954,
p. 112.)

building should be oriented to receive maximum radiation through the year,
but if temperatures are predominantly high, building orientation should be
such as to minimize the radiation load. Olgyay next determines the average
radiation received on a vertical surface during the overheated and underheated periods for various orientations. This information is readily available
from simple tables or graphs, and can be clearly shown by means of polar
diagrams (Fig. 10-17) or a circular table (Fig. 10-18).

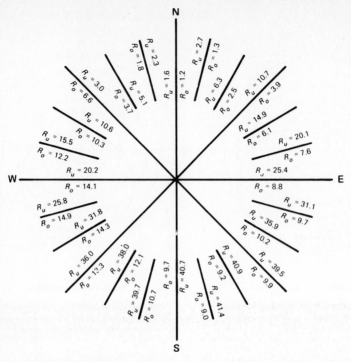

Numbers represent cal (x 10⁶)

Yearly direct average radiation received at underheated (R_u) and overheated FIG. 10-18
(R_0) periods. (*Prepared by V. Olgyay for* Housing Research, 1954, *p.* 113.)

The hourly radiation load is now added to the diurnal temperature curve for selected days through the year to provide typical curves of the sol-air index (or total temperature and radiation impact). The desired sol-air index curve is the one that is most symmetrical about noon, indicating that the heat load is well distributed through the day. Figure 10-19 shows the daily course of the total sol-air index on four selected days of the year (the solstices and equinoxes) with building orientations $17\frac{1}{2}°$ east of south, due south, and 25° west of south.

Analysis of such graphs now suggests the correct orientation for the building in the New York–New Jersey area (for which the graphs have been prepared). Figures 10-17 and 10-18 indicate clearly that the maximum of radiation received during the underheated period is found with a building orientation $17\frac{1}{2}°$ east of south. At the same time, the maximum of radiation during the overheated period is received on a vertical wall oriented $17\frac{1}{2}°$ south of west. Since these two surfaces are at right angles to one another, it suggests simply that a rectangular building with the long side oriented $17\frac{1}{2}°$ east of south (which will then have the short side $17\frac{1}{2}°$ south of west) should provide the maximum surface for radiant heating in cold seasons and the minimum surface for heating in warm seasons in the New York–New Jersey

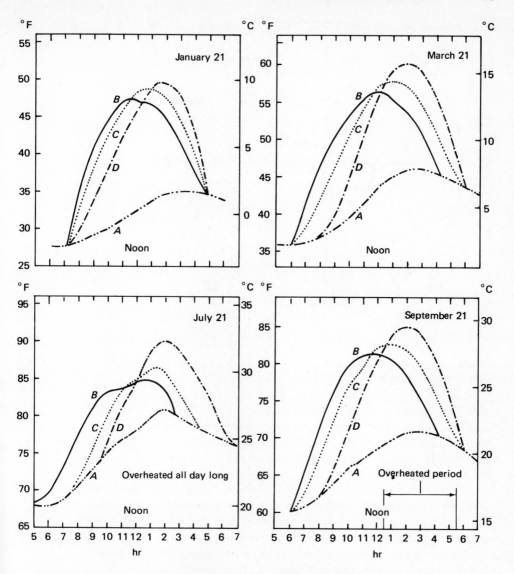

A curve indicates the daily drybulb temperature change
B curve indicates the solair impact on a vertical surface oriented 17 1/2° east from south
C curve indicates the solair impact on a south vertical surface
D curve indicates the solair impact on a vertical surface oriented 25° west from south

Heat effect for solstice and equinox periods. (*Prepared by V. Olgyay for* FIG. 10-19 *Housing Research, 1954, p.* 114.)

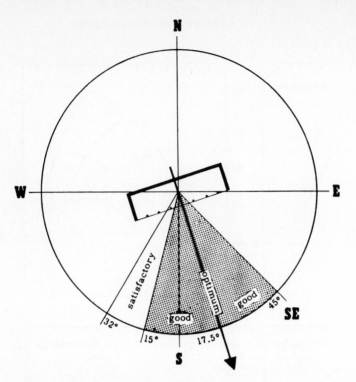

House orientation chart for New York–New Jersey area. (*Prepared by V.* FIG. 10-20 *Olgyay for* Housing Research, 1954, *p.* 113.)

area. This result is somewhat unique because the two directions are at right angles to one another. If they were not, some compromise in orientation would be necessary in order to achieve the most satisfactory distribution of total heat receipts in all seasons. From the sol-air index (Fig. 10-19), we again see that an orientation $17\frac{1}{2}°$ east of south provides the most symmetrically distributed curve around noon for all four selected days. At the same time, with that orientation heat loads are less in the two overheated periods, and on September 21 the sol-air index is decreasing just at the beginning of the overheated period while the indexes for other orientations are still increasing. The sol-air analysis reinforces the conclusions reached from the radiation analysis alone (if this had not been the case, further adjustments of orientation would have been necessary) and provides us with the most desirable orientation for a building in the New York–New Jersey area (Fig. 10-20). Similar analyses for other areas would suggest other preferred orientations depending on radiation and temperature conditions during the overheated and underheated periods.

Olgyay made several assumptions in working out the foregoing example. First, the example used data that referred only to the New York–New Jersey area. The answers obtained are representative of a wide area of conditions in midlatitudes, but the conclusions concerning orientation and heating loads should not be used in different climatic regions.

Second, values of total direct radiation were used. It would have been more correct to use values of total direct and diffuse radiation. However, Olgyay points out that his purpose was not to determine quantitative values but rather to obtain a qualitative understanding of the variation in radiation impacts with direction. The sol-air values are really index numbers which are entirely satisfactory in defining orientation.

Third, various house types need to be tested. For simplicity, Olgyay has assumed a simple "unilateral" house with all living areas facing only one direction. Often the designer is faced with a "bilateral" building in which the living areas are back to back or in completely opposite directions. Some areas must then be oriented in less-desirable directions. More judgment is necessary on the part of the designer to decide the actual location of rooms within the building. There is a greater range of orientation possible for "bilateral" building types, but many of these types are not well suited to tropical locations.

HOUSE SHAPES

From his radiation calculations in different environmental situations, Olgyay feels that a square house is not the optimal shape anywhere. However, shapes elongated in a north-south direction are less efficient in both winter and summer than square houses. He concludes that in every case the most satisfactory shape is one in which the house is elongated in some general east-west direction (Olgyay, 1963, p. 88).

His analyses show that a square shape is most nearly optimum in cool climate areas because of the general need to conserve heat. The greatest internal area can be contained within a cube—the larger the cube, the smaller the surface-to-volume ratio which improves the heat-retention characteristics of the building. A two-story structure is quite desirable.

Considerable freedom in house shapes is available in temperate areas (for example, the New York–New Jersey area) because of the lack of any real climatic stress in either summer or winter. While an elongated east-west design is desirable, it is possible to construct buildings with wings in other directions without seriously impairing environmental relationships.

Winter conditions in hot, arid regions would permit an elongated house design, but the severe heat stresses in summer would argue for a square house to minimize surface area. Thus a compromise is the square building with an interior courtyard containing shade, or a generally cubical building with a portion of one wall cut out, the area being grassed and containing trees, a pool, an evaporative cooling device, or a trellis with vegetation. If the outside wall is continued around the whole building, the small garden environment (similar to the old courtyard plan) provides a cool area within the house structure while minimizing the solar radiation impact on the outside walls of the house.

The hot-humid environment emphasizes the need for shade and elimination of radiation conditions on the east and west walls. This suggests marked east-west elongation of the building with small east and west walls. If protective shade is available, considerable freedom is possible in house shapes. Orientation to take advantage of ventilation effects can then become as important as orientations to minimize radiation effects.

WIND AND VENTILATION

Movement of air from outside to inside a building may be both desirable and undesirable, depending on its temperature in relation to the interior temperature. In cold areas, such movement is minimized as much as possible through the use of storm windows and doors along with other weather-sealing devices on all openings. In hot regions, if the air temperature is below the temperature in the building or below body temperature, airflow is encouraged in order to promote cooling by evaporation and a feeling of comfort with air movement. Air speed is important since it is the rate of movement of the air column that brings most of the relief from the hot conditions. In hot, humid areas, airflows of 2 m/sec are entirely desirable. In hot, dry climates, especially if the air temperature is near body temperature or above, reduction of daytime ventilation to a minimum is desirable to prevent additional heating of the interior. Ventilation is only desirable during the warm nighttime period.

Givoni (1969) has listed three objectives for ventilation: (1) for health, (2) for thermal comfort, and (3) for structural cooling. In the first case, ventilation for health is necessary to remove high levels of odors and carbon dioxide, to provide necessary oxygen for breathing, and to keep the products of combustion within safe health levels. Thermal-comfort ventilation removes water vapor, thus promoting evaporative cooling, and provides comfortable indoor temperature conditions. Ventilation for structural cooling is necessary because of the low heat capacity of air; without ventilation the temperature of the indoor air rapidly equals that of the surrounding window surfaces.

Air movements through a building result from the establishment of pressure gradients due both to different external wind effects and to different temperature conditions between inside and outside. If the temperature inside is not the same as the temperature outside, air densities differ and vertical pressure gradients inside and outside are different. With only one opening, air pressure quickly equalizes across the opening and no air movement can result in spite of the temperature difference. If, however, two openings exist, one above the other, an airflow is established with air moving outward through the upper window and inward through the lower window as long as higher temperatures occur inside the building. With higher temperatures inside, air density varies less with height than outside so that at the higher opening the pressure inside should be above that outside. With higher temperatures outside, the reverse flow should occur. Air movement will always be from the area of higher pressure to the area of lower pressure.

Pressure and Suction Zones

As the wind blows against a structure, air piles up on the windward side, creating an area of high pressure. With the air flowing around the building, an area of reduced pressure is created directly downwind of the structure. Thus a pressure difference exists between the windward and leeward side in such a way that air will move through the structure (if openings exist) from the high-pressure side to the low-pressure side (or from the pressure zone to the suction zone as it is also called). If air is blowing perpendicularly to one wall of a building, lower pressure or suction will be found to occur on the other three walls. This means that windows need not be placed opposite one another to create airflow through a structure. Wind obliquely hitting a building will result in pressure zones on the two upwind sides and suction zones on the two downwind sides. Wind stress on roofs depends on the degree of slope. With a fairly flat or gently sloped roof, suction will occur on both the windward and leeward slopes, while with a steeply pitched roof, a pressure zone will occur on the windward side and a suction zone on the leeward side.

It is the combined thermal and pressure forces that produce the final air movement into a building. The gradient existing across a window opening will be the algebraic sum of the pressure differences generated by the thermal and wind forces separately. These forces may reinforce one another or act in opposite directions (Givoni, 1969, p. 258).

The Effect of Partitions

The actual airflow through or around a room results from (1) the pressure difference that exists around the structure itself and (2) the inertia of the moving air. As a result of inertia, the path followed by air within a room depends primarily on its initial direction upon entering the room. When the room is divided by means of a partition (or several rooms occur together with inlets and outlets separated by doors or halls), the air changes direction and speed as it passes through the room. This, in general, reduces air movement, although by creating a turbulent, circulating movement of air within the room it may result in effective ventilation of more of the room area. Givoni (1969) has suggested directions of flow (Fig. 10-21) and airspeeds (Fig. 10-22) with various window and partition locations. In these cases he has considered only a wind perpendicular to the opening.

Velocities are lowest when the partition is close to the inlet window so that the air has to change direction rapidly upon entering the room. This leads to the conclusion that in multiroom apartments with openings on opposite sides, the larger room should be in the prevailing upwind direction to maximize air movement within the apartment complex.

On the basis of many investigations into ventilation problems, Givoni (1962, 1968) concludes that better ventilation conditions often occur within a room when the outside airflow arrives obliquely at the window. This is especially true if the outlet window is directly opposite the inlet and results from the turbulent circling motion induced in the airstream by the angle of the entering air. This result does not occur when the outlet is in a side wall so

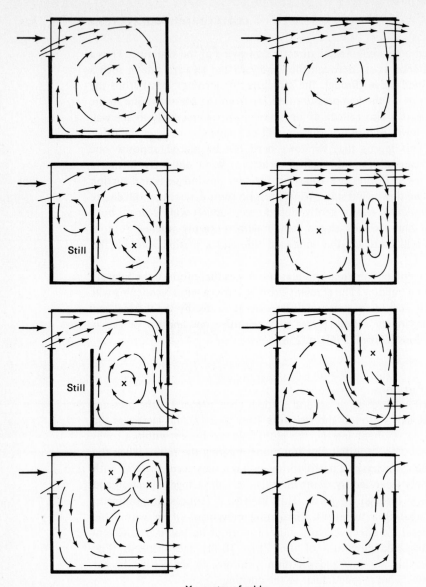

X center of eddy

Flow patterns in room models with different subdivisions of the internal FIG. 10-21
space. (*With permission of the Elsevier Publishing Company, London, from
Givoni*, 1969, *p.* 274.)

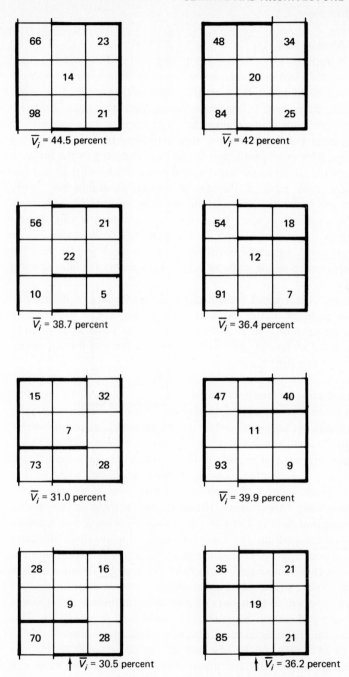

Effect of subdivision of a room on the distribution of internal airspeeds. FIG. 10-22
(*With permission of the Elsevier Publishing Company, London, from Givoni,*
1969, p. 273.)

that the obliquely entering air moves directly to the outlet window in a fairly narrow, straight stream.

The Effect of Window Size

Window size is important if windows are on one wall and the wind is at an oblique angle to the window. The larger the opening the greater the average air velocities within the room. Pressure differences will exist along the wide window, allowing air to enter in one region and to leave in another. When the wind is directed perpendicularly against the single opening, no such areas of pressure differences exist and window size is unimportant in determining inside velocities. With cross ventilation (at least one opening in the pressure zone and one in the suction zone without specifying which walls) an increase in window size does increase average air movement within the room. However, inlet and outlet openings must be increased together. If only one of the openings is increased in size, the average air movement is only slightly affected although the maximum velocity at a point within the room can be increased significantly, especially if the outlet size is much greater than the inlet size. In this case a strong jet of air is found just within the inlet opening (Fig. 10-23).

The average indoor speed depends generally on the size of the smaller opening, either inlet or outlet, while the maximum indoor speed varies inversely with the ratio of inlet to outlet size. Since the higher air movement is limited to a small section of the room, it does not increase the average air movement in the room significantly.

Controlling Ventilation by Design Features

When a room is not cross ventilated, the average inside air movement is quite low. Givoni (1969) has suggested certain design features to improve interior air movement even though openings are only possible on one wall of a room. For example, he suggests making two smaller openings on the same wall and inducing pressure and suction zones along the wall by adding a single vertical projection from each window (located on opposite sides of the windows). With an oblique wind, this creates a pressure zone in front of the upwind opening and a suction zone in front of the downwind opening and improves circulation conditions within the room markedly. Figure 10-24 indicates quantitatively the changes in interior air movement possible with different window arrangements on a single exterior wall. V_i in each case is the average internal air movement expressed as a percentage of the exterior airspeed. If projections are provided on both sides of a window, the whole effect is lost since the pressure and suction zones created cancel each other exactly. This can be a problem in an apartment complex when projections from the windows in one apartment may be adversely influenced by similar projections from the windows of the next apartment that is located on the same wall surface.

Since the direction of the entering air has an important influence in its subsequent movement through the room due to inertia, it is possible to direct the airflow within the room by the use of jalousie windows, louvers of

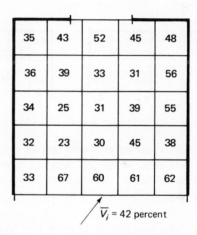

36	24	24	28	84
31	26	25	24	93
29	24	27	39	78
30	27	27	107	28
24	28	71	152	29

$\overline{V}_i$ = 44 percent

35	43	52	45	48
36	39	33	31	56
34	25	31	39	55
32	23	30	45	38
33	67	60	61	62

$\overline{V}_i$ = 42 percent

Effect of inlet and outlet size on speed of air movement (percent of external   FIG. 10-23 speed) in room models. (*With permission of the Elsevier Publishing Company, London, from Givoni,* 1969, *p.* 264.)

different types, or center-pivot hung windows. These window types will all affect the wind direction within the room although they will have little influence on the average internal speed.

Controlling Ventilation by Vegetation

The use of vegetation around high structures has very little influence, but around low structures the position and size of vegetation can have significant effect. We have already mentioned the use of vegetation for shading from solar radiation. Trees, shrubs, walls, and fences can also create pressure and suction zones in the wind flow and so influence the movement of air over and around low structures.

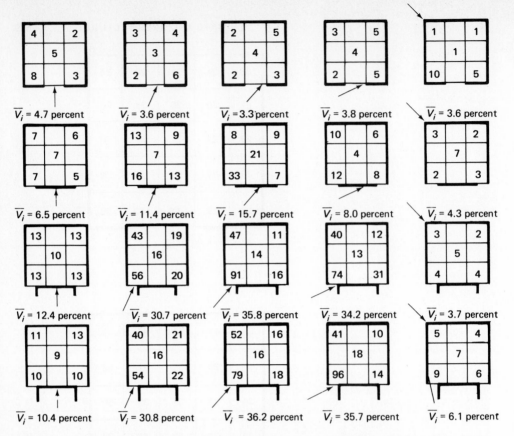

4		2
	5	
8		3

$\overline{V_i}$ = 4.7 percent

3		4
	3	
2		6

$\overline{V_i}$ = 3.6 percent

2		5
	4	
2		3

$\overline{V_i}$ = 3.3 percent

3		5
	4	
2		5

$\overline{V_i}$ = 3.8 percent

1		1
	1	
10		5

$\overline{V_i}$ = 3.6 percent

7		6
	7	
7		5

$\overline{V_i}$ = 6.5 percent

13		9
	7	
16		13

$\overline{V_i}$ = 11.4 percent

8		9
	21	
33		7

$\overline{V_i}$ = 15.7 percent

10		6
	4	
12		8

$\overline{V_i}$ = 8.0 percent

3		2
	7	
2		3

$\overline{V_i}$ = 4.3 percent

13		13
	10	
13		13

$\overline{V_i}$ = 12.4 percent

43		19
	16	
56		20

$\overline{V_i}$ = 30.7 percent

47		11
	14	
91		16

$\overline{V_i}$ = 35.8 percent

40		12
	13	
74		31

$\overline{V_i}$ = 34.2 percent

3		2
	5	
4		4

$\overline{V_i}$ = 3.7 percent

11		13
	9	
10		10

$\overline{V_i}$ = 10.4 percent

40		21
	16	
54		22

$\overline{V_i}$ = 30.8 percent

52		16
	16	
79		18

$\overline{V_i}$ = 36.2 percent

41		10
	18	
96		14

$\overline{V_i}$ = 35.7 percent

5		4
	7	
9		6

$\overline{V_i}$ = 6.1 percent

Internal airspeeds in room models with vertical projections of different depths, FIG. 10-24 compared with values in models without projections. Window width is one-third of wall width. (*With permission of the Elsevier Publishing Company, London, from Givoni, 1969, p. 267.*)

For example, a large tree upwind of a building will deflect the wind over, under, and around itself. The airspeed under the tree may be greater than it otherwise might be. If the tree is too close to the building, unsatisfactory airflow into the building can result. Similarly, hedges near buildings will deflect airflow and may result in unfavorable movement of air into the buildings. Movement of the hedge line 3 to 6 m from the building can produce a much more satisfactory distribution of airflow in the room. Hedges extending outward (perpendicular to the wind) from the upwind edges of a building will provide most unsatisfactory ventilation effects, while hedges extending outward from the leeward edge will divert the wind into the building from the sides by means of turbulent motion. The brief sketches from Olgyay (Fig. 10-25) represent some of the conclusions of R. F. White (1945), who undertook a study on the effect of vegetation on ventilation for the Texas Engineering Experiment Station. Clearly vegetation arrangement must ultimately be based on a knowledge of the magnitude of the respective winter and summer stresses.

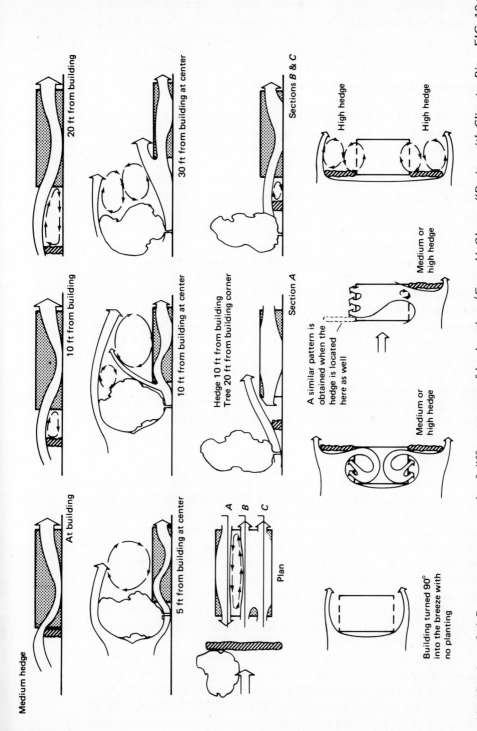

Modification of airflow pattern as a result of different types of landscaping. (*From V. Olgyay, "Design with Climate: Bio-climatic Approach to Architectural Regionalism," fig. 200, p. 102, copyright © 1963, by Princeton University Press.*)   FIG. 10-25

Screens

Screens on windows, of course, markedly reduce interior air movements. Light winds are more effectively stopped by screens than stronger winds. A 16-mesh, 30-gage, wire screen decreases a $1\frac{1}{2}$ mph wind by 60 percent while it reduces a 10 mph wind by only 25 percent (Van Straaten, 1967). The screen also has a particularly noticeable effect on an oblique wind, which seems to slip over the screen without penetrating it effectively.

THE PROBLEM OF GLASS WINDOWS

The wall surface of a building or house provides partial support for the building, as well as protection and privacy for those inside. It also serves as a barrier to the movement of environmental conditions either inward or outward. The materials of the wall are most important in determining inside conditions; some materials prevent rapid transmission of heat by radiation, conduction, and convection while other materials offer hardly any barrier at all. The overall performance of any wall depends on the transmission characteristics of all the materials in the wall.

With a greater use of glass materials in walls at present (large picture windows and glass doors, whole walls made of glass bricks, more glass areas in skylights or clerestorys), we need to understand more about the heat-transmission characteristics of glass materials and how they can be modified by shading either on the inside or the outside. For example, a heat-absorbing glass has been developed to replace regular window glass. This glass is able to absorb over 40 percent of the impinging radiant energy. While this is highly desirable in summer (even though some of this later enters the room by conduction and convection), the same characteristics also operate in winter even though increased transmission of radiation might then be desirable. Thus shading devices, some possibly movable, might be more desirable to keep out the sun's rays in the overheated period and to allow them to enter during the underheated period.

Olgyay has determined the so-called "shading coefficient" of various glass and shade combinations to illustrate the range of possible shading devices available for windows (Table 10-5). The shading coefficient is the ratio of the total solar heat gain (by transmitted, absorbed, and reradiated energy) for the particular shade and glass combination to the total solar heat gain through a single unshaded pane of common window glass. The coefficients must be considered as only approximate since the color, thickness, material, shape, and orientation of the shading device, the sun's angle, and the amount of diffuse radiation are not fully specified. These factors all influence the amount of transmission. The coefficients can, however, be considered as reasonable indicators of the characteristics of different glass and shade combinations.

As expected, light colors reflect more sunlight than dark colors; for example, aluminum venetian blinds have a shading coefficient some 30 percent less than dark venetian blinds, both exposed inside. Probably more significant is the location of the shading device. Inside shading devices are handicapped by the fact that the short-wave solar radiation has already passed the window surface before impinging on the shade. Short-wave reflection from the shading device can pass outward through the window but most of the long-wave reradiation from the shading device is trapped by the glass. It adds heat to the room by reradiation. Outside shading devices intercept the short-wave radiation before it passes the glass surfaces; any long-wave reradiation from the device is kept outside the glass window. The table shows an average increase in protection of about 35 percent using outside rather than inside shading devices. Yet most window-shading devices are still exposed on the inside of the window. Such a technique either puts an added burden on the mechanical cooling system in the summer or results in less comfortable inside conditions.

Fitch (1950) has sketched a window-shading device which is quite architecturally pleasing and climatically effective especially against the heating load of an afternoon sun. Figure 10-26 shows an outside, secondary sash fitted with heat-absorbent glass. The sash can slide back and forth or be removed in the winter. The arrows indicate the possibility of air circulation behind the outside glass sash to keep temperatures moderate. With overhead shade and the heat-absorbent glass, the shading coefficient approaches about 30 percent yet the view is not restricted by heavy shades.

TABLE 10-5

Shading coefficient for various window and shade combinations (ratio of total solar heat gain by window and shade combination to total solar heat gain by single unshaded glass window)*

SHADING COEFF.	WINDOW AND SHADE COMBINATION
1.00	Regular single double-strength window glass
0.91	Regular glass with inside dark roller shade, half-drawn
0.81	Regular glass with inside medium roller shade, half-drawn, or dark shade, fully drawn
0.75	Regular glass with inside dark venetian blind, fully drawn
0.71	Regular glass with inside light roller shade, half-drawn
0.65	Regular glass with inside medium venetian blind, fully drawn
0.62	Regular glass with inside medium roller shade, fully drawn
0.56	Regular glass with inside white venetian blind, fully drawn
0.60–0.50	Regular glass with tree outside giving light shade
0.45	Regular glass with inside venetian blind, reflective aluminum
0.43	Regular glass with outside venetian blind awning, two-thirds drawn
0.41	Regular glass with inside white roller shade, fully drawn
0.28	Regular glass with outside aluminum-painted shading screen
0.25	Regular glass with outside medium or dark canvas awning
0.23	Regular glass with outside bronze-painted dense shading screen
0.25–0.20	Regular glass with tree outside giving heavy shade
0.15	Regular glass with outside cream-white venetian blinds, or venetian blind awnings
0.15–0.10	Regular glass with outside movable horizontal louvers

* From V. Olgyay, "Design with Climate: Bioclimatic Approach to Architectural Regionalism," copyright © 1963, Princeton University Press, adapted from pp. 68–71.

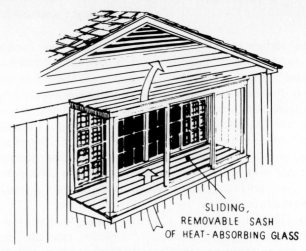

Solar sunshade of heat-absorbent glass. (*"Weather and the Building Industry,"* FIG. 10-26
Proc. BRAB Conf. Rept. No. 1, *Building Research Advisory Board, National
Academy of Science–National Research Council,* 1950, *from Fitch, p.* 96.)

New developments in glass technology are providing architects with
even more tools for control of radiation without the need for exterior or
interior shading. For example, a glass wall that is curved or set at an angle
from the vertical can reduce reflections from inside light sources and, thus,
give an undisturbed view outward. Special glass is now available that will filter
out unwanted wavelengths of radiation or refract light, thus permitting some
control over how much radiation is admitted or where it falls. A new glass,
coated on the inside with a thin film of metal, acts much like a one-way
mirror. The view outward is unimpaired, while from the outside it appears
that the glass is a polished mirror surface. Curtains are not needed for privacy
and the mirror surface is effective in reducing the entry of heat energy.
Photochromic glass even has the odd ability to be darkened by ultraviolet
radiation, yet as the amount of ultraviolet is reduced the glass regains its
transparent characteristics. Thus the glass is able, within reason, to transmit a
more constant intensity of sunlight through the day.

A new light-polarizing material made of many layers of special plastic is
under development. It has advantages over regular polarized glass in that it is
effective in all directions, does not absorb as much light, need not be tinted,
and can be frosted to soften the light. It reduces the glow of reflected light
from an object that tends to blur colors and textures. While not weatherproof
at present it still has interior uses which are beyond the role of the applied
climatologist.

CONCLUSIONS

We can no longer afford the luxury of site development or housing design
without due recognition of the existence of the environment. Nor should we
have ever sought such an expensive luxury. Houses and towns designed to
work with nature, to take advantage of the climate potentials, and to mitigate

severe climatic stresses by utilizing readily available building materials and emphasizing a healthy regionalism in style and construction methods—these are the bases for a real architecture.

The present chapter has discussed many varied aspects of architecture in relation to climate. The point has been stressed that in the design and building of a house the layout of the whole property and the influence of the surrounding environmental features are primary inputs. Selection of building materials, of finishes and surfaces, of colors and textures must all be part of the design plan. Shading of window and wall areas, length of overhangs, and wind and ventilation effects must also be included in the planning, for it is only when all aspects function together that maximum adjustment to the environmental demands becomes possible. The achievement of comfort in housing through harmony with nature should be the goal of a revitalized architectural climatology.

CLIMATE, 11
INDUSTRY,
COMMERCE,
AND ENGINEERING

Climate plays a significant role in many aspects of commerce and industry [technoclimatology is the name Landsberg (1960) has employed] just as it does in health, clothing, housing, and hydrology. But while climate's influence in these latter fields may be easily identified, the effect of climate on many industrial or engineering operations is often not at all well recognized. Management may be so concerned with labor, raw material, transportation, or sales problems that they fail to recognize the role that climate can and does play at many stages of their production, distribution, and sales programs.

In developing a balance sheet of "climatic costs" in industrial production, management must include the effect of climate on such things as (1) heating and cooling requirements; (2) water supplies; (3) air and water pollution abatement activities; (4) warehousing, storage, and transportation of raw or finished products; (5) weathering or deterioration of stockpiled items such as fuels; (6) health, efficiency, morale of workers; and (7) all outside activities. While it would be possible, given our technological competence, to design, build, and operate a manufacturing plant under the most rigorous climatic conditions experienced on earth, it would not be economically feasible. Thus climate must be considered an important factor in site selection plans as well as in the later stages of the manufacturing process.

SIGNIFICANT CLIMATIC VARIABLES

Russell (1957) has listed five meteorological variables that impair industrial operations, ranked generally in order of decreasing importance, as follows:

1   Extreme temperature (especially low temperature and including permafrost effects)

2   Snow, sleet, and ice

3   High wind

4   Heavy rainfall

5   Other factors: high humidity, poor visibility, and other weather elements

The experience of management and workers in responding to any climatic challenge will be important in determining the limiting nature of these meteorologic factors. Usually several of these factors occur together, thus compounding the effect. For example, low temperatures may be associated with the snow, ice, and high winds of a winter blizzard. Poor visibility can often accompany snow or heavy rains. While it is more efficient to discuss the influence of each of these factors separately, it might be more realistic to consider their combined impact.

1   *Extreme temperature* Although a major limitation of low temperature results from the decreased efficiency of labor, certain industrial operations are severely hampered or stopped altogether by a reduction in the temperature even though it may remain above freezing. For example, painting is usually restricted by temperatures below 16°C (60°F), and a temperature below 4°C (40°F) will usually terminate all outside painting operations; precision machining is often difficult at temperatures below 16°C, while Russell (1957) reports that up to one-quarter of the parts used in airframes cannot be produced satisfactorily when the temperature drops below 4°C. Figure 9-1 indicated the rather marked change in labor efficiency as the temperature departed either upward or downward from 60°F (16°C). At 30 or 95°F (-1 or 35°C) efficiency has fallen to about 75 percent of maximum, while at 20 or 105°F (-7 or 41°C), it only approximates 50 percent of maximum. At -10 or 110°F (-23 or 43°C), labor efficiency is about 25 percent of maximum, and, finally, below -20°F (-29°C) or above 120°F (49°C) practically all the effort of an individual must be devoted to maintaining life. Labor efficiency is essentially zero. These temperatures assume still-air conditions, for with low temperatures and any appreciable air movement the windchill effect (Fig. 8-6) will reduce labor efficiency even further. These are only average values so that experience (or lack of it) on the part of the worker results in considerable variations in reactions to the same cold or hot temperatures. For instance, Russell points out that the efficiency

of workers normally accustomed to inside work will drop to about 65 percent of normal at temperatures of 40°F (4°C).

When temperatures drop below freezing, a new set of temperature problems seems to develop. Vehicles that are not prepared for such conditions may be frozen or, at least, difficult to start, many plastics become unworkable, machine parts may become brittle, and tolerances established at normal temperatures may no longer be satisfactory. Outside assembly production in the airframe industry is effectively curtailed at 20°F (-7°C). However, industries such as the railroads, power, iron and steel, and petroleum refining will experience only minor inconveniences with temperatures between 0 and 32°F (-18 and 0°C).

Below 0°F (-18°C), practically all industrial operations are affected. Those which are housed and used to warmer temperatures are, of course, completely curtailed at these temperatures. Even many of the essentially outdoor activities that are little influenced by temperatures between -18 and 0°C develop problems. Russell (1957) notes that trains lose about 5 percent of their tractive power at -18°C; by -29°C tonnages must be reduced by 15 percent, and by -40°C tonnages are reduced by some 40 percent. At -40°C, rails are brittle and only emergency maintenance can be undertaken. Practically all industries indicate that -40 to -45°C is the lower limit for any type of operations whatsoever.

Large temperature ranges during the day may result in some problems in certain industries, especially those requiring close tolerances for machining and assembly of parts. However, seldom is fine machine work conducted in areas that might undergo such wide swings of temperature through the day.

Permafrost, or permanently frozen ground, resulting from constantly low temperatures, can create significant engineering problems. Construction of houses, roads, towers, bridges, factories, or other structures requiring foundations may result in heat conduction into the ground. If the permafrost is melted, the meltwater may have no place to go. A thick gooey substance, often referred to as "slud," can develop under the structure, the building or road may begin to settle into the melted permafrost; water and slud may be forced up into the building. There is no real cure for this situation except to reestablish the permafrost. This can be done but it is easier to prevent melting in the first place by insulating the pilings and footings so that heat is not conducted into the ground, or by using cooling coils around the pilings to keep the permafrost frozen. Techniques are available to permit construction of roads and buildings in permafrost areas but costs are significantly increased. Water and pipes that are usually buried in the ground must be insulated or heated to prevent freezing in a permafrost area. The added cost and engineering required argue against the development of large metropolitan centers in permafrost areas.

2   *Snow, sleet, and ice* The impact of snow, sleet, and ice on industrial operations is difficult to evaluate because it is influenced by the experience of the individual and of management. A small amount of snow in a city unprepared and without experience in handling it can cause considerable

havoc. Cars become stalled on highways, automobile accidents increase, snow removal becomes a major operation, and routine activities just about come to a standstill. Industrial operations suffer because of the immobility of both workers and materials. In regions where winter snows often reach 100 to 200 mm at one time, there may be hardly a pause in the work schedule because of a moderate snow storm. Roads are cleared, cars have chains or snow tires, drivers are experienced at driving in snow, and alternate means of transportation may be available to take workers and materials to and from the factory. Heavy snows occurring with high winds result in rapid drifting which can, of course, cause problems even in areas that are prepared. City streets, even though frequently plowed, may be blocked by drifts that will stall traffic. Snow does hamper outside activities, cover stockpiled materials, and result in a slowdown in such industries as construction, utilities, and basic metals as well as in transportation and communications.

What has been said for snow can also be generally applied to sleet and ice, although these weather factors create some unique problems of their own. The major effects of sleet and ice are to immobilize workers and to stop the movement of raw materials and manufactured goods. Only small accumulations of ice are needed; just a fraction of an inch of ice can bring most vehicular traffic to a halt. Ice accumulating on railroad tracks or on power lines and poles can quickly slow or stop railroad operation and bring power and telephone lines down. Both snow and ice have only little effect on the processes of production in most cases, but without workers, power lines, or communications most industries must still curtail production programs.

3   *Wind* Most buildings and other engineering works can withstand all but the most severe winds. However, the secondary effects of the wind, in blowing dust or snow, or in making the effect of low temperatures even more severe, can place limitations on many industrial operations. In certain instances, even a wind of 15 to 20 mph (6.7 to 8.9 m/sec) can curtail a painting operation if there is blowing dust or other debris. Russell (1957) has suggested that generally up to 40 mph (18 m/sec) wind causes very little impairment in industrial operations. Between 40 and 60 mph (18 to 27 m/sec) there will be increasing difficulties in outside operations especially if there is loose debris that might be moved. Construction at any height above the ground becomes extremely hazardous and is usually terminated. Above 60 mph (27 m/sec) practically all outside activities are stopped, and structural damage may occur to billboards, signs, wires, and poles, as well as other industrial items that are not secured.

4   *Heavy rainfall* Rainfall, in general, and heavy rainfall, in particular, play important direct and indirect roles in industrial operations. Heavy rainstorms can cause immediate flooding of low-lying or flat areas where drainage is inadequate. Workers may be hampered in getting to or from work and the movement of materials may be temporarily interrupted. Even portions of the work area or factory may be inundated if drainage is

particularly poor. Most outside activities are, of course, influenced by rain since few workers relish work under such conditions except in emergencies. Thus rain of almost any intensity above the lightest drizzle will usually force a modification of scheduled outside work programs.

A second and more indirect effect of rainfall concerns the water supply or effluent removal operations that are necessary for the proper functioning of any industrial enterprise. Depending on the location of the factory and its industrial process, it may be necessary to depend on the annual rainfall to supply processing water or to provide adequate dilution of factory effluent. As state and federal regulations make stream disposal of effluent less likely, this latter aspect may decrease in significance.

5   *Other weather factors* The last items in the list of limiting weather factors are of more limited applicability. While important in certain operations they may have no more than nuisance value for other operations or whole industries. High humidities have generally a twofold effect: (*a*) they can lead to deterioration, mildew, and rotting of certain stored products or raw materials or corrosion of metals; or (*b*) they can produce feelings of discomfort on the part of workers especially when accompanied by high temperatures. Low visibility is mainly important in restricting the movement of workers and materials. Usually it does not persist so there is only a temporary interruption of activities. A heavy fog will temporarily suspend many outside activities in the aircraft, painting, trucking, and shipping industries.

Electrical storms or lightning have only limited influence by themselves. When accompanied by the heavy rainfall or high wind of a thunderstorm, however, they may result in a significant curtailment of industrial operations.

Stored items are often damaged by intense radiation so that it may be necessary to provide some shading if long cloudless periods with nearly vertical sun are expected. Some materials will lose strength or disintegrate completely under strong sunlight conditions.

APPLICATION OF CLIMATIC INFORMATION IN SPECIFIC INDUSTRIES

1   *Construction.* The Environmental Science Services Administration (ESSA), as part of its responsibility to provide information to all who may be affected by weather, has undertaken a detailed survey of the weather-sensitive aspects of the construction industry (ESSA, 1966). About 45 percent of the total money spent annually in the United States on construction ($88 billion, or more than 10 percent of the gross national product at that time) occurs in areas of construction that are influenced by weather conditions (Russo, 1966). Further, losses in the construction industry due to weather range between $3 and $10 billion a year. Weather and climate clearly play an important economic role in various phases of construction. Improved short-range weather forecasts will significantly reduce these staggering weather-related losses; application of climatic knowledge in the planning of

construction or the siting of the actual structure can result in additional savings.

The ESSA study identifies some 43 operations in construction that are weather sensitive to some degree. The study includes a table (Table 11-1) listing the 43 operations—in approximate time sequence from beginning of construction to the final paving and fencing—and indicating at what point different weather conditions or combinations of conditions become critical in the operation.

The results strongly support the earlier analysis of individual weather factors by indicating the limiting nature of temperature, rain, snow, sleet, and high winds on many operations in the construction industry. The list is oriented primarily to short-period weather phenomena although it is possible to utilize it climatologically as well. The engineer or designer must recognize the need to take account of the changing climatic likelihood of different conditions at different times of the year and to adjust cost figures accordingly. Knowledge of the influence of low or high temperatures, of winds, or of other weather factors can suggest possible locations for the construction that might be climatically more favorable or less costly in weather delays than other nearby sites merely because of microclimatic differences between sites. Knowing what weather conditions are limiting is basic in any attempt to reduce costs by taking advantage of future climatologic probabilities.

Maunder (1972b) reports on the recent development by Benjamin and Davis (1971) of a simulation model to provide quantitative information on the economic consequences of weather on building construction. The model should aid in the planning of construction programs to minimize delays and losses due to adverse weather. Benjamin and Davis have used the model to test the economic effect of changing the starting date for construction as well as the influence of different degrees of reliability of weather forecasts.

To operate the model, the daily weather forecasts of minimum temperature, maximum wind speed, and total precipitation are obtained and used to determine whether or not to schedule work on current building activities. The actual weather for the day is then compared with the forecasts to determine whether the correct decisions were made and to permit evaluation of the costs associated with the various correct or incorrect decisions.

Benjamin and Davis tested the model on the hypothetical construction of a two-story office building beginning on April 1 each year from 1941 to 1950, assuming the reliability of weather forecasts to be 80 and 90 percent. They determined that the actual cost of the construction, assuming no weather delays, was $76,600 and construction required 101 days. Table 11-2 gives the results of the calculated costs and length of construction for the different years utilizing the actual weather data for those years as well as the forecast data. Costs were higher and duration of construction longer when

OPERATION	RAIN	SNOW AND SLEET	FREEZING RAIN	LOW TEMPERATURES, °F
Surveying	L*	L	L	0– –10
Demolition and clearing	M	M	L	0– –10
Temporary site work	M	M	L	0– –10
Delivery of materials	M	M	L	0– –10
Material stockpiling	L	L	L	0– –10
Site grading	M	M	L	20–32
Excavation	M	M	L	20–32
Pile driving	M	M	L	0– –10
Dredging	M	M	L	0– –10
Erection of coffer dams	M	L	L	32
Forming	M	M	L	0– –10
Emplacing reinforcing steel	M	M	L	0– –10
Quarrying	M	M	L	32
Delivery of premixed concrete	M	L	L	32
Pouring concrete	M	L	L	32
Stripping and curing concrete	M	M	L	32
Installing underground plumbing	M	M	L	32
Waterproofing	M	M	L	32
Backfilling	M	M	L	20–32
Erecting structural steel	L	L	L	10
Exterior carpentry	L	L	L	0– –10
Exterior masonry	L	L	L	32
External cladding	L	L	L	0– –10
Installing metal siding	L	L	L	0– –10
Fireproofing	L	L	L	0– –10
Roofing	L	L	L	45
Cutting concrete pavement	M	M	L	0– –10
Trenching, installing pipe	M	M	L	20–32
Bituminous concrete pouring	L	L	L	45
Installing windows and doors, glazing	L	L	L	0– –10
Exterior painting	L	L	L	45–50
Installation of culverts and incidental drainage	M	L	L	32
Landscaping	M	L	L	20–32
Traffic protections	M	M	L	0– –10
Paving	L	L	L	32–45
Fencing, installing lights, signs, etc.	M	M	L	0– –10

* L indicates light; M indicates moderate.
† Indicates operation affected by this condition but critical limit is undeterminable.
‡ Indicates water freeze.
Note 1. All operations are hindered by air temperatures over 90° F (32° C) and THI over 77.
Note 2. The survey data reported in this table are furnished for purposes of illustration only and do not constitute a representation applicable to particular construction or projects.

Critical limits of weather elements having significant
influence on construction operations (from ESSA, 1966)

HIGH WIND, MPH	DENSE FOG	GROUND FREEZE	DRYING CONDITIONS	TEMPERA- TURE INVERSION	FLOODING ABNORMAL
25	x†	–	–	–	–
15–35	x	x	–	x	–
20	x	x	–	–	–
25	x	–	–	–	–
15	x	–	–	–	–
15–25	x	x	x	–	–
35	x	x	x	x	–
20	x	x	–	–	x
20	x	x‡	–	–	x
25	x	x	x	–	x
25	–	x	–	–	–
20	–	x	–	–	–
25–35	x	x	x	x	–
35	x	x	–	–	–
35	–	x	x	–	–
25	–	x	x	–	–
25	–	x	x	–	–
25	–	x	–	–	–
35	x	x	x	–	–
10–15	x	–	–	–	–
15	–	–	–	–	–
20	–	x	x	–	–
15	–	–	–	–	–
15	–	–	–	–	–
35	–	–	–	–	–
10–20	–	–	x	–	–
35	–	x	–	–	–
25	–	x	x	–	–
35	x	x	x	–	–
10–20	–	–	–	–	–
15	x	–	x	–	–
25	–	x	x	–	x
15	x	x	x	–	–
15–20	x	x	–	–	–
35	x	x	x	–	–
20	x	x	–	–	–

forecast reliability was only 80 percent, as would be expected because of the
additional missed forecasts. Costs were as much as 13 percent higher than
estimated with the 80 percent reliable forecasts and 8 percent higher than
estimated with the 90 percent reliable forecasts. Use of the model with other
starting dates permitted evaluation of alternate decisions and would permit
the builder to obtain some feeling for the possible cost of weather delays at
various times of the year to use in his construction bids.

2 *Petroleum and chemical industries.* Chemical industries, which include
petroleum refining, synthetic rubber, coke chemical, and synthetic ammonia
industries, have many common areas of operations. They are most frequently
conducted in unhoused or unenclosed units. Thus they are subject to all the
vagaries of climate and, if located in areas of severe conditions, require more
costly shelter or weather protection. This type of industry operates most
efficiently in a cool, dry climate with fairly light, steady winds for ventilation
purposes. Temperatures between 50 to 60°F (10 to 15°C) are favored, in
areas without heavy rains or high humidities.

Booth (1957) has made a fairly comprehensive survey of the climatic
problems of these industries. His conclusions concerning the difficulties that
might result from both high or low temperature, high precipitation, and high
winds are presented in tabular form in Table 11-3.

In each of the industries, there is the common low-temperature
problem of possible freezing materials flowing in pipelines or of water in
cooling towers. Ice around the intake to the tower may reduce its size and
make cooling less efficient. In cold areas, cooling towers need not be as large
as in warm areas, but in areas with significant seasonal temperature contrasts,
the small winter cooling towers may not be of sufficient size to permit
adequate cooling in summer.

In all these industries, flooding will interrupt operations. In areas of
high moisture, it will be necessary to provide good insulation for all exposed

TABLE 11-2

Cost and duration of building two-story office building starting
on April 1, using the weather forecasts with different degrees of reliability*

| | RELIABILITY 80% | | RELIABILITY 90% | |
YEAR	COST, $	DURATION, DAYS	COST, $	DURATION, DAYS
1941	86,500	135	81,800	130
1942	84,500	135	81,600	118
1943	84,000	152	80,700	121
1944	82,800	121	80,200	114
1945	83,100	142	81,400	133
1946	83,600	124	80,800	111
1947	85,000	132	82,100	119
1948	86,300	130	81,400	119
1949	86,200	138	83,000	126
1950	84,400	128	81,000	115

* *With permission of New Zealand Meteorological Service, after Maunder,
1972b, table 5, abridged from Benjamin and Davis, 1971.*

TABLE 11-3

Effect of climate on operation of petroleum and chemical industries*

	LOW TEMPERATURE	HIGH TEMPERATURE	PRECIPITATION	WIND
Petroleum refineries	Below −7°C critical. Housing of exchangers, pumps, etc., can present safety problem if combustible vapors are present without ventilation. Excessive cooling of pipelines restricts flows, might cause water in gas to freeze or hydrates to form	Cooling water must be between 15–32°C. Above 32°C cooling towers must be used. Reduces maximum recovery of light oil since increased temperature affects partial pressure of gases	Causes corrosion and deterioration of units. Old, worn insulation wets easily, contacts give trouble. Oil collectors may become clogged when moisture seeps into unit	Tall units, cracking, fractionating, cooling towers susceptible to high wind speeds. Gasoline refinery, fires, or fired heaters should be upwind to prevent gas or vapor being blown toward them
Coke chemical	50% of normal efficiency at 0°C. At −18°C, all weather-sensitive operations must be housed	High ambient temperatures prevent maintenance of proper oven coking temperature due to reduced upstack draft—requires blowers to supplement draft	Fifty-mm rain may shut down all operations. Rain may cool coke ovens and cause shutdown or inefficient operation	Coal bridge may be damaged by high winds. Wind of 30–40 mph may cool coke ovens too rapidly, affect efficiency of recovery and quality of light oils. Need winds over 4 mph for proper ventilation
Synthetic ammonia	Scrubbing towers subject to freezing. Solubility of scrubbing agent decreases with temperature and so does efficiency. More difficult to remove impurities	High temperature lowers the efficiency of production; 28% conversion of gas to ammonia at water temperature of 15°C, while at 35°C only 12% conversion occurs. Water scrubbers do not remove as much carbon monoxide. High temperatures cause high vapor pressures which result in loss of stored ammonia	Stored ammonium sulfate degraded by moisture; complete housing required for protection	

* With permission of American Meteorological Society, adapted from Booth, 1957.

wires and protection for all motors and contacts. Snow is not a significant problem for many of these industries; the large amounts of heat generated during operations will melt much of the snow that falls locally, making possible outside operations long after other areas are hampered by the accumulation of snow. The problem of movement of men and goods to and from the plant through snow still remains, however.

3   *Iron and steel industry.* The iron and steel industry is little influenced by most weather or climatic conditions found in habited regions of the globe. While optimum operating conditions exist in fairly cool (-18 to 4°C), dry (but with adequate supplies of water) climates with light breezes, it is also clear that only the most extreme climatic conditions create any serious problems. Operating efficiencies drop somewhat under high temperatures and humidities but these are hardly limiting. Very cold temperatures and especially permafrost conditions can result in problems because of (*a*) lack of adequate water, (*b*) possible melting of permafrost, and (*c*) increasing brittleness of iron and steel products. Any summary of climatic influences (e.g., Waters, 1957*b*) reveals that the iron and steel industry is not basically weather sensitive.

4   *Electric utilities.* The influence of weather or climate on electric power utilities may take at least two different and distinct forms: (*a*) the effect of climate on the maintenance of equipment, lines, power sources, and buildings, and the efficiency of workers; (*b*) the effect of climate on power demands and the application of climatic information to the problem of anticipating future loads. In addition, the role of atmospheric pollution or potential pollution on the location of generating stations is becoming an area of concern for climatologists. Many utilities employ private meteorologists to aid them in adjusting their operations to the vagaries of the local weather.

(*a*) Weather effects on equipment and power generation. Several sources for the production of power must be considered since they respond differently to severe weather conditions. Waters (1957*a*), who has surveyed several electric utilities operating in the north-central part of the country, has considered steam, hydro-, and diesel power production plants. The climatic limitations he identified are summed briefly in Table 11-4.

Waters makes clear that low temperature is the single most limiting factor in the operation of electrical utilities. Three effects are quite significant: (1) the freezing of water supplies and water intake systems, (2) the freezing of coal, and (3) the difficulty of starting diesel motors. Secondary temperature effects include possible engineering limitations due to permafrost or ground frost and the effect of cold on the efficiency of workers. The engineering problems of permafrost will be of concern to all three types of power-producing plants; careful study and more expensive design and installation of foundations, towers, pole lines, piping, dams, and roads are needed in such areas.

Low temperatures are often associated with snow or ice conditions, which are the greatest hazards to poles, towers, and transmission lines. The combination of these three factors also creates real problems in terms of

TABLE 11-4

Weather hazards to electric power utilities and distribution lines*

	LOW TEMPERATURE	HIGH TEMP.	RAIN	SNOW, SLEET, ICE	HUMIDITY	WIND	LIGHTNING
Steam	Below 0°C, freezing of water supplies and water intake systems, freezing of coal in railroad cars, freezing of wet conveyor belts, cooling tower problems. Lignite coal freezes easily and creates difficult winter handling problems	None	None	Slows unloading and handling of coal	High humidity reduces efficiency of cooling towers	Might blow water from intake opening	None
Hydro	Temperature 0°C critical because of ice problems, decrease in supply of water runoff. Possible frost damage to hydroplant dams. Winter water supply must be continuous to keep intake equipment free of ice.	None	None	Where snowmelt is source of water, annual changes in snow volume can result in changing surface water supplies	None	Might blow water from intake opening	None
Diesel	Diesel fuel must be heated prior to motor use. Starting diesel engines with temperatures below −18°C not normally possible because of cold lubricants	None	None	None	None	None	None
Distribution and transmission lines	Maintenance work hindered due to effect on repair crews	None	None	Lines become coated and poles and lines can be broken by weight. Critical if accompanied by high wind	None	High winds knock down poles and lines	Transformers, poles can be damaged if struck

* *With permission of American Meteorological Society, adapted from Waters, 1957a.*

repair and maintenance of a wide network of lines at a time when the movement of repair crews may be limited. Frost on transmission lines results in the need to reduce the electric power load which, in turn, can lead to the interruption of service to marginal customers.

(b) Weather influences on load dispatching. Load dispatching is a particularly weather-sensitive area of operations. Many weather-related conditions affecting load dispatching are of a short-range nature, involving weather forecasts of a few hours to a day; others, of a longer-term nature, involve more of a climatic forecast or estimate. One such example is found in the operation of a combined steam-hydroelectric generating plant. Few hydroplants rely entirely on water to produce all power needs; most either use the hydroplant to satisfy the peak power loads or have steam-generating capacity for use when water flow is insufficient to provide adequate power.

For most efficient operation of such combined steam-hydro generating plants, it is desirable to maximize the use of the less expensive hydrofacility, thus reducing the need to burn expensive fuel. Water still stored in reservoirs at the end of the dry season represents wasted power just as does water that overtops the dam due to runoff into an already full reservoir. Operations that bring the reservoir to its lowest permissible level just at the end of the dry season so that increased rains will have a whole reservoir to fill will make maximum use of the total water power stored in the reservoir. Thus some estimate of seasonal climatic conditions, the outlook for above- or below-normal precipitation and temperature, can be of help in planning the proper seasonal mix of fossil fuel and water-power generation to minimize costs. Even on a short-range basis, knowledge of an impending heavy storm with significant amounts of runoff will allow the utility to draw down the reservoir storage more than normal in anticipation of later reservoir recharge. This permits maximum conservation of fossil fuel supplies.

Steam-generating plants have a great need for water for cooling purposes. Thus, hydrometeorologic information has considerable value to ensure adequate surface or groundwater supplies throughout the year. Dry spells and their persistence for several years in humid or subhumid areas can be very serious if resulting water supplies are not adequate for cooling purposes. Any estimate of the nature of deviations of moisture supplies from normal over an extended period can prove helpful in being prepared for future eventualities.

Johnson, McQuigg, and Rothrock (1969) have shown a direct relation between electric power consumption and daily temperatures during summer months. Based on this relation, they have calculated power production costs as well as the savings with a reduction in daily air temperature of 2 to 3°C. Using data for 1962, the authors calculate that for each of 14 large utilities in the midwestern United States the average difference in hourly production costs using actual and modified temperatures would range between $405 and $529. Daily differences can be estimated by multiplying by 12 since daily load curves are at or near their peaks from 10 A.M. to 10 P.M. during hot summer weather. Their research provides basic information on the daily influence of climate (especially air temperature) on power demands and shows how accurate forecasts on both a short-range and longer-period basis can aid in more efficient load dispatching. Their study suggests the possibility that utilities might try to lower excessive summer temperatures by creating a cloud layer (by means of aircraft contrails, for example). Such a suggestion has most serious implications far beyond the significant savings that might accrue to the power companies. It is to be hoped that there will be careful consideration of all the possible results of such action before any implementation of the suggestion is attempted.

The load dispatcher must be aware of local weather in all parts of his distribution area, especially as far as severe storms, icing conditions, and the possibility of sudden local power demands are concerned. This latter situation might develop if a heavy cloud bank passes over a portion of the distribution area, forcing residents to turn on lights in greater numbers than usual. Severe storms may cause breaks in electric lines and require the rerouting of power through other lines in order to restore service as rapidly and efficiently as possible. With icing of lines, the local dispatcher must know when to send additional current through lines causing them to heat internally and to remain free of ice longer. Such heating techniques are most effective when started just before icing and, of course, should not be carried out after icing ends to keep costs down. Here accuracy in timing and in identifying the area of severe icing conditions is important.

Lightning strikes may damage poles, transformers, and transmission towers and, if nothing more, result in short-period "trip-outs" until the lightning arrester begins to function. Since problems are complicated if the line is heavily loaded at the time of a lightning strike, line loads are often reduced in regions of moderate to heavy thunderstorm activity.

Recent tie-in arrangements between large utilities reduce some of the need for reserve power supplies to handle emergencies caused by weather, unexpected breakdowns, or even shutdowns for routine maintenance. However, such tie-ins require knowledge of conditions over a wide area, covering several systems in size, since the load dispatcher must know conditions not only in his own dispatch area but also in nearby areas in the event of power transfers resulting from weather emergencies.

Benner (1958), in a comprehensive review of the role of weather on load dispatching, divides the system load of an electric utility into three components: (1) the fixed load, reflecting the general business activity and customer demand; (2) the variable load superimposed by the vagaries of the weather and climate; and (3) a "mass human activity" load component resulting, for instance, when a large number of households all (a) turn on TV for a particular show of national importance, (b) reduce demand by leaving home for Christmas shopping, or (c) increase demand by displaying decorative Christmas lights. For most efficient operation, the dispatcher must be fully aware of all three components of his daily or hourly load as well as how they may vary over the next few hours.

Benner has summarized the effects of various weather parameters in a brief table (Table 11-5). Effective temperatures are often used in place of dry bulb temperature in an effort to express the combined influence of temperature, wind, and humidity in a single index. While the weather factors listed generally refer to the short-range weather influence, they are equally important to the applied climatologist, for they help him understand longer-period relations as well.

5   *Railroad industry.* Hay (1957) has surveyed the effects of weather on all phases of railroad construction, operation, and maintenance (see Table 11-6). He indicates that any weather event that will wash out or block tracks

TABLE 11-5

Effects of weather parameters on electric load*

WEATHER PARAMETER	EFFECT ON LOAD
Primary factors: Dry bulb temperature	1 Two percent change in load for each 2.8°C change from 18°C (Philadelphia area) (Dryar, 1949) 2 Load saturation produced by heavy use of home heating systems and auxiliary appliances during protracted and/or extreme low temperatures† (Dryar, 1944) 3 Increased loads in hot weather by air conditioning and poor line transmission equipment characteristics‡ (Corey, 1949)
Cloud cover	1 Increased load during variable cloudiness and for several hours thereafter (Drummond, 1949) 2 No prediction problem for cloudless days or for cirrus cover (Dryar, 1944; Neuberger and Stephens, 1948) 3 Increased load for low and middle clouds; shift of load pattern as a result of clouds effectively changing the "sunrise" and "sunset" hours (Thomas and Drummond, 1953)
Secondary factors: Wind vector	1 Increased load demand in cold weather as a result of increased heat losses from buildings 2 Decreased load by aiding cooling in summer‡ (Dryar, 1944, 1949) 3 Can increase load demand if electric traction vehicles are supplied by the power system‡ 4 Destructive velocities may cause outages and/or rerouting (Corey, 1949)
Moisture	1 Increased load demand as a result of human discomfort in summer‡ 2 Presence creates possibility of sudden load surges from electrical arc-over of duty insulators (Neuberger and Stephens, 1948) 3 Presence may shift the load peak if human habit schedules are altered by precipitation (Dryar, 1944)

* *With permission of American Meteorological Society, after Benner, 1958, p. 518.*
† *Personal communication, R. T. Purdy, 1957.*
‡ *Personal communication, W. Beckwith, 1957.*

TABLE 11-6
Weather effects on railroad operations*

1 Roadway and track

(a) Large temperature changes: causes excessive rail expansion or contraction; "sun kinks" which can bend rails out of shape, keep drawbridges from closing

(b) Dry weather: increases danger from fires

(c) Low temperature: frost heaving of track and roadbed. Ice and freezing may damage bridge piers. Steel rails become brittle at very low temperatures. Switches may freeze

(d) Snow: switches, joints, crossings, etc., can become impacted with snow and rendered inoperative. Drifting of snow or snow slides can block open tracks especially in defiles.

(e) Wind: only significant when it results in blowing snow or sand, or windchill effect (except for hurricane or tornado velocities)

(f) Moisture: excessive humidity can cause rusting and corrosion. High rainfall can cause soft spots in roadway or slippage of subgrade materials, landslides. Floods and washouts can occur

2 Signals, communications

(a) Temperature: large range can cause breaking of wires with repeated expansion and contraction. Moving parts can freeze

(b) Snow: signals covered by drifting snow, lines destroyed by ice and snow coating. Snow slides can carry away signals or wires. Signals and lines especially susceptible to ice storms; communications are disrupted by downing of lines and poles

3 Yards and terminals

(a) Temperature: cars roll more freely in hump yards with high temperatures, more slowly with low temperature. May require two different humps. Coal may freeze with low temperature. Diesel fuel oil needs heating when below $-18°C$ to flow more readily

(b) Snow: heavy falls as well as drifting can disrupt yard just as it stops movement over open track

(c) Switches, rails, roadbed have same weather problems as listed in 1 above.

4 Locomotives and rolling stock

(a) High temperature: increased incidence of hot boxes in cars. Perishable materials in shipment must be handled in refrigerator cars. Domes needed in liquid cargo cars so liquids can expand and gases can collect and escape

(b) Low temperature: frost or ice on rails causes loss of adhesion (coefficient of friction of 0.15 for slippery rails vs. 0.25 for dry rails would mean 40–50 percent reduction in traction and in hauling capacity). Diesels likely to freeze in very cold temperatures should be housed at $-40°C$ or below. Lubrication less effective, resulting in overheated journals and hotboxes

TEMP. RANGE, °C	DIESEL TONNAGE REDUCTION, %
−1 to −18	0
−18 to −29	15
−29 to −40	20
−40 to −51	40

(c) Wind: blowing sand may block tracks, clog air filters, increase wear on bearings

(d) Rainfall: maximum depth of only 75 mm over rails is permitted for diesels in order to keep moisture out of motors

* *With permission of American Meteorological Society, adapted from Hay, 1957.*

(floods, snow storms, dust storms) or disrupt signals or communications (snow, ice, floods) will halt railway service until the track is repaired, the right-of-way is cleared, and signals are returned to operation. Under most conditions, weather does not cause a significant delay, but with severe flooding or snow storms sometimes as much as a week or more delay may be encountered on certain sections of track. Fortunately, it may be possible to bypass the particular section of track and to move materials and passengers by a more circuitous route.

The railroad industry is well equipped and very experienced in meeting weather hazards. Thus, when snow or floods are anticipated, crews are already at work to protect the right-of-way and to keep signals functioning. It is uncommon for serious weather delays to occur.

Based on a survey of operation of the Chicago, Milwaukee, St. Paul and Pacific Railroad in two different months, Hay listed six major causes for winter interruption of railroad service (Table 11-7). More than 90 percent of all delays involved snow or ice problems in one way or another. The table suggests the rather specific nature of weather hazards to railroad operation.

6   *Other transportation industries.* Aeronautical meteorology has been an extensively studied branch of weather science. Comprehensive surveys of the field are included in many books on meteorology and climatology. Rather than recounting how knowledge of weather results in improved in-flight operations, this section will only seek to identify selected areas in which climatic knowledge can aid in terminal selection and operation, flight scheduling, and aircraft safety.

Climatology can play an important role in the selection of an airport site. Especially in hilly terrain, it is important to have a considerable period of observations along the slopes as well as in the valley bottom. Site selection based only on location of flat ground or nearness to urban centers may ultimately result in increased weather hazards or fewer operating days. A site further removed from the urban center or even on rolling terrain, involving higher initial cost for construction, might ultimately prove more desirable by

TABLE 11-7

Major causes of winter interruption of railway service
(Chicago, Milwaukee, St. Paul and Pacific RR)*

CAUSE	PERCENT OF TOTAL WEATHER DELAYS	
	JANUARY	MARCH
Thawing of steam lines	4.8	
Snow and ice in switches	3.8	1.4
Heavy, drifting snow	11.9	27.2
Blizzard conditions	64.5	37.3
Snowplow delays	10.4	31.3
All other weather delays	4.6	2.8
	100.0	100.0

* *With permission of American Meteorological Society, from Hay, 1957, p. 31.*

having more operating days due to weather or providing greater weather safety.

Prevailing upslope winds may increase the frequency of cloudy conditions with the possibility of fogs, icing conditions near the ground level, precipitation, or other generally unfavorable flying conditions. Across the valley, in an area of prevailing downslope winds, adiabatic heating of the air may result in dissipation of clouds and fog, less icing, and possibly more flying days all within the same general climatic area.

Extensive wind roses may not always be necessary for determining the layout of runways, since modern planes can land with reasonable safety with steady, light to moderate, crosswinds. It is the gusty or high-speed winds that need to be avoided. This information is usually not available from the average wind rose.

Actual aircraft operation is much more influenced by weather than by climate factors. However, climatology can contribute in terms of identifying logical alternate airports in case the main destination is closed or in suggesting most appropriate flight paths or altitudes. Such knowledge, based on climatic records, must be modified for the particular weather situation but, at least, the climatological planning will provide logical starting points. For example, in case a particular airport is closed by weather, it is desirable to know the probability of adverse weather conditions also closing other nearby airports. Such information is readily available from the record of existing weather conditions at possible alternate airports at the time when limiting conditions exist at the selected destination. Knowledge of possible alternate airports will aid in planning fuel loads for aircraft operation and in focusing the attention of the forecaster on other areas of possible concern as he prepares his forecast for the flight.

Water transport is not as dependent on weather as it once was, but still temperatures and winds are important factors in planning ocean voyages. Wind direction and speed are significant in determining both fuel consumption and smoothness of the voyage. The pressure of the wind on the vessel varies with the square of the wind speed so that doubling the wind speed increases the pressure of the wind four times. Moving against a headwind, the real wind effect on the ship is the sum of the wind speed itself and the speed of the ship relative to the surface. As a result, a longer route can often result in greater economy of operation, less elapsed time en route, and a smoother voyage.

The whole field of ship routing was previously the responsibility of climatologists who prepared seasonal charts of prevailing winds, frequencies of storms, and ocean currents for the use of navigators. It is now the meteorologist who helps the navigator plan his course on the basis of weather forecasts and keeps him informed of weather developments along his route so that adjustments of course can be made en route.

In a discussion of ship routing, White (1971) reported that one tanker company sailing regularly across the Pacific Ocean considered that a time saving of just $\frac{1}{2}$ hr for the trans-Pacific voyage would pay the weather-routing fee for a 95,000 ton dead weight tanker. He noted, however, that time saving was only one of the economic benefits from weather routing; smoother voyages result in less damage to cargo and ships and reduce insurance costs.

The main objective in ship weather-routing is to select the optimum track for oceanic crossings. It is a complex undertaking for it requires an understanding of meteorological, nautical, and economic factors and it needs precise knowledge of the weather and sea conditions for a week or 2 weeks in advance. The reason that weather routing works well even without all needed information is that, at present, it does not rely greatly on detailed features of the weather and waves but rather on the general circulation and its variation with time and latitude. Often it is sufficient to know whether a ship's path should be to the north or south of the main tracks of cyclones or anticyclones in the area (Kruhl, 1971). The West German Seewetteramt, in Hamburg, investigated 32 successive winter trans-Atlantic crossings and found an average of 19 hr of traveling time was saved per crossing by ship-routing techniques. Another study of 200 Atlantic crossings in winter showed that weather routing saved an average of 10.5 hr per crossing. The time savings were more than enough to pay for the ship-routing service itself. In this latter study it was found that 77 percent of the recommendations to sail either north or south around Great Britain on the trip from Germany to Canada were correct. Kruhl (1971) points out that some figures on time savings are questionable since it is not always certain what particular route the shipmaster would have followed had the ship-routing forecast not been available.

Fog occurrence is a second important weather factor in shipping. The dual problems of possible ship damage due to collision in the fog and the time lost due to immobilization by fog are the critical factors. Use of radar greatly reduces both of these problems although ship collisions still can and do occur in periods of reduced visibility.

Temperature also has a significant impact on shipping. First, there is the effect of low temperatures on ice formation, or the existence of icebergs themselves. Ice patrols, as well as aircraft, and satellite observations of icebergs are important in order to route ships out of those areas where danger from icebergs is high. Ice cover in harbors may prevent ships from entering or leaving. The date of opening of the Great Lakes to shipping, of importance especially to the iron ore carriers, is closely correlated with the air temperature over the Great Lakes area during the winter period. Forecasters have found the best relation to be between the opening date of the lakes and the February temperature—the lower the February temperature, the later the opening of the lakes.

There is also a relation between environmental temperature and conditions within the hold of the ship where cargo is stored. The upper cargo layers in the hold and any deck cargo will be at or near air temperature (unless in direct sunlight), while the temperature of the cargo near the bottom of the hold will approach water temperature. The color of the ship can influence the temperature of the upper portion of the cargo through the amount of energy absorbed or reflected. Some care must be exercised in placing cargo within a ship, especially on long voyages through regions of changing temperature, if the cargo is temperature sensitive.

Humidity conditions within the hold can be quite variable. Many materials have an optimum humidity for storage and will react adversely when humidities are either too high or too low. In wood, paper, and leather products, high humidity leads to warping and swelling, while low humidity results in shrinking and possibly cracking. Condensation on materials stored in the holds can become a major problem, especially when a ship moves into colder waters and the shipments are not in vapor-tight containers. If holds must be maintained at fairly uniform temperature and humidity conditions, some knowledge of the variation in outside air and water conditions is required to ensure proper regulation of the ship's ventilating or heating equipment. Climatology can suggest the control equipment needed on the ship for each cargo and route to prevent loss or deterioration of cargo while en route.

Highway transportation has been mentioned several previous times in connection with the operation of various industries as being highly weather sensitive. The urban and suburban transportation system and its relation to weather and climate need special mention, however, because of the significant role the system plays in the lives of so many North Americans and Europeans. Many examples could be cited of the location of superhighways or turnpikes through hazardous climatic areas even though the shifting of the highway by just a few miles might have eliminated the hazard. This is especially true in the case of fog. Study of the distribution and frequency of fog along several proposed rights of way before purchase or construction is undertaken might well result in considerable later savings in lives as well as in increased operating hours for the turnpike.

Lowered visibility, icy or wet surfaces, and gusty winds all greatly increase driving hazards as can be seen by any survey of accident statistics. Maunder (1970) has summarized the data for all traffic accidents in British Columbia in 1966. While 53 percent of the 44,187 accidents occurred during generally clear weather, a significant 26 percent of the accidents occurred during rainy weather, nearly 5 percent in snow, and just under 2 percent in fog or mist. The percentage of accidents during adverse weather is much higher than would be anticipated as a result of the normal expectancy of such weather conditions. Robinson (1965) reported just over 4,000 casualty road accidents on 161 rainy days in Melbourne in 1960 as compared with just under 4,000 accidents on the 205 nonrainy days. This results in a 30 percent increase in accidents on rainy days.

Because of the weather-sensitive nature of highway travel and the high speeds possible with modern cars and turnpikes, it is imperative that climatology be considered in the problem of highway design, layout, and construction. Final location of high-speed arteries must be in the areas of least possible weather delays from fog, ice conditions, drifting snow, and flooding. Climatologists can provide information on the distribution and frequency of these hazards; such information will also prove useful in suggesting surface materials to use for best trafficability, possible road colors for most rapid melting of snow and ice, the need for drainage facilities or storm sewers, and the desirability of reflective signs or lines to increase driving safety in hazardous areas. The applied climatologist has not been utilized effectively in most highway transportation problems.

7   *Advertising.* Advertising, whether it be a nationwide campaign by a major industry member or the decision of a single store owner to feature a particular product by moving it to a prominent position in the store, is often climate dependent. The most obvious example would be in the advertising of some product such as antifreeze for automobiles. Motorists will not consider the need for antifreeze until the forecast for the first real frost of the fall occurs. There is no advantage in featuring antifreeze in the spring or early summer for sales will be minimal. Rather, in any campaign to sell antifreeze, the seller must be aware of the dates of first fall frost in each area and have supplies and ads on hand in each locality at these times. Advertising that begins just before the actual date of first frost and continues for a period thereafter will be most effective.

Picnic supplies, bathing and boating items, clothing, and soft drinks all might fit into this same climate-dependent category although, in some instances, advertisers have successfully changed buyer attitudes so that off-season advertising is successful. In this regard, the recent trends toward advertising cold drinks or "thirst quenchers" all year around or bathing suits early in spring or late in the fall at reduced prices is often effective (especially now that many people take vacations to other climatic regions where such products are actively in use). With more mechanical control over climate, with heated or inside swimming pools, and with air conditioned homes, buildings, and cars, long-standing clothing, eating, drinking, and, thus, advertising habits are being subjected to considerable change.

8   *Retail sales.* Some examples have appeared in the literature of the relation between retail sales of certain products, such as raincoats, cold drinks, ice cream, overcoats, and bathing suits, and weather factors, The Federal Reserve Board, which compiles the figures of retail sales and compares them with sales in similar weeks of previous years, often notes that year-to-year changes result from the occurrence of favorable or unfavorable weather events such as widespread blizzards, hurricanes, or sunny skies that cover large portions of the nation. Figures for retail sales over the nation for a particular month are adjusted for seasonal climatic changes, but when the weather in a particular year is either extremely favorable or unfavorable, the seasonal correction factors cannot adjust sufficiently for the extreme deviation from normal.

Steele (1951) studied the effect of weather on daily sales in three departments of a large store in Des Moines, Iowa, from 1940 to 1948. Using only the 7 weeks prior to Easter, Steele concluded that retail sales were weather sensitive in four different ways: (1) the weather might physically prevent people from getting to stores (heavy snows and drifts), (2) the weather might be disagreeable enough to discourage some people from shopping, (3) the weather might produce certain psychological effects that would change shopping habits, and (4) the weather might be such as to make certain items more saleable (e.g., raincoats and umbrellas during rainy periods). He found that when Easter was early, the depth of snow was the most significant weather factor, while with a late Easter, the amount of precipitation during the day was most significant. Other factors considered were daily air temperature, wind speed, and amount of sunshine. Some 88 percent of the variance in store sales were accounted for by the five weather factors.

Linden (1959) studied the effect of weather on customer traffic and sales in a single downtown store as well as in several suburban stores in the New York City area and found a number of weather-related events. Customer traffic was 8 percent less on wet days than on cloudless days. However if the duration of the rain was 3 hr or less, sales volume decreased only about 5 percent as compared to an 8 percent decrease for a 4- to 6-hr rain and more than a 15 percent decrease if the wet period lasted all day. Intensity of rain was not as significant as duration in its relation to sales although the time of occurrence was important. Rain before noon hurt sales more than afternoon or evening rains (even with night store hours). Rains on Monday and Saturday were less effective in cutting sales than rains on the other business days. Sales were better on the second of two consecutive rainy days than on the first. Linden also found that sales lost to bad weather were often not made up with later increased buying.

Maunder (1972a) has attempted to achieve a weekly nationwide weather index to relate to weekly trade statistics. While he based his weather index on departure of weekly mean temperature and precipitation from normal at 147 cities across the United States, he recognized that people react to weather rather than to the individual elements themselves, and that other elements such as sunshine, humidity, and precipitation duration would be valuable additions to any index if these data were readily available. The individual weather indices were "weighted" on the basis of a "buying power index" for each of the 147 cities selected. The buying power index (E_i) is

$$E_i = \frac{5I + 3R + 2P}{10}$$

where I = percent of U.S. effective buying power
R = percent of U.S. retail sales
P = percent of U.S. population in the city

Summation of the various weighted weekly weather indices provided a nationwide index for each week that could be related to retail sales. To eliminate certain problems associated with Christmas buying, the study covered only the weeks from February to November inclusive in each of 3 years, April 1966 to March 1969.

Several significant conclusions resulted from the analysis of the 3 years of data. For example, Maunder found that drier than normal conditions were generally associated with above-average retail sales in late winter and early fall while wetter than normal conditions were often associated with above-average retail sales in late spring and early summer. At the same time colder than normal falls led to increased retail sales as did warmer than normal spring and early summer periods. Maunder warned that the econoclimatic associations achieved varied in their significance within the different retail-trade groups studied. Many of the groups did not involve particularly weather-sensitive items, while the weather index itself needs to be refined to reflect more accurately those aspects of weather most actively associated with sales.

The Federal Reserve Bank of Chicago has undertaken brief studies of the relation between retail sales in Chicago and weather. Petty (1963) reports on a study comparing temperature and precipitation data with dollar values of sales. Retail sales volume is significantly related to temperature and to a smaller extent to precipitation. Normal temperatures are not as desirable as what might be called "better than normal" temperatures. Thus warmer than normal spring temperatures or cooler than normal fall temperatures (especially early in their respective seasons) are related to above-normal sales of spring and fall apparel. However, cold springs or warm falls often depress sales; many individuals seem to go directly from winter to summer or vice versa, foregoing purchases of spring or fall clothes. This does not result in delayed purchases, but rather, the sales are lost forever. Fairly high summer temperatures and low winter temperatures both act as depressants on sales, as do severe storms. Petty concludes that relating weather or climate to retail sales is a complex undertaking involving many different aspects of weather, as well as type of goods being sold, price lines, and even the day of the week. Understanding the relationship however, should result in more efficient selling. While many economic activities (such as sales, rates of personal income, housing starts, or transfer payments) are related to weather factors, there are also many other factors not related to weather that may mask the weather relations. Using the seasonal pattern of personal income in each of the four southern states of Florida, Georgia, Louisiana, and Mississippi, McLeary (1968) shows that each has its own distinctive seasonal pattern related to the source from which the income in each area is derived. In each state, the proportion of the income derived from agriculture (ranging from 3 to 11 percent of the total income) is the most variable—in part because of the weather-sensitive nature of agriculture. Recognition of the seasonal nature of economic activity, and especially understanding the possible variability of this activity due to weather abnormalities, should be a prime concern of the banking institutions in any area. Seasonal demands for loans or credit that may result from such weather anomalies seldom coincide with bank deposits, and so the banking concern must be prepared to seek other sources of funds to meet both the recurring seasonal as well as the abnormal weather-related demands.

9   *Weather insurance.* One developing field of applied climatology is that of insurance against the various hazards posed by the vagaries of weather and climate. Wind and hail insurance are often routinely written into home owners' or automobile insurance policies although flood and water damage insurance is more difficult to obtain. Whatever weather coverage is included in these cases is merely additional insurance on a larger and more comprehensive policy. Only in the last several decades has any serious attempt been made to provide a comprehensive policy to insure only crops or other weather-sensitive items against the chance occurrence of destruction or damage by weather factors. Maunder (1970) has reported that crop-hail insurance purchased by farmers from private companies has increased from about $1.06 billion to $2.92 billion during the period 1950 to 1964, while purchases from federal sources have increased from $0.24 billion to $0.54 billion in the same time period. Hendrick and Friedman (1966) point out that property in the United States valued at more than $500 billion is insured against storm damage, the annual claims paid by the insurance companies for storm damage coming to $0.5 to $1.0 billion. Weather insurance is already a big business but it could become even more significant in the years ahead.

While standard crop-hail insurance, covering mainly hail and fire, has been written for many years both by the federal government and by private insurance companies, it was not until 1937 that the Congress established a crop insurance corporation in the Department of Agriculture. The Crop Insurance Act, as originally passed, extended coverage only to wheat because of the availability of experience information. Due to pressure, however, it was extended in 1939 to cover cotton as well. Many questions and problems arose, and because of large losses the program was canceled in 1943. Only about 30 percent of eligible wheat growers and 10 percent of eligible cotton producers participated in the program. World War II, then in progress, was, in part, instrumental in reinstating the program in the following year since food production was vital to our national needs and farmers needed the protection and security of insurance. The Federal Crop Insurance Act of 1944 started the program again with certain modifications to eliminate, hopefully, the problems of the earlier program. In 1947 another bill passed Congress to reduce the scope of the Federal Crop Insurance program and to place it on an experimental basis. It was recognized that the program would have to be continued for a number of years, of the order of 20 as a minimum, before enough experience was obtained to permit any real evaluation of the results.

In the mid-1950s, stock fire insurance companies began to offer an all-risk crop insurance protection program (Roth, 1955, 1956). These policies covered hail, fire, drought, flood, excessive moisture, insect infestation, plant disease, wind, tornado, frost, freeze, and other miscellaneous perils. The experimental program protected only corn and soybeans in 41 counties and tobacco in 11 counties. The multiple-peril policy devised at that time offered two levels of protection: "A" perils were covered fully while "B" perils were only covered in part with the farmer having to accept essentially one-half the loss himself. The farmer paid only one premium for the two parts of the insurance package. In the A part, he was reimbursed in the same percentage as the percentage of the crop loss. For example, if 40 percent of a particular acre of a crop was lost, the farmer received 40 percent of the amount of the insurance. If he had $100 insurance on that acre, he received $40 for his loss. Thus a farmer could have full protection for loss under the A perils if he had sufficient coverage per acre. Under the B perils, the farm was considered as a unit rather than each acre and the yield was computed for the whole farm. If the yield fell below some guaranteed amount, a proportional payment was made by the company. Thus it was true catastrophe insurance.

During the more than 20 years that the Federal Crop Insurance Corporation (FCIC) has been in operation, there has been an opportunity to study and test the whole idea of crop insurance. The major problem, of course, has been to obtain enough experience with the risks involved to permit the establishment of fair premium rates. Other concerns with crop insurance include

(a) The loss must be purely accidental and not a function of poor farming practices on the part of the farmer—the "moral hazard" must be removed

(b) Losses must be easily determined and verified

(c) Risks must be dispersed in such a way that a weather catastrophe in one area (a large drought or area of insect infestation) cannot produce insurance losses that will place undue financial strain on the insurer

(d) The probability of loss must not be so high that the risk becomes uneconomical to cover

Very seldom are all these problems and limitations clearly under control. The "moral hazard" factor can hardly ever be eliminated, and it is extremely

TABLE 11-8

Weather factors causing payment of indemnities, Montana, 1947–1967*

WEATHER FACTOR	PERCENTAGE OF INDEMNITIES PAID
Drought	61.0
Hail	27.7
Insects	4.1
Wind	2.6
Disease	1.3
Frost, freeze, cold, winterkill	1.2
Excess moisture	0.4
Flood	0.1
All other	1.6

* With permission of Montana Agricultural Experiment Station, from Myrick (1970), p. 15.

difficult at times to measure the extent of loss that is weather related. However, with extensive coverage well-dispersed geographically, some of the risks to the insurance company from unknown or undetermined factors should average out.

Myrick (1970) has listed the various weather factors causing payment of indemnities in Montana during the period 1947 to 1967 (Table 11-8).

Because drought conditions can sometimes cover large areas of the country, geographic diversification of risk may not provide adequate protection for the insurer. But since drought severity fluctuates over time in all but the arid areas, protection against widespread drought catastrophes can be achieved over time rather than by means of widespread areal coverage. This necessitates strong financial reserves on the part of the insurer.

Even the definition of what constitutes drought cannot be clearly resolved by scientists. One way to eliminate this problem is to base the payment of indemnities solely on the basis of yield. Under such a policy, the degree of drought or of excessive moisture need not be exactly known, or the likelihood of such events need not be precisely determined. Rather, only the pattern, over time, of yield per acre or per farm as influenced by all weather factors corrected for the effect of improved seeds, fertilizer programs, and farming methods must be known. Such information is not presently available for most crops, and so the applied agricultural climatologist still has much to do to aid the farmer and the insurance company to develop a sound and realistic program.

Weather insurance has been available for a number of years but the record of successful use has not been encouraging. Problems and limitations abound, not the least of which are the climatological or meteorological statistics involved. It is clear that a large potential market exists for various types of weather insurance. With the increased use of high-speed computers, with the increased availability of weather data, and with the movement of climatologists into the field of actuarial statistics, there is a real possibility for rapid development in this line of endeavor.

ENGINEERING APPLICATIONS

Heating Degree-Days

The heating industry has long used a degree-day temperature summation technique to determine fuel consumption. As established by the American Gas Association in 1927, fuel consumption is found to vary directly with the difference in the air temperature below a base temperature of 65°F (18°C). This value is the temperature above which daytime heating is not needed to keep house temperatures within the range of human comfort. As the mean air temperature falls below the base temperature, heating requirements and thus fuel consumption increase proportionally. Each degree that the mean daily temperature falls below 65° is called a "degree-day."

It is clear that temperature difference from 65°F (18°C) is not the only
factor influencing heat requirements. Certainly wind speed will influence the
rate of exchange of heat between a house and its environment as will the
presence or absence of solar radiation. A given temperature or degree-day
value will be associated with different values of fuel consumption if it occurs
with overcast or clear skies or with high wind speeds or calms.

Murphy (1960) made a detailed study of the relation between fuel
consumption and several combinations of meteorological factors in the
northeastern portion of Massachusetts, using records from the Haverhill Gas
Company for the 1951–1952 heating season. He used a linear regression
technique of the form $Y = a_0 + a_1x_1 + a_2x_2 + \cdots + a_nx_n$, where Y is the
predicted heating load, the x's are different meteorological factors, and the
a's are regression coefficients.

While it was not possible to obtain data on temperatures, wind, and
solar radiation at one spot, this is not particularly necessary since the service
area for the gas company covers 17 communities in northeastern Massachusetts, over which some variation in meteorological conditions can be
expected to exist on any given day. To obtain the best relation between
degree-days and fuel consumption, Murphy tested degree-day values from
Haverhill, from Middletown (a more centrally located station), and from
Boston somewhat to the south of the area. Several different techniques for
obtaining degree-days were investigated—first using mean daily temperatures,
then an average temperature based on five selected hourly temperature values
and, finally, an average using all 24 hourly values. Murphy also obtained
correlations between fuel consumption and the weather factors of wind
speed, and cloudiness (at Boston), daily solar radiation (at Blue Hill, the
closest recording station), and precipitation (at Middletown). Finally, he tried
various combinations of these meteorological factors, seeking the highest
correlations. Table 11-9 lists the variables tested, the size of the sample used,
and the correlation coefficients obtained.

The highest correlation coefficient occurred when a combined meteorological factor consisting of degree-days obtained from the average of 24
hourly values adjusted for wind speed (obtained by averaging five hourly
values) was related to fuel consumption. A similar high correlation was found
when a combination of degree-days based on an average of five hourly values,
corrected for wind speed and cloudiness, was used. Correlations with
degree-days alone were quite significant, but improvement was noted when
wind and cloudiness data were also included. In this respect the house is
much like an individual, producing heat internally through the burning of fuel
and losing it to the cold outside environment as a result of conduction,
convection, and radiation. Thus temperature difference between the house
and the outside, wind factors, and solar heat load (as influenced by
cloudiness) are all related to heat loss and fuel consumption and are necessary
inputs in any meteorologic evaluation of heating requirements.

TABLE 11-9

Relation between heating load and various meteorological
parameters, eastern Massachusetts, 1951–1952 heating season*

ITEM	INDEPENDENT VARIABLE(S)	SAMPLE SIZE N	CORR. COEFF. r	PERCENT REDUCTION PR
1	DD_{HAV}	147	0.89	79
2	DD_{MID}	147	0.90	81
3	DD_{BOS}	147	0.89	79
4	$DD_{BOS}(5)$	147	0.92	84.5
5	$DD_{BOS}(24)$	147	0.925	85.5
6	V_{BOS}	147	0.41	17
7	C_{BOS}	147	−0.31	9.5
8	$I_{BL\,HLS}$	147	−0.29	8
9	P_{MID}	147	−0.04	0
10	$DD_{HAV} + V_{BOS}$	147	0.92	85
11	$DD_{MID} + V_{BOS}$	147	0.94	88
12	$DD_{BOS} + V_{BOS}$	147	0.92	85
13	$DD_{BOS}(5) + V_{BOS}$	147	0.945	89
14	$DD_{BOS}(24) + V_{BOS}$	147	0.95	90
15	$DD_{MID} + V_{BOS} + I_{BL\,HLS}$	147	0.94	88
16	$DD_{BOS}(5) + V_{BOS} + I_{BL\,HLS}$	147	0.945	89
17	$DD_{BOS}(5) + V_{BOS} + C_{BOS}$	147	0.95	90

Location subscripts:

HAV: Haverhill; BOS: Boston; MID: Middletown; BL HLS: Blue Hill Observatory

Key to independent variables:

DD_{HAV} — Haverhill degree-days, computed by using a mean temperature obtained by averaging the maximum and minimum on consecutive days (due to the time of gas sendout). For example, Sunday's average temperature would be obtained by averaging Sunday's maximum and Monday morning's minimum.

DD_{MID} — Middletown degree-days, computed as was DD_{HAV}

DD_{BOS} — Boston degree-days, computed as was DD_{HAV}

$DD_{BOS}(5)$ — Boston degree-days, computed by using a mean temperature obtained by averaging the temperatures of the synoptic observations, 0700, 1300, 1900, 0100, and 0700 local time

$DD_{BOS}(24)$ — Boston degree-days, computed by using a mean temperature obtained by averaging the hourly temperatures, 0800 to 0700, inclusive

V_{BOS} — Boston wind speed, computed by averaging the speeds at the synoptic observations, 0700, 1300, 1900, 0100, and 0700 local time

C_{BOS} — Boston mean sunrise-to-sunset cloudiness

$I_{BL\,HLS}$ — Total daily solar radiation received at Blue Hill Observatory, modified to take into account the variation of surface conductance with wind speed

P_{MID} — Middletown daily precipitation amount

* With permission of American Meteorological Society, from Murphy, 1960, p. 67.

Recent concern over degradation of the environment has resulted in much confusion in the field of power generation. For example, large amounts of water are needed for cooling during the course of power production, and this heated effluent must be discharged to the environment. Many conservationists argue against the discharge of any heated water to receiving streams or lakes. Cooling towers have been used to remove heat from the effluent, but there is no real knowledge of the effect that cooling towers may have on the environment. Do they remove heat from the water without creating some other problem through the discharge of either heat or moisture to the atmosphere? Some preliminary studies have been attempted but no positive answers have been achieved.

Cooling towers can be of the natural-draft or mechanical-draft type. The former relies on a tall chimney to induce air movement up the tower, while the latter uses fans to force air up the tower. Wet-type or evaporative cooling towers result in heat dissipation by the evaporation of the heated moisture into the air while dry-type towers lose heat to the air by conduction and convection with less evaporative cooling. Based on the design of three 55-m wide, 76-m tall wet-type towers, with a discharge velocity of 5.6 m/sec, Huff and his associates (1971) estimate an average discharge of 2.1×10^{12} cal/hr to the atmosphere with a water loss to the atmosphere of 11,000 gal/min in winter and 14,700 gal/min in summer from each tower. They estimate the average yearly output of heat from the urban area of St. Louis at 13.9×10^{12} cal/hr. Thus at peak load each tower (actually designed for a nuclear facility in Zion, Illinois, on Lake Michigan) will put into the air about 16 percent of the heat produced by the St. Louis urban area—a significant amount of heat and water and possibly of environmental concern.

While quantitative information on the environmental influence of cooling towers is lacking, most researchers feel that some changes in frequency or severity of fog, icing, clouds, precipitation, and possibly severe weather such as thunderstorms may result. But little more than rough estimates are available.

Increased frequency of fog and icing conditions do appear to be the most likely weather events to develop from the use of mechanical-draft wet-type cooling towers. The taller, natural-draft towers may result in plumes that do not sink to the ground so that only clouds develop. With shorter (23 to 30 m) mechanical-draft towers, McVehil (1970) feels that, at Zion, Illinois, there might be about 90 hr/year of tower-produced fog at a point (highest frequency point located 2.4 to 4.0 km north of towers) as compared with the present 160 to 260 hr/year of natural fog. The natural-draft towers might result in only 14 hr/year of tower-produced fog, with the highest frequency point located some 11 km north of the towers. Maximum occurrence for tower-produced fog would be between 3 and 9 A.M. in the winter season. With winter temperatures often below freezing in the Zion area, icing would be likely to accompany the fog.

Various literature sources suggest the possibility of increased cloud formation downwind of a cooling tower. Both stratiform and cumuloform clouds have been suggested; some authors believe the heat and moisture might trigger cumulonimbus build-up and increase thunderstorm activity. While this is possible, there seems to be little evidence yet that such severe storms have developed from cooling tower operation. Aynsley (1970) indicates that humidity increases can be found many kilometers downwind of cooling towers. It is apparent that materials put into the air by large industrial complexes and by cooling towers may have significant downwind effects on the energy and moisture balance factors. Each instance is different and must be investigated with care.

While Huff and his associates (1971) conclude that sufficient information is not now at hand to evaluate in quantitative terms the atmospheric effects of releases of large amounts of heat and moisture by means of cooling towers, they are of the opinion that meteorological effects will be less as a result of lake discharge of heated effluent than through atmospheric release by means of cooling towers. This conclusion seems to be reasonable although it says nothing of the overall effects of the heated effluent on the ecology of the lake.

Cooling Ponds

Thackston and Parker (1971) attempt one answer to the question of the influence of heated effluent on temperatures in cooling ponds or small lakes by calculating conditions in standard-sized lakes in various parts of the country receiving a similar quantity of heated effluent. They employ an energy budget approach using the effluent from a standard 1,000-MW capacity power plant with a cooling water flow of 38 m^3/sec and with a temperature rise across the condensers of 8°C. The temperature of the pond is determined by months, for ponds of different surface areas, under average and extreme climatic conditions. Each pond is considered to be 4.6 m in depth and completely mixed although the authors point out that the deeper ponds will differ very little in response.[1] They provide information to determine results if there is "plug flow" (no mixing at all—each parcel of influent will follow a similar path across the pond, utilizing the whole cross-sectional area without mixing with the parcels ahead or behind).

Present experience suggests that between 1 and 2 acres of pond surface is needed per megawatt of generating capacity or that 20 to 40 cal of heat loss per hour is possible per square centimeter of surface area. Clearly, these figures are quite general; regional or local weather and climate conditions significantly influence the rate of heat exchange and consequently the size of the cooling pond surface required. Thus Thackston and Parker opt for the application of the energy budget, identifying all inputs and losses of heat, as the only rational way to determine cooling rates and efficiencies. These

---

[1] The authors indicate a 67 percent increase in pond depth results in a change of less than 0.1°C in pond equilibrium temperature, a difference within the rounding errors of the computer program.

authors calculate the equilibrium temperature of the cooling pond when exposed to constant meteorologic conditions and with the given input of heated effluent. They supply figures of so-called "net plant temperature" (actually, 8°C minus the cooling achieved in the pond) for different-sized pond surfaces, always assuming a 4.6-m deep pond, well mixed. They assume that the intake temperature to the plant is the equilibrium temperature of the pond.

The authors find a marked seasonal difference in pond cooling—of the order of 1°C difference between summer and winter—due to the increase in the heat exchange coefficient during the summer. All sizes of ponds showed approximately 1°C more cooling in summer than winter, although, of course, pond temperatures were higher in summer than winter. Pond size also significantly influences cooling with an 800-acre pond providing 3.6°C of cooling in June and a 1,500-acre pond giving 4.7°C cooling.

The results are presented as a series of maps (based on calculations at 88 stations) of (1) the geographical distribution of equilibrium temperatures under average and extreme weather conditions for 4 months of the year, (2) time in days that monthly average equilibrium temperatures are above specified values, (3) heat-exchange coefficients for 2 months, (4) net temperature rise for two different-sized ponds for January and July, and finally, (5) the surface area of ponds necessary for 50 percent cooling (4.2°C temperature rise) for January and July under average and extreme conditions.

Most satisfactory cooling conditions are found in the southern Great Plains centered over the Texas area and generally east of the Rockies, with a second area favorable for the use of cooling ponds along the Gulf and Atlantic coasts northward to Cape Cod. Poor pond-cooling conditions are found in the Appalachian and Rocky Mountain areas and across the northern part of the country. The maps show variations on a particular day over the United States of as much as 30°C in equilibrium temperature, and 100 percent in heat-exchange coefficients resulting in at least a 50 percent difference in heat loss from a given-sized pond, and, thus, a 200 percent difference in the water area needed for similar cooling effects.

MOVEMENT OF VEHICLES OVER OFF-ROAD SURFACES

Characteristics of Surface Materials

The tractionability of the surface over which vehicles move is determined by four characteristics of the surface materials that, in turn, are influenced by weather and climate. The four characteristics are bearing capacity, shearing strength, surface friction coefficient, and stickiness. The term "traffic-ability" is sometimes used in relation to vehicle movement over off-road surfaces but actually it only applies to the first of these four characteristics (Thornthwaite, 1950).

Soils vary widely in bearing power, shearing strength, surface friction, and stickiness, and thus in tractionability. In all soils tractionability is influenced in some way by soil moisture content although not all soils respond to the same extent, or even in the same direction, to changing moisture content. Since soil moisture is largely determined by precipitation which adds water and evapotranspiration which removes it, the basic factors of tractionability common to all soils are climatic.

With respect to their response to moisture and, therefore, to tractionability, Thornthwaite (1950) divided soils into two main groups, plastic or cohesive and nonplastic or cohesionless. Plastic soils can be molded when wet and on drying firmly retain any shape given to them by molding. Nonplastic soils will not retain any particular form on drying. If water is added when plastic soils are being molded, a point is reached at which the soil begins to stick to the fingers. Up to a certain point, the stickiness increases with an increase in moisture; it ceases at saturation when free water appears at or near the surface. Stickiness is a property of plastic soils only. Bearing capacity and shearing strength of plastic soils both decrease with increase in soil moisture. Conversely, in nonplastic soils, both bearing capacity and shearing strength increase with moisture content. In a dry state the shearing strength of nonplastic soils is very low while the bearing capacity is fairly high, especially if the material is confined.

The percentage of clay in a soil determines whether it is plastic or nonplastic. With a sandy soil having only minute amounts of clay, the reaction is characteristically nonplastic; when about 15 percent clay is present, a soil begins to react as a plastic material. A clay content of 15 percent is used as the dividing line between plastic and nonplastic soils (Thornthwaite, 1950).

Classes of Ground Conditions

On nonfrozen soils, five mutually exclusive classes of ground conditions, based on the soil moisture content in the significant soil depth for tractionability (considered to be 2 ft), have been identified (Thornthwaite et al., 1958). An interpretation of their significance in terms of tractionability is given below. The moisture percentages used are only approximate.

1   *DD (very dry)*   The average soil moisture content in the top 2 ft is less than 33 percent of the moisture content at field capacity. The moisture content at the soil surface is essentially 0, while at the depth of 2 ft it can vary between 0 and 66 percent of field capacity. Within the range of soil moisture covered by *DD*, tractionability of plastic soils is excellent; all parts of the relief except swampy areas are sufficiently hard and firm to support vehicles of any kind. When soil moisture remains in this class for extended periods, heavy traffic may grind the surface to fine dust, which may impede fast-moving equipment. The tractionability of sandy soils and particularly of deep shifting sands is very poor except in seeps and depressions where some moisture may have accumulated.

2   *D (dry)*   The average moisture content of the 2 ft layer of significance for tractionability is between 33 and 75 percent of field capacity. At the soil

surface the moisture content ranges from 0 to 50 percent of field capacity, while at the 2 ft depth it ranges from 66 percent of field capacity up to field capacity itself. Within this class, tractionability of plastic soils remains high and that of sands remains poor. Plastic soils tend to soften somewhat but not enough to impede movement of even the heaviest equipment. No appreciable improvement is noted in the tractionability of sandy soils except immediately after heavy showers. The surface of plastic soils becomes definitely slippery during and just after such showers.

3   *M (moist)* The average moisture content of the top 2 ft of soil is between 75 and 120 percent of field capacity. At the soil surface the moisture content will vary from 50 percent of field capacity to field capacity, while at a depth of 2 ft it will vary from field capacity to 133 percent of field capacity. Within the range of moisture represented by *M* conditions, the tractionability of plastic soils deteriorates rapidly. The soil becomes softer and offers considerable difficulty for heavy, wheeled vehicles. Track-laying vehicles have no trouble except after heavy showers when the surface is muddy and slippery. Tractionability of sands and other nonplastic soils is good and continues to improve with additional moisture.

4   *MM (very moist)* The average soil moisture content in the top 2 ft ranges from 120 to 160 percent of field capacity. At the soil surface the moisture content varies from field capacity to the liquid limit (about 150 percent of field capacity), while at a depth of 2 ft the moisture ranges from 133 to 167 percent of field capacity. *MM* conditions represent very poor tractionability on plastic soils. Adhesion of plastic soils to foreign objects becomes greater than the cohesion between soil particles. Traffic of practically every kind is extremely difficult on bare soil but men and light animal-drawn vehicles can travel over grassed areas without too much difficulty until the surface is cut up. The tractionability of sands and nonplastic sandy soils is excellent in this class of soil moisture.

5   *W (wet)* The average moisture content ranges from 160 percent of field capacity to the maximum water-holding capacity (about 200 percent of field capacity). At the surface, the moisture content is between the liquid limit and the maximum water-holding capacity, while at a depth of 2 ft it ranges between 167 percent of field capacity and the maximum water-holding capacity. *W* conditions represent a range of moisture content where free water begins to appear at or near the surface. While the stickiness of plastic soils is greatly diminished, the bearing strength and friction coefficient are so greatly reduced that the movement of power-driven transport is nearly impossible. The tractionability of most sandy soils remains fair although this is the moisture range in which sands with certain grain size exhibit the characteristics of quicksand.

When frozen soil is considered, a number of additional classes of ground conditions such as deep snow, permafrost, ice caps, and glaciers need to be included (Thornthwaite et al., 1958). As fairly dry soils freeze, ice is formed around the soil particles, the ice taking up more space than the water did previously. However, since there is only a small amount of water, expansion by freezing still does not fill all the pore spaces. The very thin films of water surrounding the basic soil grains do not seem to freeze even at quite low soil temperatures. As a result, tractionability of fairly dry soils is relatively unaffected by the freezing of the soil. Thus tractionability of frozen, dry, plastic soils remains good, while fairly poor tractionability is found on frozen nonplastic dry soils.

At moisture contents approaching saturation, thick water films surround the soil particles and fill nearly all the pore spaces. This water expands as it freezes so that the whole mass of water and soil will ultimately freeze solidly together. Tractionability of both plastic and nonplastic soils should improve markedly. As the soil moisture content drops to some value between field capacity and saturation, freezing will fill some, but not all, of the pore spaces with ice. Thus, certain solid ice and soil masses may exist in the soil, separated by soil particles with unfilled pore spaces. The effect of freezing of this soil on tractionability is quite difficult to determine since it depends on soil type, moisture content, and the exact nature of the resulting frozen soil and ice mass.

Unfrozen ground under a deep snow cover will usually remain unfrozen while frozen ground under such a cover usually will not remain frozen since snow is a good insulator. Cooling by radiation from the snow surface has little effect on the ground, which will slowly be warmed to near freezing by heat from soil layers below the surface. The tractionability of a deep snow surface is quite poor and entirely independent of the type of soil beneath. Ice surfaces have their own special tractionability problems beyond the scope of applied climatology.

To determine the existing tractionability class for any unfrozen soil, all that is necessary is knowledge of the moisture content in the upper 2 ft of the profile. This information can be obtained from direct field measurements or from calculations of the climatic water budget. From this information, it is a simple matter to determine which of the nonfrozen classes of tractionability exists at the particular time.

In appraising the significance of the several classes of ground conditions, the important influences of exposure and topography on soil moisture must always be kept in mind. As finally mapped, the moisture classes refer to average conditions in an area. Steep slopes and southern exposures are represented by the lower soil moisture values included in the range for the class, while the upper values within the range would apply to conditions in depressions, to areas with impaired drainage, and to northern exposures.

Plotting the data on soil moisture content as determined at a number of stations and for different time periods results in a series of generalized maps of the average distribution of tractionability with time. An example for the southeastern United States is shown in Fig. 11-1.

In March over the southeastern United States, *MM* conditions are found in a belt through eastern and central Tennessee and northern Georgia,

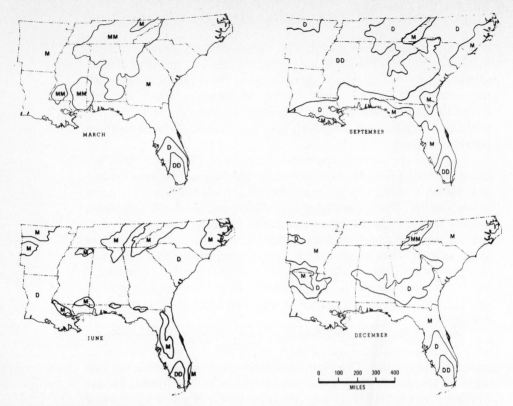

Average ground moisture conditions in the southeastern United States. (*From*   FIG. 11-1
Thornthwaite and Mather, 1955b, *fig. 5.7.*)

Alabama, and Mississippi. Isolated patches of *MM* conditions occur in
southern Mississippi and Alabama. The rest of the southeastern part of the
country experiences *M* conditions except for southern Florida where the low
precipitation and high evapotranspiration result in *D* and *DD* conditions.

By June, increased evapotranspiration results in the occurrence of *D*
conditions over most of the southeast with the exception of isolated areas of
M conditions in the mountains of North Carolina and eastern Tennessee as
well as along the Gulf Coast. These conditions of dryness intensify through
the summer so that by September a wide band of *DD* tractionability
conditions is found through the mid-South. Only in the mountainous regions
and in isolated patches along the Gulf Coast is there sufficient precipitation
to result in *M* soil moisture conditions.

With the decrease in evapotranspiration in fall, the soil moisture
content increases, and by December most of the mid-South area is once more
M. A small band of *D* conditions is found through south-central Georgia and
Alabama and again in southern Florida, while very moist (*MM*) conditions
exist in the mountains of Tennessee and North Carolina.

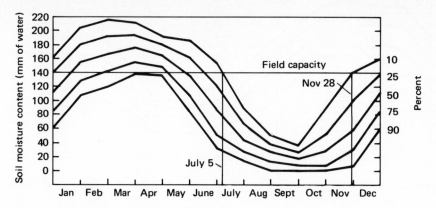

Expectancies of soil moisture in 2-ft depth of soil, Marked Tree, Ark. (*From   FIG. 11-2
Thornthwaite and Mather, 1955b, fig. 5.8.*)

    In addition to maps of the geographic variation of soil moisture values
with time, the climatic approach also makes it possible to determine the
expectancies of different values of soil moisture through the year. Such
calculations for a large number of places permit the preparation of maps or
tables of the geographic distribution of the expectancies of different levels of
soil moisture by month, week, or even day of the year.

    A 20-year record of monthly values of soil moisture has been obtained
from the climatic data at two stations in the southeastern United States.
Using a graphical method for determining probabilities developed by Hazen in
his work on flood flows, the soil moisture contents that can be expected to
be equaled or exceeded different percentages of the time have been
determined. The seasonal course of different levels of probabilities of soil
moisture based on determinations at the end of each month are given in Figs.
11-2 and 11-3. The distributions are only approximate since the determination of probabilities for each day of the year would undoubtedly result in
certain changes in the patterns shown.

    Plastic soils should be tractionable if the moisture content in the active
layer for tractionability is 140 mm or less. To ensure success of an operation,
it is desired that the chances of having a soil moisture content above this
value are 1 in 10 or less. On Figs. 11-2 and 11-3, the 10 percent probability
line intersects the 140-mm soil moisture line on July 5 and November 28 for
Marked Tree, Arkansas, and on June 17 and December 26 for Louisville,
Kentucky. During those periods, the soil moisture content should be below
the critical value more than 90 percent of the time. Local heavy showers will
bring the value of soil moisture in the top layer temporarily above the critical
value but this effect is transitory. The probabilities shown have been
determined from a limited past record, and there is no assurance that future
conditions will be exactly similar.

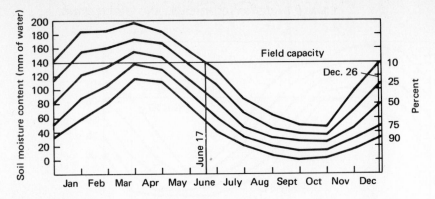

Expectancies of soil moisture in 2-ft depth of soil, Louisville, Ky. (*From* FIG. 11-3
Thornthwaite and Mather, 1955b, fig. 5.9.)

CONCLUSIONS

This chapter has been too brief to do more than touch on a few of the areas
where climatic knowledge can be profitably applied to industrial operations.
Many present examples emphasize the value of weather forecasting more than
climatic analysis. As we learn more of the role climate plays in industrial
production, manufacture, storage, and distribution, additional examples of
the value of applied climatology will be forthcoming. Given time, it is to be
hoped that an increasing number of industries will see the advantages of
employing applied meteorologists or climatologists as part of their regular
operating staffs.

BIBLIOGRAPHY

Abbe, C. (1905): A First Report on the Relations between Climates and Crops, *U.S. Dept. Agr., Weather Bur. Bull.* 36.

Adolph, E. F. (1947): "Physiology of Man in the Desert," John Wiley & Sons, Inc., New York.

—— and G. W. Molnar (1946): Exchanges of Heat and Tolerances to Cold in Men Exposed to Outdoor Weather, *Am. J. Physiol.*, vol. 146, pp. 507–537.

Albrecht, J. C. (1971): A Climatic Model of Agricultural Productivity in the Missouri River Basin, *Publ. Climatol.*, Laboratory of Climatology, vol. 24, no. 2, 107 pp.

Allmaras, R. R., W. C. Burrows, and W. E. Larson (1964): Early Growth of Corn as Affected by Soil Temperature, *Proc. Soil Sci. Soc. Am.*, vol. 28, no. 2, pp. 271–275.

Ångström, A. (1922): Note on the Relation between Time of Sunshine and Cloudiness in Stockholm, 1908–1920, *Arch. Mat. Astron. Physik.*, vol. 17, no. 15.

Arndt, K. A., D. H. K. Lee, and M. M. Key (1969): Skin—the Interface between Man and Tropical Environments, *Intern. Rev. Trop. Med.*, vol. 3, pp. 187–217.

Aronin, J. E. (1953): "Climate and Architecture," Reinhold Publishing Corporation, New York, 304 pp.

ASHRAE (1972): "Handbook of Fundamentals," Physiological Principles, Comfort and Health, Chap. 7, pp. 119-150, New York.

Aynsley, E. (1970): Cooling-Tower Effects: Studies Abound, *Elec. World*, May, pp. 42-43.

Bagnouls, F., and H. Gaussen (1957): Les climats biologique et leur classification, *Ann. Geograph.*, vol. 66, no. 355, pp. 193-220.

Baier, W. (1967): Recent Advancements in the Use of Standard Climatic Data for Estimating Soil Moisture, *Ann. Arid Zone*, vol. 6, no. 1, pp. 1-21.

—— (1969): Concepts of Soil Moisture Availability and Their Effect on Soil Moisture Estimates from a Meteorological Budget, *Agr. Meteorol.*, vol. 6, pp. 165-178.

——, D. Z. Chaput, D. A. Russelo, and W. R. Sharp (1972): Soil Moisture Estimator Program System, *Tech. Bull. 78 Agrometeorol. Sec., Plant Res. Inst., Can. Dept. Agr., Ottawa*, 55 pp.

—— and G. W. Robertson (1965): Estimation of Latent Evaporation from Simple Weather Observations, *Can. J. Plant Sci.*, vol. 45, pp. 276-284.

—— and —— (1966): A New Versatile Soil Moisture Budget, *Can. J. Plant Sci.*, vol. 46, pp. 299-315.

—— and —— (1967a): Estimating Supplemental Irrigation Water Requirements for Climatological Data, *Can. Agr. Eng.*, vol. 9, pp. 46-50.

—— and —— (1967b): Estimating Yield Components of Wheat from Calculated Soil Moisture, *Can. J. Plant Sci.*, vol. 47, pp. 617-630.

—— and —— (1968): The Performance of Soil Moisture Estimates as Compared with the Direct Use of Climatological Data for Estimating Crop Yields, *Agr. Meteorol.*, vol. 5, pp. 17-31.

—— and D. A. Russelo (1968): A Computer Program for Estimating Risks of Irrigation Requirements from Climatic Data, *Tech. Bull. 59, Agrometeorol. Sec., Plant Res. Inst., Can. Dept. Agr., Ottawa*, 48 pp.

Barger, G. L. (ed.) (1960): "Climatology at Work—Measurements, Methods and Machines," U.S. Department of Commerce, Weather Bureau, 109 pp.

Beaumont, A. (1940): Potato Blight and the Weather (Abstract), *Trans. Brit. Mycol. Soc.*, vol. 24, p. 266.

—— (1947): Dependence on the Weather of the Dates of Outbreak of Potato Blight Epidemics, *Trans. Brit. Mycol. Soc.*, vol. 31, pp. 45-53.

Bedford, T. (1948): "Basic Principles of Ventilation and Heating," H. K. Lewis and Co., Ltd., London.

Belding, H. S. (1949): Protection against Dry Cold, in L. H. Newburgh (ed.), "Physiology of Heat Regulation and the Science of Clothing," pp. 351-367, W. B. Saunders Company, Philadelphia, 457 pp.

—— (1967): Resistance to Heat in Man and Other Homeothermic Animals, in A. H. Rose (ed.), "Thermobiology," pp. 479-510, Academic Press, Inc., New York.

Benjamin, N. B. H., and C. S. Davis (1971): "Impact of Weather on Construction Planning," paper presented at Annual Meeting and

National Meeting on Environmental Engineering, American Society Civil Engineers, St. Louis, Oct. 21.

Benner, W. (1958): Engineering Meteorology: Weather Influences on Electrical Utility Load Dispatching, *Bull. Am. Meteorol. Soc.*, vol. 39, no. 10, pp. 514–520.

Benton, G. S., and R. T. Blackburn (1950): A Comparison of Precipitation from Maritime and Continental Air, *Bull. Am. Meteorol. Soc.*, vol. 31, no. 7, pp. 254–256.

———, R. T. Blackburn, and V. O. Snead (1950): The Role of the Atmosphere in the Hydrologic Cycle, *Trans. Am. Geophys., Union*, vol. 31, no. 1, pp. 61–73.

Black, J. N., C. W. Bonython, and J. A. Prescott (1954): Solar Radiation and the Duration of Sunshine, *Quart. J. Roy. Meteorol. Soc.*, vol. 80, pp. 231–235.

Blodget, L. (1857): "Climatology of the United States," J. B. Lippincott Company, Philadelphia, 536 pp.

Blum, H. F. (1945): The Solar Heat Load; Its Relationship to Total Heat Load and Its Relative Importance in the Design of Clothing, *J. Clin. Invest.*, vol. 24, no. 5, p. 715.

——— (1959): "Carcinogenesis by Ultraviolet Light," Princeton University Press, Princeton, N.J.

Booth, A. W. (1957): "Petroleum Refining and Selected Chemical Industries," in Industrial Operations under Extremes of Weather, *Meteorol. Monographs*, vol. 2, no. 9, pp. 104–111.

Bourke, P. M. A. (1953): Potato Blight and the Weather: A Fresh Approach, *Tech. Note* 12, *Irish Meteorol. Serv.*

——— (1955): The Forecasting from Weather Data of Potato Blight and Other Plant Diseases and Pests, *Tech. Note* 10, *World Meteorol. Org.*, no. 42, TP 16, Geneva, 48 pp.

——— (1957): Department Industry and Commerce Meteorological Service, (Eire), *Tech. Note* 12, pp. 1–11.

Bowen, I. S. (1926): The Ratio of Heat Losses by Conduction and by Evaporation from any Water Surface, *Phys. Rev.*, vol. 27, pp. 779–789.

Brezowsky, H. (1964): Morbidity and Weather, in S. Licht (ed.), "Medical Climatology," pp. 358–399, vol. 8, Physical Medicine Library, Waverly Press, Inc., Baltimore, 753 pp.

Bridges, E. L. (1949): "Uttermost Part of the Earth," E. P. Dutton & Co., Inc., New York, 558 pp.

Brody, S. (1956): Climatic Physiology of Cattle, *J. Dairy Sci.*, vol. 39, pp. 715–725.

Brooks, C. E. P. (1951): "Climate in Everyday Life," Philosophical Library, Inc., New York, 314 pp.

——— and T. Hunt (1930): The Zonal Distribution of Rainfall over the Earth, *Mem. Roy. Meteorol. Soc.*, vol. 3, no. 28, pp. 139–158.

Brooks, F. A. (1959): An Introduction to Physical Micrometeorology,"
Davis, Calif., Assoc. Students Store, 264 pp. Illus.

Budyko, M.I. (1955): "Atlas of the Heat Balance," Gidrometeoizdat., Leningrad.

—— (1956): "The Heat Balance of the Earth's Surface," Gidrometeoizdat.,
Leningrad (translated by N. A. Stepanova, Office of Climatology, U.S.
Weather Bureau, 1958), 255 pp.

—— (1963): "Atlas Teplovogo Balansa," Gidrometeoizdat., Moscow.

——et al. (1962): The Heat Balance of the Surface of the Earth, *Soviet
Geograph. Rev. Transl.*, vol. 3, no. 5, pp. 3–16.

Buechley, R. W., J. van Bruggen, and L. E. Truppi (1972): Heat Island=Death
Island? *Environ. Res.*, vol. 5, no. 1, pp. 85–92.

Burton, A. C. (1943): "Insulation by Reflection and the Development of a
Reflecting Cloth," *Natl. Res. Council Rept.* C2464, *Ottawa.*

——and O. G. Edholm (1955): "Man in a Cold Environment," Monographs
of the Physiological Society, no. 2, Edward Arnold (Publishers), Ltd.,
London, 273 pp.

Businger, J. A. (1955): Nachtvort-bestrjding door middel van besproeiing
(Frost Protection by Sprinkling), *Mededel. Directeur Tuinbouw*, vol.
18, no. 19, pp. 21–34.

Carter, D. B. (1954): Climates of Africa and India According to Thornthwaite's 1948 Classification, *Publ. Climatol.*, Laboratory of Climatology, vol. 7, no. 4, pp. 453–474.

——and J. R. Mather (1966): Climatic Classification for Environmental
Biology, *Publ. Climatol.*, Laboratory of Climatology, vol. 19, no. 4, pp.
305–395.

Cassedy, James H. (1969): Meteorology and Medicine in Colonial America:
Beginnings of the Experimental Approach, *J. Hist. Med. Sci.*, vol. 24,
no. 2, pp. 193–204.

Chang, Jen-Hu (1968): "Climate and Agriculture, an Ecological Survey,"
Aldine Publishing Company, Chicago, 304 pp.

——, R. B. Campbell, and F. E. Robinson (1963): On the Relationship
between Water and Sugar Cane Yield in Hawaii, *Agron. J.*, vol. 55, no.
5, pp. 450–453.

Changnon, S. A. (1968): The La Porte Anomaly—Fact or Fiction, *Bull. Am.
Meteorol. Soc.*, vol. 49, pp 4–11.

Clarke, J. F. (1972): Some Effects of the Urban Structure on Heat Mortality,
Environ. Res., vol. 5, no. 1, pp. 93–104.

Coffin, J. H. (1853): Winds of the Northern Hemisphere, *Smithsonian
Contrib. Knowledge*, no. 6, Washington, 197 pp.

——(1875) (with contributions by S. J. Coffin and A. Woeikof): The Winds
of the Globe or the Laws of Atmospheric Circulation over the Surface
of the Earth, *Smithsonian Contrib. Knowledge*, no. 268, Washington,
756 pp.

Corey, C. P. (1949): The Effects of Weather upon the Electric Power
Systems, *Bull. Am. Meteorol. Soc.*, vol. 30, pp. 239–241.

Cornwall, C. J., and P. A. B. Raffle (1961): Bronchitis-sickness Absence in
London Transport, *Brit. J. Ind. Med.*, vol. 18, pp. 24–32.

Craigie, J. H. (1945): Epidemiology of Stem Rust in Western Canada, *Sci. Agr.*, Ottawa, vol. 25, no. 6, pp. 285-401.

Crawford, T. V. (1965): Moisture Transfer in Free and Forced Convection, *Quart. J. Roy. Meteorol. Soc.*, vol. 91, pp. 18-27.

C. W. Thornthwaite Associates (1954): "The Cropmeter, the Modern Means of Calculating Plant Maturity," Elmer, N.J.

Dalton, J. (1802): Experimental Essays on the Constitution of Mixed Gases; on the Force of Steam or Vapour from Water and Other Liquids in Different Temperatures, both in a Torricellian Vacuum and in Air; on Evaporation; and on the Expansion of Gases by Heat, *Manchester Lit. Phil. Soc. Mem.* 5, pp. 535-602, illus.

Davies, J. A. (1965): Estimation of Insolation for West Africa, *Quart. J. Roy. Meteorol. Soc.*, vol. 91, pp. 359-363.

———(1967): A Note on the Relationship between Net Radiation and Solar Radiation, *Quart. J. Roy. Meteorol. Soc.*, vol. 93, pp. 109-115.

——— and P. H. Buttimor (1969): Reflection Coefficients, Heating Coefficients and Net Radiation at Simcoe, Southern Ontario, *Agr. Meteorol.*, vol. 6, pp. 373-386.

Davis, F. K., Jr. (1957): Study of Time-Height Variations of Micrometeorological Factors during Radiation Fog, *Publ. Climatol.*, Laboratory of Climatology, vol. 10, no. 1, 51 pp.

Deacon, E. L., and W. C. Swinbank (1958): Comparison between Momentum and Water Vapour Transfer, *Proc. UNESCO Symp. Arid Zone Climatol.*, pp. 38-47.

DeCandolle, A. (1855): "Géographie botanique raisonnée, ou exposition des faits principaux et des lois concernant la distribution géographique des plantes de l'époque actuelle," 2 vol., V. Masson, Paris.

Denevan, W. M. (1961): The Upland Pine Forests of Nicaragua, *Publ. Geograph., Univ. Calif.*, vol. 12, no. 4, pp. 251-320.

Desjardins, R. L., and D. Siminovitch (1968): Microclimatic Study of the Effectiveness of Foam as Protection against Frost, *Agr. Meteorol.*, vol. 5, pp. 291-296.

DeWit, C. T. (1958): Transpiration and Crop Yields, *Inst. Biol. Scheik. Onderzoek Landbouwgewassen, Wageningen, Mededel.*, no. 59.

de Zeeuw, J. W. (1958): Mogelijkheden van nachtvorstwering (Ground Frost Prevention), *Mededel. Lanbouw inst. Lab. Tuinbouw., Wageningen*, vol. 21, no. 7, pp. 430-433.

Dohan, F. C., and E. W. Taylor (1960): Air Pollution and Respiratory Disease: A Preliminary Report, *Am. J. Med. Sci.*, vol. 240, pp. 337-339.

Dove, H. W. (1846): Uber den Zusammenhang der Temperaturveränderungen der Atmosphäre und der oberen Erdschichten mit der Entwickelung der Pflanzen, *Ann. Phys. Chem.*, Ser. 2, 68, pp. 224-235.

Drummond, J. J. (1949): "Correlation of Electric Loads on Consolidated Edison of N.Y., Inc., System with Weather Variables," October Minutes Electrical Equipment Committee, Edison Electric Institute.

Dryar, H. A. (1944): The Effect of Weather on the System Load, *Am. Inst. Elec. Eng.*, vol. 63, pp. 1006–1013.

—— (1949): Load Dispatching and Philadelphia Weather, *Bull. Am. Meteorol. Soc.*, vol. 30, pp. 159–167.

Duckham, A. N. (1967): Weather and Farm Management Decisions, in J. A. Taylor (ed.), "Weather and Agriculture," pp. 69–80, Pergamon Press, Symposium Publishing Division, London, 225 pp.

Dutt, R. H., and E. C. Simpson (1957): Environmental Temperature and Fertility of Southdown Rams Early in the Breeding Season, *J. Animal Sci.*, vol. 16, no. 1, pp. 136–145.

Dyer, A. J., and F. J. Maher (1965): Automatic Eddy-flux Measurement with the Evapotron, *J. Appl. Meteorol.*, vol. 1, pp. 471–473.

Edholm, O. G. (1966): Problems of Acclimatization in Man, *Weather*, vol. 21, no. 10, pp. 340–350.

Ellis, F. P. (1972): Mortality from Heat Illness and Heat-aggravated Illness in the United States, *Envir. Res.*, vol. 5, no. 1, pp. 1–58.

——, F. E. Smith, and J. D. Walters (1972): Measurement of Environmental Warmth in SI Units, *Brit. J. Ind. Med.*, vol. 29, pp. 361–377.

ESSA (1960): Climatography of the United States, Summary of Hourly Observations, 1951–60, "Decennial Census of the United States Climate," U.S. Department of Commerce, Weather Bureau.

—— (1966): "Weather and the Construction Industry," U.S. Department of Commerce, 11 pp.

—— (1967): "The National Weather Records Center, Asheville, N.C.," U.S. Department of Commerce, Environmental Data Service, 33 pp.

—— (1968): "Climatic Atlas of the United States," U.S. Department of Commerce, Environmental Data Service, 80 pp.

—— (1969): "Climates of the World," U.S. Department of Commerce, Environmental Data Service, 28 pp.

Eyre, S. R. (1963): "Vegetation and Soils, a World Picture," Aldine Publishing Company, Chicago.

Fanger, P. O. (1967): Calculation of Thermal Comfort: Introduction of a Basic Comfort Equation, *ASHRAE Trans.*, vol. 73, pt. II, p. III.4.1.

Fitch, J. M. (1950): Buildings Designed for Climatic Control, in "Weather and the Building Industry," *Proc., Res. Advisory Board, BRAB Conf. Rept. 1, Nat. Acad. Sci.-Nat. Res. Coun.*, Washington, pp. 91–99.

Fonseca, G. F., J. R. Breckenridge, and A. H. Woodcock (1959): Wind Penetration of Fabric Systems, *Tech. Rept. EP-104 QMRE Comm.*, Natick.

Forbes, W. H. (1949): Definitions, Miscellaneous Figures and Conversion Factors, in L. H. Newburgh (ed.), "Physiology of Heat Regulation and the Science of Clothing," pp. 443–450, W. B. Saunders Company, Philadelphia, 457 pp.

Forry, S. (1842): "The Climate of the United States and its Endemic Influences," J. and H. G. Langley, New York, 378 pp.

Foster, E. E. (1948): "Rainfall and Runoff," The MacMillan Company, New York, 487 pp.

Francis, W. J. (1967): Prediction of California Quail Populations from Weather Data, *The Condor*, vol. 69, no. 4, pp. 405-410.

Freeman, R. G., and J. M. Knox (1965): The Factor of Temperature in Ultraviolet Injury, *Arch. Environ. Health*, vol. 11, pp. 477-483.

Fritz, S., and T. H. MacDonald (1949): Average Solar Radiation in the United States, *Heating and Ventilating*, vol. 46, pp. 61-64.

Gagge, A. P., A. C. Burton, and H. C. Bazett (1941): A Practical System of Units for the Description of the Heat Exchange of Man with His Environment, *Science*, vol. 94, pp. 428-430.

———, J. A. J. Stolwijk, and Y. Nishi (1971): An Effective Temperature Scale Based on a Simple Model of Human Physiological Regulatory Response, *ASHRAE Trans.*, vol. 77, pt. I.

Gangopadhyaya, M., and R. P. Sarker (1965): Rainfall Distribution and the Yield of Wheat Crop, *Agr. Meteorol.*, vol. 2, pp. 331-350.

Gates, D. M. (1962): "Energy Exchange in the Biosphere," Harper & Row Biological Monographs, New York, 151 pp.

Gaussen, H. (1955): Les climats analogues a l'echelle du monde, *Compt. Rend. Acad. Agr. France*, vol. 41.

Geiger, R. (1965a): "The Atmosphere of the Earth" (12 wall maps and text), Justus Perthes, Darmstadt, Germany.

——— (1965b): "The Climate Near the Ground," Harvard University Press, Cambridge, Mass. (translated from fourth German edition, published 1961, by Friedrich Vieweg & Sohn, Brunswick, Germany), 611 pp.

Givoni, B. (1962): Basic Study of Ventilation Problems in Housing in Hot Countries, *Res. Rep. Ford Found.*, Building Research Station, Technion, Haifa, Israel.

——— (1963): Estimation of the Effect of Climate on Man: Development of a New Thermal Index, *Res. Rept to UNESCO*, Building Research Station, Technion, Haifa, Israel.

——— (1968): Ventilation Problems in Hot Countries, *Res. Rep. Ford Found.*, Building Research Station, Technion, Haifa, Israel.

——— (1969): "Man, Climate and Architecture," Elsevier Architectural Sciences Series, Elsevier Publishing Company, London, 364 pp.

——— and H. S. Belding (1962): The Cooling Efficiency of Sweat Evaporation, *Proc. 1st Intern. Congr. Bio-Meteorol.*, Pergamon Press, New York, pp. 304-314.

——— and E. Berner-Nir (1967): Effect of Solar Radiation on Physiological Heat Strain in Relation to Work and Clothing, *Res. Rept U.S. Public Health Serv.*, Building Research Station, Technion, Haifa, Israel.

Glendinning, D., J. A. MacDonald, and J. Grainger (1963): Factors Affecting the Germination of Sporangia in *Phytophthora infestans*, *Trans. Brit. Mycol. Soc.*, vol. 46, pp. 595-603.

Gover, M. (1938): Mortality during Periods of Excessive Temperature, *Public Health Rept. (U.S.)* 53, p. 1122.

Grainger, J. (1955): The 'Auchincruive' Potato Blight Forecast Recorder, 380
Weather, vol. 10, pp. 213-222.

—— (1956): Host Nutrition and Attack by Fungal Parasites, *Phytopathology,* vol. 46, pp. 445-456.

—— (1962): The Host Plant as a Habitat for Fungal and Bacterial Parasites, *Phytopathology,* vol. 52, pp. 140-150.

—— (1967): Meteorology and Plant Physiology in Potato Blight Forecasting, in J. A. Taylor (ed.), "Weather and Agriculture," pp. 105-113, Pergamon Press, Symposium Publishing Division, London, 225 pp.

—— (1968): C_p/R_s and the Disease Potential of Plants, *Hort. Res.,* vol. 8, pp. 1-40.

—— and A. A. Rutherford (1963): Rapid Determination of Host Receptivity in Potato Blight Forecasting, *Europ. Potato J.,* vol. 6, pp. 258-267.

Grisebach, A. (1866): Die Vegetations-Gebiete der Erde, ubersichtlich zusammengestellt, *Petermanns Mitt.,* vol. 12, pp. 45-53.

Hagstrom, R. M., H. A. Sprague, and E. Landau (1967): The Nashville Air Pollution Study: VII. Mortality from Cancer in Relation to Air Pollution, *Arch. Environ. Health,* vol. 15, pp. 237-248.

Halstead, M. H. (1954): "The Fluxes of Momentum, Heat, and Water Vapor in Micrometeorology," pp. 326-358, in C. W. Thornthwaite and M. H. Halstead (eds.), Micrometeorology of the Surface Layer of the Atmosphere, Final Report, *Publ. Climatol.,* Laboratory of Climatology, vol. 7, no. 2, pp. 233-361.

Hare, F. K. (1954): The Boreal Conifer Zone, *Geograph. Studies,* vol. 1, no. 1, pp. 4-18.

—— (1972): "The Observed Annual Water Balance over North America South of 60°N and Inferred Convective Heat Exchange," in J. R. Mather (ed.), Thornthwaite Memorial Volume II, Papers on Selected Topics in Climatology, *Publ. Climatol.,* Laboratory of Climatology, vol. 25, no. 3.

—— and J. E. Hay (1971): Anomalies in the Large-scale Annual Water Balance over Northern North America, *Can. Geograph.,* vol. 15, pp. 79-94.

Harrold, L. L., and F. R. Dreibelbis (1951): Agricultural Hydrology as Evaluated by Monolith Lysimeter, *U.S. Dept. Agr., Tech. Bull.* 1050, 149 pp.

Hawaiian Sugar Planters' Association (1963): *1963 Ann. Rept.,* Experiment Station.

—— (1969): *1969 Ann. Rept.,* Experiment Station, pp. 16-20.

Hay, W. W. (1957): "Effects of Weather on Railroad Operation, Maintenance and Construction," pp. 10-36, in Industrial Operations under Extremes of Weather, *Meteorol. Monographs,* vol. 2, no. 9.

Hendrick, R. L., and D. G. Friedman (1966): "Potential Impacts of Storm Modification on the Insurance Industry," in W. R. D. Sewell (ed.), Human Dimensions of Weather Modifications, *Univ. Chicago, Dept. Geograph., Res. Paper* 105, pp. 227-246.

Henschel, A., L. Burton, L. Margolies, and J. E. Smith (1968): An Analysis of Heat Deaths in St. Louis during July 1966, Environmental Control Administration, Cincinnati, Ohio, 22 pp.

Hidore, J. J. (1971): The Effects of Accidental Weather Modification on the Flow of the Kankakee River, *Bull. Am. Meteorol. Soc.*, vol. 52, pp. 99–103.

Higgins, J. J. (1952): Instructions for Making Phenological Observations of Garden Peas, *Publ. Climatol.*, Laboratory of Climatology, vol. 5, no. 2, 11 pp.

——and A. M. Decker (1971): *Tephrosia* Leaf Development: Index of Crop Yield and Bioclimate, *Univ. Maryland, Agr. Expt. Sta. Bull* 774, 61 pp.

Hilgard, E. W. (1892): A Report on the Relations of Soil to Climate, *U.S. Dept. Agr., Weather Bur. Bull.* 3.

Hodgson, T. A., Jr. (1970): Short-term Effects of Air Pollution on Mortality in New York City, *Environ. Sci. Tech.*, vol. 4, no. 7, pp. 589–597.

Hogg, W. H. (1967): The Use of Upper Air Data in Relation to Plant Disease, in J. A. Taylor (ed.), "Weather and Agriculture," pp. 115–127, Pergamon Press, Symposium Publishing Division, London, 225 pp.

Holdridge, L. R. (1959): Simple Method for Determining Potential Evapotranspiration from Temperature Data, *Science*, vol. 130, Sept. 4, p. 572.

Holmes, R. M., and G. W. Robertson (1959): A Modulated Soil Moisture Budget, *Monthly Weather Rev.*, vol. 87, pp. 101–106.

Holzman, B. (1937): Sources of Moisture for Precipitation in the United States, *U.S. Dept. Agr., Tech. Bull.* 589, 41 pp.

——and H. C. S. Thom (1970): The La Porte Precipitation Anomaly, *Bull. Am. Meteorol. Soc.*, vol. 51, pp. 335–337.

Houghten, F. C., and C. P. Yaglou (1923): Determining Lines of Equal Comfort, *ASHVE Trans.*, vol. 29, p. 163.

Housing Research (1954): Application of Climatic Data to House Design, Housing and Home Finance Agency, U.S. Government Printing Office, 152 pp.

Hudson, H. E., Jr., G. E. Stout, and F. A. Huff (1952): Studies of Thunderstorm Rainfall with Dense Raingage Networks and Radar, *Illinois State Water Surv., Rept. Invest.* 13, 30 pp.

Hudson, J. (1962): Irrigation Water Use in the Utah Valley, Utah, *Res. Paper* 79. Department of Geography, University of Chicago, 249 pp.

Huff, F. A., R. C. Beebe, D. M. A. Jones, G. M. Morgan, Jr., and R. G. Semonin (1971): Preliminary Report: Effect of Cooling Tower Effluents on Atmospheric Conditions in Northeastern Illinois, *Illinois State Water Surv., Circ.* 100, 37 pp.

Huntington, E. (1945): "Mainsprings of Civilization," John Wiley & Sons, Inc., New York, 660 pp.

Hurst, G. W. (1967): Honey Production and Summer Temperatures, *Meteorol. Mag.*, London, vol. 96, pp. 116–120.

——(1970): Temperatures in High Summer, and Honey Production, *Meteorol. Mag.*, London, vol. 99, pp. 75–82.

——and L. P. Smith (1967): Grass Growing Days, in J. A. Taylor (ed.), "Weather and Agriculture," pp. 147–155, in Pergamon Press, Symposium Publishing Division, London, 225 pp.

Huschke, R. E. (1959): "Glossary of Meteorology," American Meteorological Society, Boston, 638 pp.

Jacobs, W. C. (1947): Wartime Developments in Applied Climatology, *Meteorol. Monographs*, vol. 1, no. 1, 52 pp.

——(1959): Meteorology Applied Is a Social Affair, *Bull. Am. Meteorol. Soc.*, vol. 40, no. 4, pp. 179–181.

Jensen, M. E., and H. R. Haise (1963): Estimating Evapotranspiration from Solar Radiation, *Proc. Am. Soc. Civil Engrs., J. Irrigation Drainage Div.*, vol. 89, IR-4, pp. 15–41.

Johnson, S. R., J. D. McQuigg, and T. P. Rothrock (1969): Temperature Modification and Costs of Electric Power Generation, *J. Appl. Meteorol.*, vol. 8, no. 6, pp. 919–926.

Jorgenson, D. L. (1949): An Objective Method of Forecasting Rain in Central California During the Raisin Drying Season, *Monthly Weather Rev.*, vol. 77, pp. 31–46.

Kibler, H. H., and S. Brody (1954): Environmental Physiology with Special Reference to Domestic Animals: XXVII. Influence of Wind on Heat Exchange and Body Temperature Regulation in Jersey, Holstein, Brown Swiss, and Brahman Cattle, *Missouri Agr. Expt. Sta. Bull.* 552, 39 pp.

Klarmann, E. G. (1949): Sunburn and Suntan, *Am. Perfumes Essent. Oil Rev.*, vol. 54, pp. 33–38, 126–135.

Kligman, A. M. (1964): The Biology of the Stratum Corneum, in W. Montagna and W. C. Lobitz, Jr. (eds.), "The Epidermis," pp. 387–433, Academic Press, Inc., New York.

Kolb, L. L., and R. R. Rapp (1962): The Utility of Weather Forecasts to the Raisin Industry, *J. Appl. Meteorol.*, vol. 1, no. 1, pp. 8–12.

Köppen, W. (1900): Versuch einer Klassifikation der Klimate, vorzugsweise nach ihren Beziehungen zur Pflanzenwelt, *Geograph. Z.*, vol. 6, pp. 593–611; 657–679.

Koschmieder, H. (1934): Methods and Results of Definite Rain Measurements, *Monthly Weather Rev.*, vol. 62, no. 1, pp. 5–7.

Kruhl, H. (1971): Ship-routing Activities in the Seewetteramt, Hamburg, *The Marine Observer*, vol. 41, no. 231, pp. 27–29.

Küchler, A. W. (1964): "Potential Natural Vegetation of Conterminous United States," American Geographical Society, New York.

Landsberg, H. E. (1950): "Microclimatic Research in Relation to Building Construction," in Weather and the Building Industry, *Proc. Bldg. Res. Advisory Board, BRAB Conf. Rept.* 1, *Nat. Acad. Sci.-Nat. Res. Coun.*, Washington, pp. 23–29.

——(1960): "Physical Climatology," 2d ed., Gray Printing Co., Du Bois, Pa., 446 pp.

——(1961): Solar Radiation at the Earth's Surface, *Solar Energy*, vol. 5, no. 3, July-September, pp. 95-98.

——(1969*a*): Biometeorological Aspects of Urban Climates, *Tech. Note* BN-620, Institute for Fluid Dynamics and Applied Mathematics, University of Maryland (NSF Grants GA1104 and 13353), 13 pp. (processed).

——(1969*b*): "Weather and Health, An Introduction to Biometeorology," Anchor Science Study Series, Doubleday & Company, Inc., Garden City, N.Y., 148 pp.

——and W. C. Jacobs (1951): Applied Climatology, in T. F. Malone (ed.), "Compendium of Meteorology," pp. 976-992, American Meteorological Society, Boston, 1334 pp.

Lave, L. B., and E. P. Seskin (1970): Air Pollution and Human Health, *Science*, vol. 169, no. 3947, pp. 723-733.

Lee, D. H. K. (1963): Human Factors in Desert Development, in "Aridity and Man," pp. 339-367, American Association for the Advancement of Science, Washington.

——(1964): Heat and Cold Effects and Their Control, *U.S. Public Health Serv. Monograph* 72, 35 pp.

——(1967): Principles of Homeothermic Adaptation, in W. P. Lowry (ed.), "Biometeorology," pp. 113-130, Proceedings 28th Annual Biological Colloquium, Oregon State University Press, Corvallis.

Leighly, J. (1937): A Note on Evaporation, *Ecology*, vol. 18, pp. 180-198.

Lettau, H. (1969): Evapotranspiration Climatonomy: 1. A New Approach to Numerical Prediction of Monthly Evapotranspiration, Run-off, and Soil Moisture Storage, *Monthly Weather Rev.*, vol. 97, pp. 691-699.

Linden, F. (1959): Weather in Business, *The Conf. Board Business Record*, vol. 16, pp. 90-94, 101.

Linsley, R. K., M. A. Kohler, and J. L. H. Paulhus (1949): "Applied Hydrology," McGraw-Hill Book Company, New York.

Linsser, C. (1867): Die periodischen Erscheinungen des Pflanzenlebens in ihrem Verhältniss zu den Wärmeerscheinungen, *Mem. Acad. Imp. Sci. St. Pétersbourg*, ser. 7, vol. 11, no. 7.

——(1869): Untersuchungen über die periodischen Lebenserscheinungen der Pflanzen, Zweite Abhandlung, Resultate aus einer eingehenden Bearbeitung des europäischen Materials für die Holzpflanzen in Bezug auf Wärme und Regenmenge, *Mem. Acad. Imp. Sci. St. Pétersbourg*, ser. 7, vol. 13, no. 8.

Liopo, T. N. (1966): A Method for Computing Optimal Heat-regulating Properties of Clothing on the Basis of the Probability of Variation of Meteorological Factors, *Dokl. Inst. Geogr. Sibir. Dal'nego Vostoka*, no. 13, pp. 48-55 (translated in *Soviet Geogr., Rev. Transl.*, American Geographical Society, vol. 9, no. 2, pp. 95-105, 1968).

List, R. J. (1949): Smithsonian Meteorological Tables, *Smithsonian Inst. Misc. Collections*, vol. 114 (6th rev. ed.), Washington, 527 pp.

Löf, G. O. G., J. A. Duffie, and C. O. Smith (1966): World Distribution of Solar Radiation, *Univ. Wisconsin, Eng. Expt. Sta., Rept.* 21.

Longley, R. W. (1970): "Elements of Meteorology," John Wiley & Sons, Inc., New York, 317 pp.

Lowry, W. P. (1957): An Inexpensive Field Instrument for Non-selective Net Radiation Measurements, *Ecology*, vol. 38, pp. 152-155.

——(1959): The Falling Rate Phase of Evaporative Soil Moisture Loss—A Critical Evaluation, *Bull. Am. Meteorol. Soc.*, vol. 40, pp. 605-608.

McArdles, B., W. Dunham, H. E. Holling, W. S. S. Ladell, J. W. Scott, M. L. Thomson, and J. S. Weiner (1947): The Prediction of the Physiological Effects of Warm and Hot Environments: The P_4SR Index, *Med. Res. Council R.N.P.*, Report 47/391, London.

McLeary, J. W. (1968): Seasonal Income Patterns in the South, *Monthly Rev.*, Sixth Federal Reserve District, November, pp. 150-153.

McMillan, I. I. (1964): Annual Population Changes in California Quail, *J. Wildlife Management*, vol. 28, pp. 702-711.

McVehil, G. E. (1970): "Preliminary Report, Environmental Effect at Zion Nuclear Generating Station," prepared for Commonwealth Edison Co. by Sierra Research Corporation, Boulder, Colo., 4 pp.

Mack, A. R., and W. S. Ferguson (1968): A Moisture Stress Index for Wheat by Means of a Modulated Soil Moisture Budget, *Can. J. Plant Sci.*, vol. 48, pp. 535-543.

Major, J. (1963): A Climatic Index to Vascular Plant Activity, *Ecology*, vol. 44, no. 3, pp. 485-498.

Martin, P. S., G. I. Quimby, and D. Collier (1947): "Indians before Columbus," The University of Chicago Press, Chicago, 582 pp.

Mather, J. R. (1949): "Potential Evapotranspiration," in C. W. Thornthwaite (ed.), Micrometeorology of the Surface Layer of the Atmosphere, *Publ. Climatol.*, Laboratory of Climatology, vol. 2, no. 3, pp. 27-32, 102 pp.

——(1950): Manual of Evapotranspiration, *Publ. Climatol.*, Laboratory of Climatology, vol. 3, no. 3, 29 pp.

——(1954a): "A Summary of Evapotranspiration at Seabrook, New Jersey, 1947-1953," in J. R. Mather (ed.), The Measurement of Potential Evapotranspiration, *Publ. Climatol.*, Laboratory of Climatology, vol. 7, no. 1, pp. 7-51, 225 pp.

——(1954b): The Determination of Soil Moisture from Climatic Data, *Bull. Am. Meteorol. Soc.*, vol. 35, no. 2, pp. 63-68.

——(1954c): The Effect of Climate on the New World Migration of Primitive Man, *Southwestern J. Anthro.*, vol. 10, no. 3, pp. 304-321.

——(1959a): Determination of Evapotranspiration by Empirical Methods, *Trans. Am. Soc. Agr. Engrs.*, vol. 2, no. 1, pp. 35-38, 43.

——(1959b): The Moisture Balance in Grassland Climatology, in H. B. Sprague (ed.), "Grasslands," pp. 251-261, American Association for the Advancement of Science, Washington.

——(ed.) (1962, 1963, 1964, 1965): Average Climatic Water Balance Data of the Continents, Part I, Africa; Part II, Asia (excluding USSR); Part III, USSR; Part IV, Australia, New Zealand, and Oceania; Part V,

Europe; Part VI, North America (excluding US); Part VII, United States; Part VIII, South America, *Publ. Climatol.*, Laboratory of Climatology, vol. 15, no. 2; vol. 16, nos. 1, 2, 3; vol. 17, nos. 1, 2, 3; and vol. 18, no. 2.

——(ed.) (1967): Factors of the Heat and Water Balance of Different Surfaces, Pt. I, Analysis, *Publ. Climatol.*, Laboratory of Climatology, vol. 20, no. 2, pp. 173–273.

——(1968*a*): Irrigation Agriculture in Humid Areas, in A. Court (ed.), "Eclectic Climatology," pp. 107–122, Yearbook, Association of Pacific Geographers, vol. 30.

——(1968*b*): Meteorology and Air Pollution in the Delaware Valley, *Publ. Climatol.*, Laboratory of Climatology, vol. 21, no. 1, 136 pp.

——(1969*a*): The Average Annual Water Balance of the World, *Proc. Symp. Water Balance in N. Am.*, American Water Resources Association Proceedings Series 7, Banff, Canada, June 23–26, pp. 29–40.

——(1969*b*): Factors of the Climatic Water Balance over the Delmarva Peninsula, *Publ. Climatol.*, Laboratory of Climatology, vol. 22, no. 3, 129 pp.

—— and G. A. Yoshioka (1968): The Role of Climate in the Distribution of Vegetation, *Ann., Assoc. Am. Geograph.*, vol. 58, no. 1, pp. 29–41.

——, F. J. Swaye, Jr., and B. J. Hartmann (1972): The Influence of the Climatic Water Balance on Conditions in the Estuarine Environment, *Publ. Climatol.*, Laboratory of Climatology, vol. 25, no. 1.

Maunder, W. J. (1970): "The Value of Weather," Methuen & Co., Ltd., London, 388 pp.

——(1972*a*): National Econoclimatic Models: Problems and Applications, *Tech. Note* 208, *New Zealand Meteorol. Serv.*, Wellington, 17 pp.

——(1972*b*): A Review of Research into the Economic Impact of Weather on the Building and Construction Industry, *Symp. Meteorol. Bldg. Ind., Paper* 8, New Zealand Meteorological Service, October 4, 33 pp. (processed).

Meher-Homji, V. M. (1963): Les bioclimats du sub-continent Indien et leurs types analogues dans le monde, *Trav. Sec. Sci. Tech.*, Institute Francais de Pondichery, tome 8, fasc. 1, pp. 13–23.

Meinardus, W. (1934): Die Niederschlagsverteilung auf der Erde, *Meteorol. Z.*, September, pp. 345–350.

Meinzer, O. E. (1949): "Hydrology," Physics of the Earth: IX. Dover Publications, Inc., New York, 712 pp.

Mell, P. H. (1893): Report on the Climatology of the Cotton Plant, *U.S. Dept. Agr., Weather Bur. Bull.* 8.

Middleton, W. E. K., and F. G. Millar (1936): Temperature Profiles in Toronto, *J. Roy. Astron. Soc. Can.*, vol. 30, p. 265.

——, and A. F. Spilhaus (1953): "Meteorological Instruments," 3d ed., University of Toronto Press, Toronto, 286 pp.

Miller, D. H. (1968): A Survey Course: The Energy and Mass Budget at the Surface of the Earth, *Publ. 7 Comm. Coll. Geograph.*, Association American Geographers, Washington, 142 pp.

Missenard, H. (1948): Equivalences thermiques des ambiences: equivalences de passage; equivalences de sejour, *Chaleur Ind.*, July–August.

Mitchell, J. M., Jr. (1962): "The Thermal Climate of Cities," in Air over Cities Symposium, *Robert A. Taft Sanitary Eng. Center, Tech. Rept.* A62-5, pp. 131-145, 290 pp.

Momiyama, M., and K. Katayama (1966): A Medico-climatological Study in the Seasonal Variation of Mortality in the United States of America, *Papers Meteorol. Geophys.*, vol. 17, no. 4, pp. 279-285.

Monin, A. S., and A. M. Obukhov (1954): Principal Law of Turbulent Mixing in the Air Layer near the Ground, *USSR Akad. Nauk. Geophys. Inst.*, 24 (Translated from Russian, Office of Technical Services, U.S. Department of Commerce, 1959).

Morton, F. I., and H. B. Rosenberg (1959): Hydrology of Lake Ontario, *J. Hydraulics Div., Proc. Am. Soc. Civil Engrs.*, vol. 85, HY-5, 2017.

Muller, R. A. (1966): The Effects of Reforestation on Water Yield: A Case Study using Energy and Water Balance Models for the Allegheny Plateau, New York, *Publ. Climatol.*, Laboratory of Climatology, vol. 19, no. 3, pp. 251-304.

Munn, R. E. (1961): Energy Budget and Mass Transfer Theories of Evaporation, *Proc. 2nd Hydrol. Symp.* (Toronto), pp. 8-30.

Murphy, Allan H. (1960): "Engineering Meteorology: Meteorology and Heating Load Requirements," pp. 65-68, in Topics in Engineering Meteorology, *Meteorol. Monographs*, vol. 4, no. 22, 98 pp.

Myrick, D. H. (1970): All Risk Crop Insurance: Principles, Problems, Potentials, *Montana State Univ., Agr. Expt. Sta., Bull.* 640, 42 pp.

Neuberger, H. H., and F. B. Stephens (1948): "Weather and Man," Prentice-Hall, Inc., Englewood Cliffs, N.J.

Newsweek (1949): Invisible Drought, vol. 34, no. 6, Aug. 8, pp. 43-44.

Nicholas, F. W. (1971): The Changing Form of the Urban Heat Island of Metropolitan Washington, *Tech. Papers*, American Congress on Surveying and Mapping, Annual Meeting, March 7-12, Washington, D.C.

Nix, H. A., and E. A. Fitzpatrick (1969): An Index of Crop Water Stress Related to Wheat and Grain Sorghum Yields, *Agr. Meteorol.*, vol. 6, pp. 321-337.

Olgyay, V. (1963): "Design with Climate: Bioclimatic Approach to Architectural Regionalism," Princeton University Press, Princeton, N.J., 190 pp.

Ollerenshaw, C. B. (1967): Climatic Factors and Liver Fluke Disease, in J. A. Taylor (ed.), "Weather and Agriculture," pp. 129-135, Pergamon Press, Symposium Publishing Division, London, 225 pp.

Palmer, W. C. (1965): Meteorological Drought, *Res. Paper* 45, U.S. Weather Bureau, U.S. Department of Commerce, 58 pp.

—— (1968): Keeping Track of Crop Moisture Conditions Nationwide: The New Moisture Index, *Weatherwise*, vol. 21, no. 4, pp. 156-161.

Parsons, J. J. (1955): The Miskito Pine Savanna of Nicaragua and Honduras, *Ann. Assoc. Am. Geograph.*, vol. 45, no. 1, pp. 36–63.

Pathak, M. A., F. C. Riley, and T. B. Fitzpatrick (1962): Melanogenesis in Human Skin following Exposure to Long-wave Ultraviolet and Visible Light, *J. Invest. Dermatol.*, vol. 39, pp. 435–443.

Paulhus, J. L. H. (1965): Indian Ocean and Taiwan Rainfalls Set New Records, *Monthly Weather Rev.*, vol. 93, no. 5, pp. 331–335.

Pelton, W. L., K. M. King, and C. B. Tanner (1960): An Evaluation of the Thornthwaite and Mean Temperature Methods for Determining Potential Evapotranspiration, *Agron. J.*, vol. 52, pp. 387–395.

Penman, H. L. (1948): Natural Evaporation from Open Water, Bare Soil and Grass, *Proc. Roy. Soc., Ser. A.*, vol. 193, pp. 120–145.

—— (1956a): Evaporation: An Introductory Survey, *Netherlands J. Agr. Sci.*, vol. 4, no. 1, pp. 9–29.

—— (1956b): The Physical Bases of Irrigation Control, *13th Intern. Hort. Conf.*, pp. 913–924.

Peterson, J. T. (1969): The Climate of Cities: A Survey of Recent Literature, *U.S. Public Health Serv., Natl. Air Pollution Control Admin. Publ.* AP-59, 48 pp.

Petty, M. T. (1963): Weather and Consumer Sales, *Bull. Am. Meteorol. Soc.*, vol. 44, no. 2, pp. 68–71.

Pratt, R. L. (1952): Effect of Solar Heat, *Environ. Protect. Branch, Rept.* 185, O.Q.M.G., U.S. Army, Washington.

Priestley, C. H. B. (1967): Microclimates of Life, *Sci. J.*, London, vol. 3, no. 3, pp. 67–73.

Réaumur, R. A. F. de (1735): Observations du thermomètre, faites à Paris pendant l'année 1735, comparées avec celles qui ont été faites sous la ligne, à l'Isle de France, à Alger et en quelques-unes de nos isles de l'Amérique, *Paris Mem., Acad. Sci.*

Reifsnyder, W. E., and H. W. Lull (1965): Radiant Energy in Relation to Forests, *U.S. Dept. Agr., Forest Serv. Tech. Bull.*, 1344, 111 pp.

Riehl, H. (1972): "Introduction to the Atmosphere," McGraw-Hill Book Company, New York, 516 pp.

Robbins, C. S. (1949): Weather and Bird Migration, *The Wood Thrush*, vol. 4, no. 4, pp. 130–144.

Robinson, A. H. O. (1965): Road Weather Alerts, *What Is Weather Worth*, Australian Bureau of Meteorology, Melbourne, pp. 41–43.

Rohles, F. H., and R. G. Nevins (1971): The Nature of Thermal Comfort for Sedentary Man, *ASHRAE Trans.*, vol. 77, part I.

Roth, R. J. (1955): Crop-Hail Insurance Actuarial Association, *Bull. Am. Meteorol. Soc.*, vol. 36, no. 8, pp. 409–411.

—— (1956): Modern Crop Protection, *The Casualty and Surety J.*, September, 7 pp.

Russell, J. A. (1957): "The Problem, Method, and Conclusions," in Industrial Operations under Extremes of Weather, *Meteorol. Monographs*, vol. 2, no. 9, pp. 1–9.

Russo, J. A. (1966): The Economic Impact of Weather on the Construction Industry of the United States, *Bull. Am. Meteorol. Soc.*, vol. 47, pp. 967–972.

Sanderson, M. (1966): A Climatic Water Balance of the Lake Erie Basin, *Publ. Climatol.*, Laboratory of Climatology, vol. 19, no. 1, pp. 1–87.

Savinov, S. I. (1933): Concerning Formulas for Direct and Scattered Radiation and Dependent on Cloudiness, *Meteorol. Vestn.*, nos. 5–6.

Schott, C. A. (1872): Tables and Results of the Precipitation in Rain and Snow in the United States and at Some Stations in Adjacent Parts of North America, and in Central and South America, *Smithsonian Contrib. Knowledge*, no. 18, Washington, 178 pp.

——(1876): Tables, Distributions and Variations of the Atmospheric Temperatures in the United States and Some Adjacent Parts of America, *Smithsonian Contrib. Knowledge*, no. 21, Washington, 360 pp.

Schultz, H. B. (1962): The Interaction of the Macro- and Micro-climatic Factors Contributing to the Success of Wind Machines for Frost Protection in Southern California, in S. W. Tromp (ed.), "Biometeorology," pp. 614–629, Pergamon Press, New York.

Seabrook, J. M. (1953): Applied Climatology at Seabrook Farms, *Weatherwise*, vol. 6, no. 2, pp. 35, 36, 59.

Selirio, I. S., D. M. Brown, and K. M. King (1971): Estimation of Net and Solar Radiation, *Can. J. Plant Sci.*, vol. 51, pp. 35–39.

Sellers, W. D. (1965): "Physical Climatology," University of Chicago Press, Chicago, 272 pp.

Sen, A. R., A. K. Biswas, and D. K. Sanyal (1966): The Influence of Climatic Factors on the Yield of Tea in the Assam Valley, *J. Appl. Meteorol.*, vol. 5, no. 6, pp. 789–800.

Shantz, H. L., and R. Zon (1924): Natural Vegetation, in "Atlas of American Agriculture," U.S. Department of Agriculture.

Singer, I. A., and M. E. Smith (1953): Relation of Gustiness to Other Meteorological Parameters, *J. Meteorol.*, vol. 10, no. 2, pp. 121–126.

Siple, P. A. (1944): Intra Clothing Climate, *Trans. Am. Geophys. Union 1944, Pt. III, Sec. Meteorol.*, pp. 412–417.

——and C. F. Passel (1945): Measurements of Dry Atmospheric Cooling in Subfreezing Temperatures, *Proc. Am. Phil. Soc.*, vol. 89, pp. 177–199.

Smith, G. W. (1959): Solar Radiation and the Duration of Sunshine in Trinidad, West Indies, *Quart. J. Roy. Meteorol. Soc.*, vol. 85, pp. 421–423.

Smith, L. P. (1956): Meadow Hay Yields, *Outlook in Agr.*, vol. 3, no. 5, Summer 1956, p. 219.

——(1968): Forecasting Annual Milk Yields, *Agr. Meteorol.*, vol. 5, pp. 209–214.

Soil Conservation Service (1964): "National Engineering Handbook Section 4. Hydrology," U.S. Department of Agriculture.

Staple, W. J., and J. J. Lehane (1954): Weather Conditions Influencing Wheat Yields in Tanks and Field Plots, *Can. J. Agr. Sci.*, vol. 34, pp. 552–564.

Starr, V. P., and J. P. Peixoto (1958): On the Global Balance of Water Vapor and the Hydrology of Deserts, *Tellus*, vol. 10, no. 2, pp. 188-194.

———and———(1964): The Hemispheric Eddy Flux of Water Vapor and Its Implications for the Mechanics of the General Circulation, *Arch. Meteorol. Geophys. Bioklimatol.*, vol. 14, pp. 111-130.

———,———, and A. R. Crisi (1965): Hemispheric Water Balance for the IGY, *Tellus*, vol. 17, no. 4, pp. 463-472.

Steadman, R. G. (1971): Indices of Windchill of Clothed Persons, *J. Appl. Meteorol.*, vol. 10, no. 4, pp. 674-683.

Steele, A. T. (1951): Weather's Effect on the Sales of a Department Store, *J. Marketing*, vol. 15, pp. 436-443.

Steward, J. H. (ed.) (1946): Handbook of South American Indians: Vol. 1, The Marginal Tribes, *Bull., Bur. Am. Ethnology*, 143, Washington.

Stout, G. E., and F. A. Huff (1962): Studies of Severe Rainstorms in Illinois, *Proc. Am. Soc. Civil Engrs., J. Hydraulics Div.*, vol. 88, no. HY-4, pt. 1.

Suomi, V. E., and P. M. Kuhn (1958): An Economical Net Radiometer, *Tellus*, vol. 10, pp. 160-163.

———, and C. B. Tanner (1958): Evapotranspiration Estimates from Heat Budget Measurements over a Field Crop, *Trans. Am. Geophys. Union*, vol. 39, pp. 298-304.

Swartz, J. R. (1963): "Selective Guide to Published Climatic Data Sources Prepared by U.S. Weather Bureau," Key to Meteorological Records Documentation No. 4.11, U.S. Department of Commerce, Weather Bureau, 84 pp.

Swinbank, W. C. (1963): Long-wave Radiation from Clear Skies, *Quart. J. Roy. Meteorol. Soc.*, vol. 89, pp. 339-348 (also vol. 90, pp. 488-493).

Tanner, C. B., and M. Fuchs (1968): Evaporation from Unsaturated Surfaces: A Generalized Combination Method, *J. Geophys. Res.*, vol. 73, no. 4, pp. 1299-1304.

Taylor, J. A. (ed.) (1967): "Weather and Agriculture," Pergamon Press Limited, Symposium Publishing Division, London, 225 pp.

Terjung, W. H. (1966): Physiologic Climates of the Conterminous United States: A Bioclimatic Classification Based on Man, *Ann. Assoc. Am. Geograph.*, vol. 56, no. 1, pp. 141-179.

Thackston, E. L., and F. L. Parker (1971): Effect of Geographical Location on Cooling Pond Requirements and Performance, *Rept. 5, Natl Center Res. Training Hydrol. Hydraulic Aspects Water Pollution Control, Dept. Environ. Water Resources Eng.*, Vanderbilt Univ., School of Eng., 234 pp.

Thomas, A., and J. J. Drummond (1953): The Role of Weather Correction in Load Forecasting, *Edison Elec. Bull.*, August.

Thompson, L. M. (1970): Weather and Technology in the Production of Soybeans in the Central United States, *Agron. J.*, vol. 62, no. 2, pp. 232-236.

Thornthwaite, C. W. (1931): The Climates of North America According to a   390
New Classification, *Geograph. Rev.*, vol. 21, pp. 633-655.

—— (1937): The Hydrologic Cycle Re-examined, *Soil Conserv.*, vol. 3, no. 4, pp. 2-8.

—— (1940): Atmospheric Moisture in Relation to Ecological Problems, *Ecology*, vol. 21, no. 1, pp. 17-28.

—— (1943*a*): Atmospheric Turbulence and the Measurement of Evaporation, *Proc. Second Hydraulics Conf., Univ. Iowa, Studies Eng., Bull.* 27, pp. 280-288.

—— (1943*b*): Problems in the Classification of Climates, *Geograph. Rev.*, vol. 33, no. 2, pp. 233-255.

—— (1945*a*): Report of the Committee on Transpiration and Evaporation, 1943-1944, *Trans. Am. Geophys. Union*, pt. V, pp. 686-693.

—— (1945*b*): Report of the Committee on Transpiration and Evaporation, 1944-1945, *Trans. Am. Geophys. Union*, vol. 26, no. 3, pp. 453-455.

—— (1947): Climate and Moisture Conservation, *Ann. Assoc. Am. Geograph.*, vol. 37, no. 2, pp. 87-100.

—— (1948): An Approach Toward a Rational Classification of Climate, *Geograph. Rev.*, vol. 38, pp. 55-94.

—— (1950): Estimating Soil Tractionability by Climatic Analysis, Rept. 167, Environ. Protect. Sec. O.Q.M.G., U.S. Army.

—— (1952*a*): Grassland Climates, *Publ. Climatol.*, Laboratory of Climatology, vol. 5, no. 6, 14 pp.

—— (1952*b*): Climate in Relation to Planting and Irrigation of Vegetable Crops, *Publ. Climatol.*, Laboratory of Climatology, vol. 5, no. 5, 12 pp.

—— (1952*c*): "Temperature Relations to Time of Maturity of Vegetable Crops," prepared for presentation 78th Annual Meeting, New Jersey State Horticultural Society, Atlantic City, December, 18 pp. (processed).

—— (1953*a*): A Charter for Climatology, *WMO Bull.*, vol. 11, no. 2, pp. 40-46.

—— (1953*b*): Operations Research in Agriculture, *J. Operations Res. Soc. Am.*, vol. 1, no. 2, pp. 33-38.

—— (1954*a*): The Place of Supplemental Irrigation in Postwar Planning, *Publ. Climatol.*, Laboratory of Climatology, vol. 6, no. 2, 29 pp.

—— (1954*b*): "A Re-examination of the Concept and Measurement of Potential Evapotranspiration," in J. R. Mather (ed.), The Measurement of Potential Evapotranspiration, *Publ. Climatol.*, Laboratory of Climatology, vol. 7, no. 1, pp. 200-209, 225 pp.

—— and F. K. Hare (1955): Climatic Classification in Forestry, *Unasylva*, vol. 9, no. 2, pp. 50-59.

—— and B. G. Holzman (1942): Measurement of Evaporation from Land and Water Surfaces, *U.S. Dept. Agr. Tech. Bull.* 817, 75 pp.

—— and others (1943): Effect of Turbulence and Sedimentation, *Proc. Am. Soc. Civil Engrs.*, vol. 69, no. 8, pp. 1300-1303.

—— and J. R. Mather (1954): Climate in Relation to Crops, *Meteorol. Monographs*, vol. 2, no. 8, pp. 1-10.

——and——(1955a): Climatology and Irrigation Scheduling, *Weekly Weather Crop Bull.*, National Summary of June 27, 2 pp.

——and——(1955b): The Water Balance, *Publ. Climatol.*, Laboratory of Climatology, vol. 8, no. 1, 104 pp.

——and——(1955c): "The Water Budget and Its Use in Irrigation," in Water, *Yearbook Agr.*, *U.S. Dept. Agr.*, pp. 346–357.

——and——(1957): Instructions and Tables for Computing Potential Evapotranspiration and the Water Balance, *Publ. Climatol.*, Laboratory of Climatology, vol. 10, no. 3, pp. 185–311.

——,——, D. B. Carter, and C. E. Molineux (1958): Estimating Soil Moisture and Tractionability Conditions for Strategic Planning, *Air Force Surv. Geophys.* 94, AFCRC-TN-58-201, Geophysics Research Directorate, 56 pp.

——, W. J. Superior, F. K. Hare, and K. R. Ono (1959): Measurement of Vertical Winds in Typical Terrain, *Publ. Climatol.*, Laboratory of Climatology, vol. 12, no. 3, pp. 97–204.

——,——, and J. R. Mather (1961): Vertical Winds near the Ground at Centerton, N.J., *Publ. Climatol.*, Laboratory of Climatology, vol. 14, no. 2, pp. 95–244.

Trewartha, G. T. (1968): "An Introduction to Climate," 4th ed., McGraw-Hill Book Company, New York, 408 pp.

Turner, J. C. (1965): Some Energy and Microclimate Measurements in a Natural Arid Zone Plant Community, *Arid Zone Res.—Methodology of Plant Ecophysiol.*, *Proc. Montpellier Symp. UNESCO*, pp. 63–70.

U.S. Air Force (1961): "Applied Military Climatology," Air Weather Service Manual 105-3, Hq., Air Weather Service (MATS), Scott Air Force Base, Ill., 26 pp.

U.S. Army Corps of Engineers (1956): "Snow Hydrology," North Pacific Division Corps of Engineers, Portland, Ore., June 30, 437 pp.

U.S. Dept. of Commerce (1970): "The National Climatic Center, Asheville, North Carolina," National Oceanic and Atmospheric Agency, 34 pp.

U.S. Weather Bureau (1955): "Climatological Services of the U.S. Weather Bureau" (processed manuscript).

——(1957): Rainfall Intensity—Frequency Regime. Part I—the Ohio Valley, *Tech. Paper* 29, prepared by Cooperative Studies Section, Hydrologic Services Division, U.S. Department of Commerce, June, 44 pp.

——(1959): "Weather Bureau Programs in Climatology," U.S. Department of Commerce, 8 pp.

van Bavel, C. H. M. (1966): Potential Evaporation: The Combination Concept and Its Experimental Verification, *Water Resources Res.*, vol. 2, pp. 455–467.

van Everdingen, E. (1926): Het Verband Tussen de Weersgesteldheid en de Aardappelziekte, *Tijdschr. Plantenziekten*, vol. 32, pp. 129–140.

van Hylckama, T. E. A. (1956): The Water Balance of the Earth, *Publ. Climatol.*, Laboratory of Climatology, vol. 9, no. 2, pp. 59–117.

van Straaten, J. F. (1967): "Thermal Performance of Buildings," Elsevier Publishing Company Ltd., London.

Viessman, W., T. E. Harbaugh, and J. W. Knapp (1972): "Introduction to Hydrology," Intext Educational Publishers, New York, 415 pp.

Waggoner, P. E. (ed.) (1965): Agricultural Meteorology, *Meteorol. Monographs*, vol. 6, no. 28, 188 pp.

———, P. M. Miller, and H. C. DeRoo (1960): Plastic Mulching: Principles and Benefits, *Conn. Agr. Expt. Sta., New Haven, Bull.* 634, 44 pp.

Wahlberg, H. E. (1950): Wind Machines Cost Analysis—Cost Records Studied for Effectiveness of Machines in Frost Protection, Operation and Maintenance, *Calif. Agr.*, vol. 4, no. 11, p. 13.

Walter, H. (1955): Die Klimagramme als Mittel zur Deurteilung der Klimaverhältnisse für ökologische, vegetationskundliche und landwirtschaftliche Zwecke. *Ber. Deut. Botan. Ges. Jahrgang,* band 68, h. 8, pp. 331–334.

———and H. Lieth (1964): "Klimadiagramm—Weltatlas," Gustav Fischer Verlag, Stuttgart.

Wang, Jen-Yu (1963*a*): "Agricultural Meteorology," Pacemaker Press, Milwaukee, 693 pp.

———(1963*b*): A Graphical Solution on Temperature-moisture Responses of Tomato Yields, *Proc. Am. Soc. Hort. Sci.*, vol. 82, pp. 429–445.

———(1972): "Agricultural Meteorology," 3d rev. ed., Milieu Information Service, San Jose, Calif., 537 pp.

Waters, J. W. (1957*a*): "Weather Limitations to Electric Power Utility Operation," in Industrial Operations under Extremes of Weather, *Meteorol. Monographs*, vol. 2, no. 9, pp. 53–65.

———(1957*b*): "Weather Limitations to Primary Iron and Steel Plant Operations," in Industrial Operations under Extremes of Weather, *Meteorol. Monographs*, vol. 2, no. 9, pp. 87–95.

White, G. A. (1971): Practical and Economic Aspects of Ship Routing, *The Marine Observer*, vol. 41, no. 231, pp. 25–26.

White, R. F. (1945): Effects of Landscape Development on the Natural Ventilation of Buildings and Their Adjacent Area, *Texas Eng. Expt. Sta., Res. Rept* 45, Texas A&M College, College Station, Tex.

Whitney, M. (1892): Some Physical Properties of Soils in Their Relation to Moisture and Crop Distribution, *U.S. Dept. Agr., Weather Bur. Bull.* 4.

Wilm, H. G., C. W. Thornthwaite et al. (1944): Report of the Committee on Evaporation and Transpiration, 1943–1944, *Trans. Am. Geophys. Union*, vol. 25, pp. 686–693.

WMO (1970): World Weather Watch: Third Status Report on Implementation, Geneva, Switzerland.

Woodcock, A. H. (1964): Clothing and Climate, in S. Licht (ed.)., "Medical Climatology," pp. 557–580, vol. 8, Physical Medicine Library, Waverly Press, Inc., Baltimore, 753 pp.

———and T. E. Dee (1950): Wet Cold: I. Effect of Moisture on Transfer of Heat through Insulating Materials, *Environ. Protect. Sec., Test Rept.* 170, O.Q.M.G., Washington.

——, R. L. Pratt, and G. F. Fonseca (1955): Comparison of Experimental Pile Clothing with the Standard Army Arctic Uniform of 1953, *Tech. Rept. EP-13, QMRE Comm.*, Natick.

Woollum, C. A., and N. Canfield (1968): Washington Metropolitan Area Precipitation and Temperature Patterns, *U.S. Weather Bureau Tech. Memo.* ER-28,Weather Bureau Eastern Regional Headquarters, Garden City, N.Y.

Wright, H. H. (1916): Certain Characteristics of the Winds at Mount Tamalpais, Cal., *Monthly Weather Rev.*, vol. 44, no. 9, Sept., pp. 512-514.

Wüst, G. (1922): Verdunstung und Niederschlag auf der Erde, *Z. Ges. Erdkunde*, pp. 35-43.

——(1954): Gesetzmassige wechselbeziehungen zwischen ozean und atmosphare in der zonalen Verteilung von Oberflachenselzgchalt, Verdunstung, und Niederschlag, *Arch. Meteorol. Geophys. Bioklimatol. Ser. A(7)*, pp. 305-328.

Zon, R. (1941): "Climate and the Nation's Forests," in Climate and Man, *Yearbook Agr., U.S. Dept. Agr.*, pp. 477-498.

INDEX